QUANTITATIVE ANALYSIS FOR PUBLIC POLICY

John Kenneth Gohagan

Associate Professor of Engineering and Public Policy
Department of Technology and Human Affairs
School of Engineering and Applied Science
Washington University, St. Louis

McGraw-Hill Book Company

New York   St. Louis   San Francisco   Auckland   Bogotá   Hamburg
Johannesburg   London   Madrid   Mexico   Montreal   New Delhi   Panama
Paris   São Paulo   Singapore   Sydney   Tokyo   Toronto

QUANTITATIVE ANALYSIS FOR PUBLIC POLICY

1234567890 DODO 89876543210

This book was set in Times Roman. The editors were Charles E. Stewart, Donald G. Mason,
and James S. Amar; the production supervisor was Leroy A. Young. The drawings
were done by Santype International Limited.
R. R. Donnelley & Sons Company was printer and binder.

Library of Congress Cataloging in Publication Data

Gohagan, John Kenneth.
    Quantitative analysis for public policy.

    (McGraw-Hill series in quantitative methods for management)
    Includes bibliographies and index.
    1. Policy sciences—Statistical methods.
    2. Public administration—Statistical methods.
    I. Title.
HA29.G633   309'.01'51   79-21525
ISBN 0-07-023570-8

To the Two Most Important
People in My Life

Annette L. Gohagan
and
Daniel S. Gohagan

CONTENTS

Preface   xiii

1 Analysis as a Learning Process   1

1.1   The Learning Process   2
1.2   More on Role of the Analyst   7
1.3   Report Writing   13
1.4   Supplemental Reading   13

Part 1   Foundations   15

2   Probabilistic Reasoning   17

2.1   Interpretations of Probability   18
2.2   Statements of Truth   19
2.3   Rules of Probability   21
2.4   Illustrative Applications   28
2.5   Problems   32
2.6   Supplemental Reading   34

3   Statistical Description   35

3.1   Averages   35
3.2   Variability   40
3.3   Skewness, or Asymmetry   41
3.4   Histograms   42
3.5   Trends   44
3.6   Problems   48
3.7   Supplemental Reading   52

Part 2 Statistical Data Analysis 53

4 Probability Distributions 55

4.1	Statements, Outcomes, and Random Variables	56
4.2	Distribution Functions	58
4.3	Expected Value and Variance	61
4.4	Probability Models	65
4.5	Probability Tables	77
4.6	Problems	77
4.7	Supplemental Reading	80

5 Statistical Comparisons 81

5.1	Confidence in or Accuracy of Statistical Estimates	82
5.2	Comparing Two Estimates	93
5.3	Comparing Many Proportions	105
5.4	Comparing Multiple Means	108
5.5	Problems	113
5.6	Supplemental Reading	117

6 Statistical Modification of Prior Judgment 118

6.1	Bayesian Modification of Prior Information	119
6.2	Initial Estimates from Prior Information	120
6.3	Revising Proportions and Likelihoods	124
6.4	Revisions for Means	130
6.5	Problems	132
6.6	Supplemental Reading	135

7 Linear Relations among Variables 137

7.1	Linear Correlation	138
7.2	Two-Variable Equations	142
7.3	Many-Variable Equations	148
7.4	Stepwise Inclusion of Variables	153
7.5	Regressions on Time	155
7.6	Problems	156
7.7	Supplemental Reading	159

8 Applications of Statistical Data Analysis and a Case for Analysis 160

8.1	Estimates of Insulin Supplies Need Improvement	160
8.2	Testing for Carcinogencity of Food Additives: The Saccharin Example	164
8.3	Locks and Dam No. 26 Replacement: A Case for Analysis	173

Part 3 Benefit-Cost Analysis 181

9 Time Value of Money 183

9.1 Discount Rates 183
9.2 Present, Annual, and Future Worth Factors 185
9.3 Special Discount Factors 191
9.4 Compounding Period and Discount Rates 193
9.5 Inflation 195
9.6 Theoretical Generalizations 196
9.7 Problems 199
9.8 Supplemental Reading 200

10 Economic Evaluation Methods 201

10.1 Measures of Economic Efficiency 201
10.2 Efficiency Benefits and Costs 202
10.3 Net Present Equivalent 205
10.4 Benefit-Cost Ratio 208
10.5 Rate of Return 211
10.6 Projects Portfolio 215
10.7 Theoretical Considerations 217
10.8 Problems 220
10.9 Supplemental Reading 222

11 Incorporating Uncertainty 223

11.1 Sensitivity Analysis 223
11.2 Contingency Studies 226
11.3 Break-Even Analysis 228
11.4 Problems 231
11.5 Supplemental Reading 233

12 Nonmonetary Considerations and Equity 234

12.1 Comparable Measurable Impacts 235
12.2 Explicit Accounting 240
12.3 Equity 241
12.4 Problems 243
12.5 Supplemental Reading 245

13 Applications of Benefit-Cost Analysis and a Case for Analysis 246

13.1 Uniformed Services University of the Health Sciences 247
13.2 Earth Resources Survey System (ERS) 251
13.3 Polio Vaccine Development: A Case for Analysis 261
13.4 Supplemental Reading 266

Part 4 Resource Allocation with Linear Programming 267

14 Elements of Linear Programming 269

14.1 Intuitive or Graphic Solutions 269
14.2 Corner Point Theorem 272
14.3 Simplex Method 274
14.4 Simplex Tableau 278
14.5 Problems 281
14.6 Supplemental Reading 285

15 Problem Formulation 286

15.1 Big M Solution Method 287
15.2 Approximately Linear Objective Functions 295
15.3 Free Variables 296
15.4 Computerization 298
15.5 Problems 299
15.6 Supplemental Reading 301

16 Sensitivity Analysis 302

16.1 Vector and Matrix Operations 303
16.2 Simplex Tableau Vectors 306
16.3 Revising Resource Limits 309
16.4 Revising Rates of Profit (Cost) or Resource Consumption 312
16.5 New Constraints or Decision Variables 314
16.6 Role of the Computer 317
16.7 Problems 317
16.8 Supplemental Reading 321

17 Application of Linear Programming and a Case for Analysis 322

17.1 Air Pollution Control for St. Louis 322
17.2 Sludge Management Optimization for the Bistate Region: A Case for Analysis 326
17.3 Supplemental Reading 331

Part 5 Formal Decision Analysis 333

18 Decision Trees and Algorithms 335

18.1 Decision Trees 337
18.2 Equivalence of Extensive and Normal Forms 342
18.3 Situational Advantages 347
18.4 Problems 351
18.5 Supplemental Reading 354

19 Value of Additional Information   355

19.1 Opportunity Loss   356
19.2 Expected Value of Additional Information   357
19.3 Problems   362
19.4 Supplemental Reading   364

20 Utility and Multiattribute Decisions   365

20.1 Risk Profiles   366
20.2 Assessing a Simple Utility Curve   368
20.3 Multiattribute Utility Functions   370
20.4 Other Approaches to Multiattribute Decisions   378
20.5 Problems   378
20.6 Supplemental Reading   383

21 Applications of Decision Analysis and a Case
       for Analysis   384

21.1 Evaluating Breast Cancer Detection Strategies   385
21.2 Pumped Storage Site Selection   398
21.3 EPSDT/Medicaid Implementation in State X: A Case for Analysis   400
21.4 Supplemental Reading   408

       Appendixes   409

1 The Federal Statistical System   411
2 Probability Distributions   414
3 Interest Tables   434

       Index   457

PREFACE

This book is addressed to present and future public managers and nascent analysts. It is designed to communicate concepts and their applicability with enough attention to technique to ensure that concepts are mastered. I have written this book primarily for students in public policy, technology and policy, and public administration programs. They may be undergraduate or graduate students in regular academic programs or they may be practicing civil servants in extensive continuing education programs. My objective is to prepare students to function as intelligent consumers of analytical studies. The book is not intended as a single basic text for students who wish to become sophisticated analysts in their own rights, although I have found that it serves as a good introductory text for juniors and seniors studying engineering and public policy.

The book is especially designed for use in an intensive two-term course sequence, with only algebra as a prerequisite. That is how I use it.

Some academic programs require four separate courses covering basically the same material. That is entirely too much time devoted to quantitative methods, unless students are training for positions as analysts; too much attention is given to technical details and intricacies. Two courses of effectively presented material provide adequate training for most public policy career tracks. The other two course slots should be used to develop other skills and extend the scope of knowledge in topical areas.

The topics covered in this text do not exhaust the menu of techniques available. I chose to limit the scope for the sake of depth. I selected topics for a number of reasons, including proven value in public sector decision making; potential for fairly in-depth treatment without the use of higher mathematics; value as a framework and basis for formulating and evaluating decision problems; and, of course, personal experience.

I have been teaching quantitative methods of analysis to nontechnical people since 1970. I began at the Massachusetts Institute of Technology, where I redesigned a course in probability and statistics for midcareer federal employees in the year-long Civil Service–sponsored Educational Program in Systematic Analysis. Participants came from a multitude of public agencies, including the Foreign Service, the Food and Drug Administration, the Veterans Administration, and the U.S. Agency for International Development. The academic backgrounds of the participants were as varied as their professions. Some had had no mathematical training beyond high school algebra 15 years prior, while a few had degrees in engineering.

Since 1970 I have taught much of the material in this book at in-house workshops for managers and project engineers with the Department of the Army; at regional workshops for state, local, and regional public administrators; and to graduate and undergraduate students of technology and public policy, public affairs, social work, business, and engineering.

One important fact I have learned over the years is that one should never underestimate the ability of students to grasp these often new and strange concepts. If the material is presented properly and not buried in esoteric language and mathematical niceties, most students will understand the concepts, develop basic skills of analysis, and carry their new knowledge to their professional careers.

Three themes carry throughout the book. First, analysis is a learning process. Its purpose is to help define options and develop insights into the possible consequences of various courses of action. It cannot be expected to identify *the best* decision. Second, although quantitative analysis can provide unique and essential insights for decision making, qualitative analysis can be just as important. Thus, the sophisticated manager (or analyst) will judiciously balance the quantitative and qualitative aspects of analysis. I have chosen to emphasize these themes in the introductions to each major part of the book before launching into the technical chapters. The applications chapters and cases for analysis reemphasize these themes and relate them to reality. By "bracketing" the technical material in this way, I can concentrate on methods and concepts in the technical chapters. This is the most efficient approach I can conceive of to develop simultaneously skills and mature perspectives on the value of analysis. Finally, decisions must be made and implemented on the basis of uncertain information and probabilistic hypotheses as to their consequences. Therefore I have made the factor of uncertainty a major theme in the text.

I have included a bit more material that I usually teach in a two-course sequence, and I have developed it in slightly more detail than I cover in class. Furthermore, I have broken the text into five major parts. Included in each part is a chapter describing one or more actual applications of the material in the public sector along with a case for analysis by teams of students using techniques developed in the text.

The first two parts of the book cover the basic elements of probabilistic reasoning and data analysis. The next two parts are devoted to benefit-cost analysis and resource allocation with linear programming. In these modes of analysis procedures for incorporating uncertainty are introduced after the basic techniques are developed. The final part develops a very important methodology called formal decision analysis. Decision analysis is presented as a methodology in its own right, and a paradigm for decision making under uncertainty, as well as a framework within which one may apply much of the material developed in earlier chapters.

Like all authors I owe a debt of appreciation to a great many people. My former professors, my colleagues, and my students have all had an influence on my development and, directly or indirectly, on the development of this book. It would be impractical to name them all in this space, so I must thank them as a group, with the exception of those mentioned below, to whom I owe a special debt.

My professional development has been influenced most by three people. Mr. George Schecter, a former supervisor, now with Analytics, Incorporated, strongly influenced the direction of my professional development in the early years of my career with the federal government. He remains to this day a very special friend and mentor. Dr. Amedeo R. Odoni of the Massachusetts Institute of Technology and Dr. Harold A. Thomas, Jr., of Harvard, two of my former professors and advisors, powerfully influenced the direction and quality of my professional training. They continue to provide models for my own performance and growth even though our paths seldom cross now. Had I not had the advantage of association with these fine people, this book might never have been written.

In the actual drafting of chapters, two former students, Mr. Thomas Helscher of Monsanto Corporation and Mr. Louis Callen, an administrator with University City, Missouri, provided objective and somewhat different perspectives which influenced the ultimate design and level of the book. They reviewed and commented on early drafts of every chapter. Mr. Helscher, originally trained in chemical engineering and economics, was at that time a corporate liason with the federal government on environmental matters. Mr. Callen, originally trained in political science and public affairs, was the zoning administrator for the city.

For technical production of the manuscript I owe special recognition to Ms. Lynn Lawson and Ms. Donna Wilson, who typed and retyped numerous drafts and managed the production of bound volumes for review by McGraw-Hill consultants and for use by students in my classes. They carried a heavy load and their performance was indeed laudable.

Finally, many thanks to the McGraw-Hill staff and consultants for their many contributions.

John Kenneth Gohagan

QUANTITATIVE ANALYSIS
FOR PUBLIC POLICY

ONE

ANALYSIS AS A LEARNING PROCESS

One's view of the purpose of analysis strongly influences one's judgment regarding the value of analytical methods in public management. The view taken in this book is that analysis is a learning process within which problems are identified and defined, responses formulated and evaluated, and courses of action selected, monitored, and revised as needed.

One must not expect analysis, quantitative or otherwise, to produce "the answer," as in 2 plus 2 is 4, or even an unequivocal recommendation for the "optimal" course of action. Such expectations are completely unrealistic except perhaps in trivial cases. Any analysts who attempt to sell their analyses on an absolute basis must be judged naive and overenthusiastic, or possibly incompetent or deceitful; any administrators who expect such miracles must also be judged naive or possibly, as Pooh would say, "of little brain."

Seasoned analysts and managers know that the purpose of analysis, particularly quantitative analysis, is to provide insight into complex situations by:

- Elucidating issues
- Exposing points of sensitivity in the balance of risks and benefits
- Identifying preference relations and opportunities for trade-offs or bargaining
- Anticipating impacts of actions, in a probabilistic sense, under various conditions
- Organizing information and data in support of reasoned decision making

Experienced analysts and administrators do not despair that there are no perfect measures for the consequences of public programs or that data bases upon which to base policy studies are typically far less than perfect. They are not put off by the facts that bold forecasts of the consequences of policy decisions have been wrong in the past or that analysts and managers alike have failed on occasion to anticipate important political factors which, if accounted for in the analysis stage, would have rendered their recommendations more useful. Instead, they take a balanced view of the history of quantitative analysis in the public sector (especially since the early 1960s, when it really came into its own) and observe that systems analysts, operations researchers, management scientists, and economists have contributed enormously to better-informed, more insightful decision making. For there can be no doubt that analysis has clarified issues and lifted much of the fog obscuring who shall win and who shall lose from public programs and policies. It is a fact that analysis has led to policy-significant and sometimes counterintuitive conclusions, often dissolving long-standing misconceptions, especially in the social policy arena. Furthermore, analysts have led the way in developing data bases where none existed before and significantly upgrading many more. These products in and of themselves have had long-term, positive impacts on fact-based decision making.

Today analysis is a way of life in the public sector. It is a major factor in the entire process of decision making and policy formulation and is likely to become even more pervasive. The reason is simple. Extraordinary insight can be gained through well-conceived and well-executed quantitative analyses. As a former student, Margaret Power, wrote on a homework paper, "Moral: if one tortures the data sufficiently, it will confess."

1.1 THE LEARNING PROCESS

David Kolb's model of the learning process is an excellent structure within which to describe the analysis process (see Figure 1.1).

Problems or needs are identified by observing reality, comparing it to a mental ideal, and noting discrepancies. Continued observation, verbal analyses or discussions, back-of-the-envelope analysis of existing data, and reflection lead to problem definition and the formulation of hypotheses or theories as to the nature of the problem. Systematic data collection and analyses ensue and the theories or hypotheses are tested against reality. Then alternative responses are formulated and analyzed; they may be tested experimentally before a final course of action is selected.

Phase one Problem formulation may continue for months or even years in the public sector. Private citizens, special interest groups, elected and appointed officials, and others may be involved and each may bring a different interpretation to the observations. Most of the analysis in this phase is likely to be verbal and based on opinion, but some participants in the process may build their arguments on rough but systematic analyses of data. The quantitative methods most applicable in this phase are descriptive statistics and retrospective economic studies.

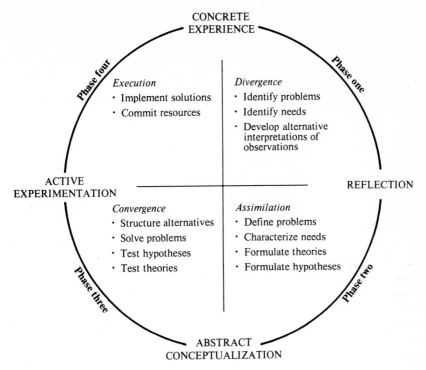

Figure 1.1 Analysis as a learning process. (Based on the work of D. A. Kolb, with modifications.)

Phase two As specific problems or needs are more clearly defined, the analytical emphasis shifts to the formulation of hypotheses or theories regarding the roots of the problem. In this phase theoretical studies and the formulation of conceptual models to explain the observed situation in terms of decision variables which might be controlled may be major efforts in preparation for data collection and analysis. Sociologists, operations researchers, economists, and policy analysts working in universities, major consulting companies (such as the Rand Corporation), and public agencies may be involved. The analytical methods applied to retrospective data are likely to include economic analyses, descriptive and inferential statistics, computer simulation, and a great deal of verbal discussion.

Phase three Hypotheses, theories, and models must be tested against experience. If the tests are to be quantitative, appropriate data must be collected and processed. This may be a major effort including carefully designed, controlled experiments to test conceptual models linking decision variables and the state of affairs. Even qualitative tests (i.e., logical evaluation without the benefit of numerical data) may entail a major effort to devise controlled situations from which logically consistent conclusions may be drawn. Statisticians, physical scientists, operations analysts, economists, and social scientists may be engaged in this activity.

Alternative courses of action to resolve a problem or meet a need are often tested on an experimental basis. In the public sector these are usually called *demonstration projects* and may be multiyear, multimillion dollar efforts. They represent experiments or field tests of theories equivalent to pilot projects in engineering or business. These projects generate data which may be analyzed by technical specialists and used by decision makers as a basis for selecting a final course of action. Demonstration projects also generate nonquantitative or even impressionistic information which may be critically important for decision making. Experimental (statistical) design is especially critical here, ensuring that the proper data are collected in as unbiased a fashion as possible to support analyses in final decision making.

Phase four Finally, all information so far obtained is evaluated and a course of action is selected. Statistical techniques, benefit-cost analysis, decision analysis, and optimization techniques such as linear programming may be employed to support decision making. Technical specialists have an important role in this decision-making phase. They function as technical advisors who can reformulate information for greater effectiveness and identify weak points in prior studies. They may even conduct short supplemental studies to help fill information voids which frequently appear at this point. But, of course, they must never attempt to dictate a course of action; the major resource commitments are made in this phase, and managerial judgment, which is typically much more broadly based than the analysts', must prevail.

At this point the process begins anew as the impacts of actions taken are observed.

One practical scheme for integrating available information of various types to illuminate options, expected consequences, and uncertainties is displayed in Figure 1.2. The rationale or philosophy behind the scheme can be summarized as follows:

1. Important policy and program choices in the public sector will inevitably be made on diffuse and uncertain information. At best, quantitative hypotheses in the form of judgmental probability distributions linking alternative actions with potential consequences can be established.
2. Predecision analyses should include:
   a. Explicit discussion of the problems, needs, goals, and objectives.
   b. Clearly specified alternatives.
   c. Quantitative hypotheses concerning impacts. These should be based on all forms of relevant information available, from statistical data to expert judgments.
   d. Carefully mapped out expected consequences and accompanying uncertainties for each viable alternative.
3. Implemented programs should be monitored and reevaluated against expectations so that corrective action can be taken as needed.

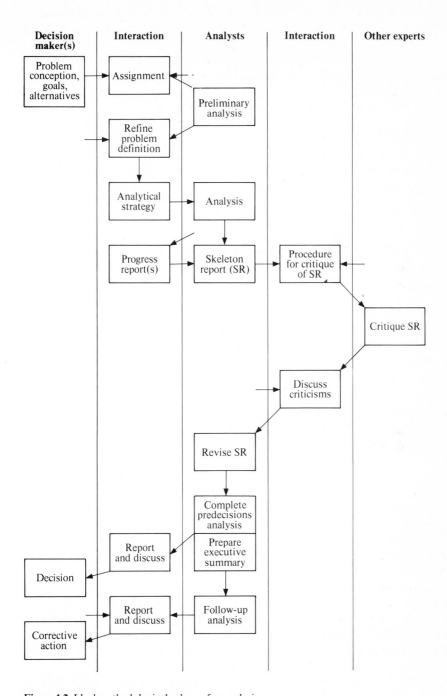

Figure 1.2 Ideal methodological scheme for analysis.

In this scheme, problem formulation and development of analysis strategies are cooperative efforts engaging both decision makers and analysts. The analyst becomes an expert on the problem and eventually generates quantitative hypotheses in the form of judgmental probabilities linking program alternatives to potential consequences. The analyst prepares a special report, a "skeleton report," which clearly presents all components of his or her initial predecision analysis, including supporting arguments, assumptions, and constraints, in skeletal form. This report serves as a vehicle for soliciting individualized criticism by recognized technical experts as well as management. Subsequently, the analyst revises his or her quantitative hypotheses as appropriate to complete the predecision analysis of the alternatives. Finally, of course, the analyst prepares a concise final report laying out the options and their expected impacts, delineating the uncertainties involved, and discussing significant disagreements between the analyst and experts, and among the experts themselves. Management and analysts discuss the results and management selects a course of action. Then, in iterative fashion, monitoring and program evaluation procedures are implemented and postdecision analyses ensue in anticipation of future decisions.

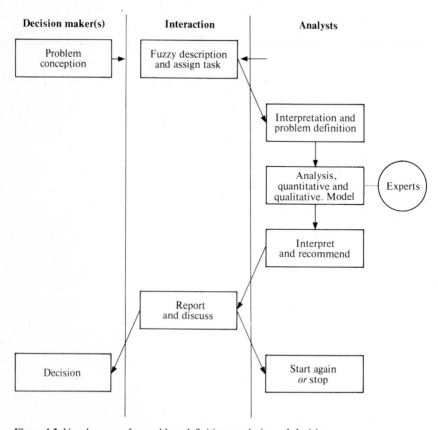

Figure 1.3 Usual process for problem definition, analysis, and decision.

A scheme which more accurately describes the process as it normally works is shown in Figure 1.3. Note the lack of communication between managers and analysts and the ad hoc use of experts. This weakness can and often does lead to wasted analytical effort and mutual dissatisfaction, a point made forcefully in the next section.

1.2 MORE ON ROLE OF THE ANALYST

Problem definition is far and away the most important phase of a policy or decision analysis. Unfortunately, this truism is too often ignored. Thus managers often complain that analysts miss the point, analyze the wrong problem, over-complicate the analysis in a mathematical sense, ignore time constraints which were an integral part of the problem, or make unrealistic recommendations because they failed to understand the problem in context. On the other hand, analysts often fault decision makers, claiming that they fail to provide a full picture of the problem in all its ramifications. Managers, they complain, describe problems in fuzzy terms, leaving analysts to complete the definition as best they can. That, they argue, is why their work sometimes results in unrealistic solutions and recommendations.

There is truth on both sides of the issue. Analysts tend to be skilled in a limited set of techniques and are prone to formulate or reformulate problems to fit these molds, occasionally by ignoring relevant information. One writer (an analyst himself, and a good one) described this phenomenon as the " screwdriver syndrome " (see Machol, 1974). He portrayed an analyst who was at home on a weekend looking for a way to help around the house. Being good with tools, he picked up a screwdriver and searched through the house for screws needing tightening. When he came upon a protruding nail, he went to the basement, got his file, which lay next to his hammer, filed a slot in the nail, and proceeded to try to reseat the nail with the screwdriver.

On the other hand, decision makers often cannot or for some reason do not convey to analysts the more subtle, but no less significant, aspects of problems. This failure to communicate is due, most likely, to the style of most managers. Gathering and mentally processing tidbits of information that are not generally available to subordinates is central to their functioning. Their information comes from multitudinous sources and often is in the form of rumors or suspicions. When conceiving a problem they engage in mental gymnastics they can seldom describe, to process very fuzzy information. Decision makers also tend implicitly to assume that analysts have a larger perspective and more background information than the analysts' positions permit.

The results of these failings on both sides are similar; analytical labors may be wasted and both parties are frustrated, each blaming the other. Better communication would naturally exist if decision makers knew more of the information requirements of good analyses and practiced a more open form of management. Analysts must consider it a part of their job to actively inquire into and understand the real needs of management and address these in their studies. As more analytically trained persons take their place at the top, and the decision-making process evolves toward more open inquiry and analysis, this difficulty may be overcome to a large extent. But there will always be a gap, which both parties must strive to overcome in the interest of more informed decision making.

A crucial fact often missed by both parties is that analysis, especially if undertaken in the spirit of cooperation illustrated in Figure 1.2, has an important role in problem definition. False analytical starts can lead rapidly to more useful problem reformulations, provided that managers and analysts keep the communication channels open and proceed as a team. In fact, this is one of the most useful consequences of analysis, because problem definition is often more than halfway to problem resolution.

Of course, not all decision problems are amenable to quantitative analyses, and those that are often have nonquantifiable aspects which may be at least as important as the quantifiable aspects. Thus, to reiterate, care must be exercised to define a problem before proceeding with its analysis. Quantitative methods should be applied to those aspects of the problem where they are appropriate and promise to yield significant results. Quantitative and qualitative analyses should proceed in concert, where both are appropriate.

The following two personal experiences illustrate clearly the admonition that effective analysis is a collaborative effort engaging both management and analysts.

The first experience demonstrates the screwdriver syndrome and the effects of poor communication between managers and analysts. In January 1973, government agencies throughout the United States received orders to reduce fuel consumption significantly. Different agencies set different goals for fuel savings, depending on what they thought they could achieve and what they thought would look good in the eyes of their parent agencies.

At one agency the goal for fuel consumption was set at 20 percent savings over the previous year. Management wanted a "nifty" mathematical "model" to predict each month what the year-end fuel savings would be so that corrective actions could be taken. The model was to be accurate to within a specified small error on each monthly forecast.

The analyst assigned to the task was a recent Ph.D. from the Massachusetts Institute of Technology. His objections that unpredictable weather and an uncontrollable heating system precluded successful completion of the task could not compete with the glorious vision of sophisticated control held by management, so he undertook a week of investigative work and drafted a memo to his chief, arguing his point again, this time with supporting evidence (see Figure 1.4). He also described what he considered a realistic plan. His chief was not satisfied and assigned two junior analysts to the job. The junior analysts worked for 4 months, November through February, on the problem and prepared a 16-page memo including heat flow equations, experimental heat loss estimates for one building, and considerations of weather factors. They concluded, in essence, that

8 January 1974

MEMO FOR RECORD

TO: Chief

I am convinced that there is no sophisticated mathematical procedure for projecting, month-by-month, expected year-end fuel savings here, or elsewhere, with any specified degree of confidence. For the current year at least, our optimal strategy is essentially the current strategy, namely, adjust thermostats downward during the day and even further during the night while investigating the peculiarities of the physical plant and heating system in search of additional ways to conserve fuel. In coming to these conclusions I have had the benefit of discussions with Dr. D. of the National Bureau of Standards and Dr. F. of Drexel University.

Dr. D. is an expert in the field of fuel conservation in buildings. (The development of technology and strategies for conserving energy is one of the major programs at the Bureau. A principal concern of that program is the development of fuel conservation measures in space heating.) He is also the Bureau's energy officer and therefore the person responsible for insuring that the Bureau itself reduces significantly its consumption of heating fuel from last year's level. Dr. D. told me that the Bureau knows of no better procedure for reducing heating fuel consumption than to lower thermostat settings. The Bureau has set thermostats at about 68° in the hope of ending the year with a fuel saving of about 7 to 15 percent. Because fuel savings are dependent on outside temperatures which cannot be projected with any degree of confidence, this range is, in Dr. D.'s words, "a guess."

Dr. F. is a well-known meteorologist. He told me there are almost certainly no models for projecting outside temperatures, with any degree of confidence, for more than one or two days in advance.

Since we are already on the short term optimal path toward fuel savings, it is my intention to take the time to study the book Dr. F. has sent me, How to Save Energy in Buildings, and analyze historical temperature data, which I received from the National Weather Service, as well as our data on fuel consumption versus temperature. This effort may provide some scientific rationale for a future energy conservation plan.

As for keeping track of heating fuel conservation, I suggest that a cumulative record of fuel consumption be kept for comparison with last year's record.

Analyst

Figure 1.4

savings could be achieved by lowering thermostats in buildings during the day and reducing them further at night (in those few buildings where thermostats existed). However, they provided no insight into the magnitude of savings to be expected over the previous year's consumption and no additional guidance on how to conserve energy. Their work was worthless. It is fortunate that an agencywide task force also had been formed at the outset to work with major organizational units to close down some buildings, lower thermostats wherever possible, and take other conservation measures, as appropriate; otherwise, almost no savings would have been realized that year.

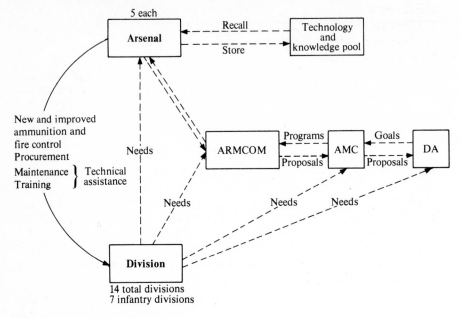

Figure 1.5 Analyst's response.

A. *R&D installation's value—attributes*
  1. New and improved technology
     • Defense:
       Weapons/systems
       Production
     • Nondefense
  2. New knowledge/concepts
  3. Economic/political (domestic)
  4. Interface with industry:
     • Preproduction engineering
     • Verification of industry systems/studies, etc.
  5. Procurement
  6. Emergency industrial base:
     • Skills reservoir
     • Tooling
     • Liaison
  7. Quick reaction (field trouble shooting/spare parts production)
B. *Division's value—attributes*
  1. Fighting
  2. Strategic
  3. Political (foreign)
  4. Manpower:
     • Employment
     • Training

5. Economic:
    - Foreign spending
    - Domestic support spending
C. *Close R&D installation?*
   Essential questions:
   1. Can the essential technical and production support levels be maintained under reconfiguration?
       - Short term
       - Long term
   2. What will be the domestic political impact?
   3. What will be the local economic impact?
       - Unemployment
       - Monetary
   4. Will essential procurement activities be significantly impaired?
       - Short term
       - Long term
   5. What budget savings can be expected under a reconfiguration which maintains the essential arsenal functions?
D. *Eliminate a division?*
   Essential questions:
   1. Will our strategic posture suffer?
   2. What foreign political implications are there?
       - Friends
       - Adversaries
   3. What will be the military effects if war breaks out?
   4. What are the economic implications?
       - Foreign
       - Domestic
   5. What are the manpower implications?
       - 12,000 more young men in civilian job market
       - Training needs for young men
       - Unemployment among the support work force
   6. Could dollars saved be better spent to improve defense/fighting/deterrent through arsenal R&D?

Figure 1.5 (continued) Analyst's written response.

Another true experience demonstrates the value of interaction between analysts and management in the problem definition stage. It also demonstrates the importance of appropriate analysis.

In the spring of 1974 the Department of Defense was considering alternative means of reducing its budget. Disparate options were under consideration. At one point a general-to-be asked his staff to evaluate the marginal return on investment to the country of a research and development organization versus a combat division. He wanted the answer in one week.

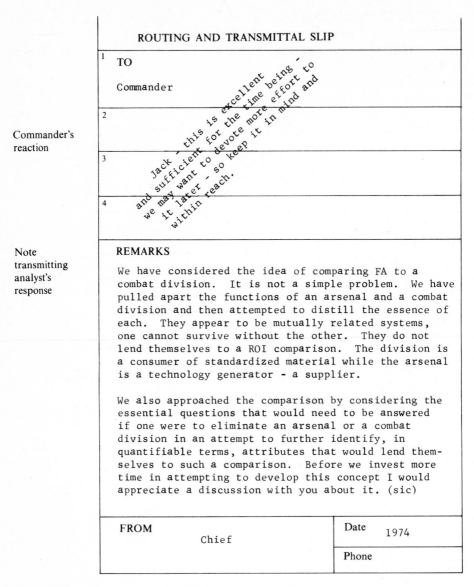

ROUTING AND TRANSMITTAL SLIP

Commander's reaction

Note transmitting analyst's response

1 TO

Commander

2

Jack - this is excellent and sufficient for the time being - and may want to devote more effort to it later - so keep it in mind and within reach.

3

4

REMARKS

We have considered the idea of comparing FA to a combat division. It is not a simple problem. We have pulled apart the functions of an arsenal and a combat division and then attempted to distill the essence of each. They appear to be mutually related systems, one cannot survive without the other. They do not lend themselves to a ROI comparison. The division is a consumer of standardized material while the arsenal is a technology generator - a supplier.

We also approached the comparison by considering the essential questions that would need to be answered if one were to eliminate an arsenal or a combat division in an attempt to further identify, in quantifiable terms, attributes that would lend themselves to such a comparison. Before we invest more time in attempting to develop this concept I would appreciate a discussion with you about it. (sic)

FROM

Chief

Date 1974

Phone

Figure 1.6 Interoffice communication: Division versus arsenal.

The analyst assigned the task considered the question absurd and chose to respond by writing a brief memo describing the attributes of both kinds of unit, the issues related to the elimination of either, and the links between units. Thus he ignored the question originally posed and instead responded with a clarification of the problem (see Figure 1.5).

His response was appreciated, as is evident in Figure 1.6. In this case a simple analysis carried the day. It often does, and that is an important point.

1.3 REPORT WRITING

Communication is central to the learning process. It is not enough that analysts understand the problem and the relative merits of alternative courses of action. Management is the consumer here and analysts have an obligation to communicate their findings completely.

Although much of the relevant information may, indeed should, be communicated to management informally, in progress reports and at formal briefings, a final written report is usually a must. The final report should be concise, complete, informative, and easy to read. Verbosity and shoddy writing are deadly. To ensure that essential results are communicated quickly, the report should begin with an executive summary describing the problem, the analysis process, assumptions, omissions, and findings or recommendations. Chapter 1 should be an introduction to the report, including a guide to readers. Succeeding chapters present the actual analysis; technical mathematical details, if there are any, should be included as appendices. Analysts should always provide drafts for comment to interested parties to help ensure the readability and effectiveness of the final draft.

1.4 SUPPLEMENTAL READING

Archibald, K. A.: "Three Reviews of the Expert's Role in Policy Making: Systems Analysis, Incrementalism, and the Clinical Approach," *Policy Sciences*, vol. 1, 1970, pp. 73–86.

Bladwin, Fred: "Evaluating the OEO Legal Services Program," *Policy Sciences*, vol. 4, no. 3, 1973, pp. 347–364.

Burnstein, Samuel, et al.: "The Problems and Pitfalls of Quantitative Methods in Urban Analysis," *Policy Sciences*, vol. 4, no. 1, 1973, pp. 29–39.

Flagle, Charles: "Operations Research in the Health Services," *Operations Research*, vol. 10, no. 4, 1962, pp. 591–603.

Goldfard, Robert S.: "Learning in Government Programs and the Usefulness of Cost Benefit Analysis: Lessons from Manpower and Urban Renewal History," *Policy Sciences*, vol. 6, no. 3, 1975, pp. 281–299.

Hoos, Ida: *Systems Analysis in Public Policy: A Critique*, University of California Press, Berkeley, 1972.

Keeney, Ralph L., and Howard Raiffa: "A Critique of Formal Analysis in Public Decision Making" in Drake, A. (ed), *Analysis of Public Systems*, MIT Press, Cambridge, Mass., 1973.

Kolb, D. A.: "On Management and the Learning Process," Sloan School Working Paper 652–73, MIT, Cambridge, Mass., 1972.

Machol, Robert: "Principles of Operations Research, The Screwdriver Syndrome," *Interfaces*, vol. 4, no. 3, May 1974.

Mantel, Samuel J., et al.: "A Social Service Measurement Model," *Operations Research*, vol. 23, no. 2, 1975, pp. 218–219.

Mintzberg, Henry: "The Manager's Job: Folklore and Fact," *Harvard Business Review*, vol. 53, no. 4, July–August 1975.

Packer, A. H.: "Applying Cost-Effectiveness Concepts to the Community Health System," *Operations Research*, vol. 16, no. 2, March 1978, pp. 227–254.

Quade, E. S.: *Analysis for Public Decisions*, American Elsevier, New York, 1971.

Rivlin, Alice M.: *Systematic Thinking for Social Action*, The Brookings Institution, Washington, D.C., 1970.

Sherwood, Clarence C.: "Issues in Measuring Results of Social Action Programs, "*Welfare in Review*," vol. 5, no. 7, 1976, pp. 13–17.

Spindler, Arthur: "Social and Rehabilitation Services: A Challenge to Operations Research," *Operations Research*, vol. 18, no. 6, 1971, pp. 1112–1129.

Staats, Elmer: "Challenges and Problems in Evaluation of Governmental Programs," *Interfaces*, vol. 5, no. 1, November 1975, p. 25.

Storey, James R.: "Systems Analysis and Welfare Reform: A Case Study of the FAP," *Policy Sciences*, vol. 4, no. 1, 1974, pp. 1–11.

Tribus, Myron: "Along the Corridors of Power," *Mechanical Engineering*, vol. 98, no. 4, April 1976, pp. 24–28.

Trinkl, Frank H.: "A Stochastic Analysis of Programs for the Mentally Retarded," *Operations Research*, vol. 22, no. 6, 1974, pp. 1175–1191.

White, Michael J., et al.: *Management and Policy Science in American Government*, Heath Book, Toronto, 1975.

Wholey, Joseph S.: "What Can We Actually Get from Program Evaluation?" *Policy Sciences*, vol. 3, 1972, pp. 361–369.

Wilson, John Oliver: "Social Experimentation and Public Policy Analysis," *Public Policy*, vol. 22, no. 3, 1974, pp. 15–37.

ONE

FOUNDATIONS

Elementary data analysis and inductive reasoning are indispensable tools of policy or decision analysis. We are all fond of saying that we deduce or infer some conclusion from the evidence. We are quite aware that most conclusions are merely plausible, not inescapable. The stronger the evidence, the more plausible the conclusion. This process of deducing from evidence is called *inductive reasoning*. It is not to be confused with *deductive reasoning*. The best we can do inductively is conclude that such and such is probably true, is very likely true, or, if we have exceptionally strong evidence, is almost certainly true. The deductive reasoning process applies only when there is no uncertainty, because the major premise includes the conclusion.

Since the real world is riddled with uncertainty, inductive reasoning is the logical basis for policy analysis and decision making. Therefore our first order of business is to master the rules of probability, which are actually the rules of inductive logic. These are presented, discussed, and illustrated in Chapter 2. Consider this material essential, for as the examples illustrate, faulty reasoning can lead ultimately to actions with serious undesirable consequences. Furthermore, these rules are the foundation upon which data analysis stands.

The second order of business is to master the elementary procedures of data summarization and first-order analysis. Very often more sophisticated methods of analysis will be unnecessary. Analysis should always begin with simple inquiries into the information content of available data, using the basic methods developed in Chapter 3.

PROBABILISTIC REASONING

The concept of probability is central to decision making and thus to policy analysis and program development. On the one hand, one cannot rationally choose among alternative courses of action unless one knows the likelihoods linking them to potential consequences, except possibly in trivial cases. For example, suppose one had a choice of action A or action B and the possible payoffs depended on some factors which could be categorized into two uncontrolled, distinct scenarios S_1 and S_2, as in Table 2.1. Choice B would be preferable, if one assumed the chance of S_2 were large. Choice A would be preferable, if the chance of S_1 were large. The question would be, How much larger must be the likelihood of one with respect to the other for it to be the best choice?

Table 2.1 Payoffs

Decisions	Scenario	
	S_1	S_2
A	10	-2
B	-6	16

On the other hand, seldom do decision makers have conclusive data on which to base decisions. They must infer likelihoods or expected outcomes from fragmentary, partially relevant information of many types and from disparate sources. As new and better information is acquired they may try to account for it and adjust their decisions. The formal rules for inductive reasoning, including integrating new with old information, are the rules of probability. Thus, understanding the concept and adhering to the rules of probability are fundamental to rational decision making.

The purpose of this chapter is to introduce the basic concepts of inductive reasoning. Different interpretations of probability are explained in Section 2.1. In Sections 2.2 and 2.3 the rules themselves are explained and illustrated.

2.1 INTERPRETATIONS OF PROBABILITY

Some people argue that there are true or objective probabilities. Others argue that a probability is an expression of ignorance or a degree of belief that is in no way objective. Both groups, however, adhere to the calculus or rules of probability.

The so-called objectivists argue that the only legitimate probabilities are those which, at least in principle, can be verified or determined experimentally. They view all probabilities in relative-frequency terms. That is, unless an experiment can be repeated under identical conditions and the relative frequencies of the outcomes determined, there is, in their view, no meaning to the statement, "The probability that the outcome will be such and such is p."

The difficulty with this view is obvious. It rules out projective uses of probability. That is, nothing can be said of the future. Only descriptive statements of the past are permissible. Prospective decision making, planning, and anticipation are rendered impossible. To illustrate the difficulty, consider tossing a coin. In principle the coin could be tossed infinitely often under identical conditions to determine whether it was biased or fair. However, if such a coin were actually tossed, say, 1000 times to check its fairness, and it landed heads 45 percent of the time, the objectivists would be constrained to the statement, "The coin was tossed 1000 times and landed heads 45 percent of the time." They could say nothing about the likelihood that the coin would land heads on the next toss. They could not even project that in the next 1000 tosses the coin would land heads 45 percent of the time, unless they made subjective judgments that the conditions of the next N tosses would be identical to those of the first 1000 tosses, including the shape of the coin.

Judgmental Interpretation

In the early seventeen-hundreds J. Bernoulli wrote of the prohibitively restrictive quality of the objectivist view of probability. He argued that probability should be interpreted as a degree of belief. P. S. de Laplace later (ca. 1825) argued that probability is really an expression of ignorance and is applicable in everyday life. A. De Morgan (ca. 1847) reinforced these arguments with his own. J. M. Keynes, H. Jeffreys, and F. P. Ramsey also viewed probability as a degree of belief. B. De Finetti, L. J. Savage (deceased 1978), and H. Raiffa are contemporary champions of this view.

According to this interpretation there is no such thing as a true probability. Instead, upon consideration of all the relevant evidence or information, one should assign a degree of belief P that a statement a is true. The only restriction is that one must abide by the rules of the calculus of probability. For example, in the coin-tossing experiment one may look at the coin; make some symmetry measurements; consider the kinds of conditions under which the coin will be tossed; anticipate possible deformations of the coin during experimentation; and conclude that the conditions of the previous 1000 tosses only approximate the conditions for the next toss, and that under the new conditions the likelihood of obtaining a head is 48 percent instead of 45 percent.

Clearly this degree of belief interpretation permits a wider range of uses of the concept of probability. It is applicable to all projective or decision situations. It provides the foundation for inductive reasoning, which proceeds by assigning probabilities (called *a priori probabilities*) as measures of the strength of simple or conditional statements or arguments.

2.2 STATEMENTS AND TRUTH

A statement is a sentence which makes a factual claim. The sentence, "The apple is red," is a statement. Complex statements are formed by combining simple statements with *logical connectives* "or," "and," and "not." Thus, the complex statement, "Joe is tall and fat," is composed of two simple statements, "Joe is tall," and, "Joe is fat," using the logical connective, called a *conjuction*.

It is convenient to write statements in symbolic form. Simple statements are denoted by lowercase letters such as a, c, d, and so forth. The logical connectives are denoted by:

or—requires the truth of *at least one*
&—requires the truth of *both*
$\bar{a}$—*denies* the truth of a

The use of parentheses in symbolic representations of complex statements is essential for ensuring clarity of meaning. For example, a & b or c is confusing; it might mean $(a$ & $b)$ or c or it might mean a & $(b$ or $c)$, which is quite different.

Example 2.1: Conjuction "All public officials are highly skilled, cost conscious, guardians of the public good," is a conjunction of three simple statements in the form a & b & c. If one component is false, the whole statement is false.

Statements may be *self-contradictory*, meaning inherently false, or they may be *tautological*, meaning inherently true. Statements which are neither are called *contingency statements*, because their truth is contingent on the truth of their components. The compound statement, "The book is entirely mine and it is entirely yours," is obviously a self-contradiction. The statement, "The project will pay for itself or it will not," is a tautology. The most interesting statements are contingencies such as, "The new road is necessary," or, "The drug will cure the patient." The truth of such statements is not obvious and may never be known for certain. Our whole purpose in gathering evidence is to estimate the likelihood that they are true.

Pairs of statements can be *logically equivalent*, which means that they make identical truth claims. Alternatively, statements may be *mutually exclusive* or inconsistent. In this case, the truth of one precludes the truth of the other. The statements, "The old firehouse is better than the new one," and, "The new firehouse is not as good as the old one," are logically equivalent. On the other hand, the two statements "The new firehouse is better than the old one," and, "The old firehouse is better than the new one," are mutually exclusive.* *Complementary* statements are by definition mutually exclusive. If one is true, the other must be false.

Truth tables are handy devices for observing the influence of simple statements on the truth of complex statements. They can be helpful in identifying logically equivalent or mutually exclusive statements. The next example demonstrates their use.

Example 2.2: Truth tables Consider any two simple statements a and b. Numerous compound statements can be formed from these. Some are indicated as column headings on the truth table below.

a	b	$(a$ or $b)$	$(\bar{a}$ & $\bar{b})$	$(a$ & $b)$	$(a$ or $b)$ or $(\bar{a}$ & $\bar{b})$	$(a$ or $b)$ & b
T	T	T	F	T	T	T
T	F	T	F	F	T	F
F	T	T	F	F	T	T
F	F	F	T	F	T	F

Regardless of the truth of its component statements, $[(a$ or $b)$ or $(\bar{a}$ & $\bar{b})]$ is true so it is a tautology. The two statements $(a$ or $b)$ and $(\bar{a}$ & $\bar{b})$ are complementary, since whenever one is true, the other is false. The statements b and $[(a$ or $b)$ & $b]$ are logically equivalent because one is true only when the other is true.

The truth of one statement may give an indication of the truth of another. There are two fundamental concepts here. We say that a statement *a is sufficient evidence for b* if whenever a is true, b is true. The significance of this concept is

---

* Of course, people may hold different opinions depending on how they interpret the situation. The idea here is that if in some specified way one firehouse is better than the other, it cannot at the same time be worse on the same measure.

Table 2.2 Necessary and sufficient evidence

Result 1

Statement a is sufficient for b if and only if b is necessary for a. The proof is obvious by substitution in the definitions.

Result 2

If a is sufficient for b, then $\bar{b}$ is sufficient for $\bar{a}$. This is clear, for the first part of the statement requires the truth of b whenever a is true. If b is not true (that is, $\bar{b}$ is true), then a cannot be true (that is, $\bar{a}$ is true). The argument works in the other direction as well.

Result 3

If a is necessary for b, then $\bar{b}$ is necessary for $\bar{a}$. This is true, because a is true whenever b is true, which means that when a is not true (that is, $\bar{a}$ is true), b cannot be true (that is, $\bar{b}$ is true). This argument goes both ways, so the proof is complete.

Result 4

If a is sufficient for b, then $\bar{a}$ is necessary for b. This is clear from Results 1 and 3. That is, whenever b is not true or $\bar{b}$ is true, $\bar{a}$ is true, which is another way of saying that $\bar{a}$ is necessary for $\bar{b}$.

Result 5

If a is necessary for b, then $\bar{a}$ is sufficient for $\bar{b}$. This, too, follows from Results 1 and 3.

clear. One may not be able to show directly that b is true, because there is no test for b; yet it may be that the means for verifying the truth of a are available. If a is sufficient for b, one simply proves the truth of b by verifying the truth of a. *The truth of a guarantees the truth of b.* For example, blood analyses are useful in early diagnoses, because finding certain compounds in the blood may indicate the presence of certain diseases even though symptoms are not yet manifest. If a compound is present in the blood only when a certain disease is present, observing it is sufficient for a diagnosis of the disease.

We say that a statement c *is necessary for d* if whenever d is true, c is true. If c is not true, d cannot be true. However, *the truth of c does not guarantee the truth of d*. For example, a particular disease always produces certain compounds in the blood. But these compounds sometimes occur in the blood for other reasons as well. If the compounds are not found in the blood (assuming the test is infallible), the disease cannot be present. However, if they are found, it is only reasonably likely that the patient has the disease; the compounds may have been produced by another mechanism.

Some important results which follow from these two fundamental concepts are listed in Table 2.2.

2.3 RULES OF PROBABILITY

For any given situation there is a family of mutually exclusive, collectively exhaustive relevant statements. There are also irrelevent statements. Statements are relevant or not, depending on their relation to the questions of interest in the context of the problem.

In the coin-tossing example there are two elementary disjoint statements: "The coin will land with heads showing," and, "The coin will not land with heads showing." From these the other two relevant statements may be constructed: "The coin will land with heads showing or it will land with heads not showing," a tautology; and, "The coin will land with heads showing and with heads not showing," a self-contradiction.

If one were interested in the outcomes of a certain treatment regimen for a specified condition, one would define a statement for each outcome or family of possible outcomes so that all relevant possibilities could be accounted for by appropriate combinations. For example, one might describe the range of all possible outcomes as:

$c \sim$ total cure
$t \sim$ up to 10% partial paralysis of legs
$f \sim 10\% <$ paralysis of legs $\leq 50\%$
$n \sim 50\% <$ paralysis of legs $\leq 90\%$
$h \sim 90\% <$ paralysis of legs $\leq 100\%$
$d \sim$ death

A tautological statement would be (c or t or f or n or h or d). Two self-contradictory statements would be (c & t) and (c & d).

In decision-making situations, one would like to know the likelihoods of truth for each relevant statement. By convention, tautologies are assigned likelihood values of one, the maximum possible value, because they are always true. This corresponds to the conventional meaning of "100 percent of the time." In the other extreme, self-contradictory statements are categorically false and are assigned the minimum possible likelihood value, namely, zero. All other statements are contingent and have likelihood values between zero and one.

So far, three rules of probability have been stated: tautologies have likelihood or probability values equal to one, self-contradictions have likelihood values equal to zero, and all other statements (contingency statements) have likelihood values between zero and one. They can be written compactly in symbolic form by defining $P(a)$ to mean, "the likelihood that a is true," and using the symbol s for all tautologies and ϕ for all self-contradictions. Thus, the first three rules of the probability calculus are

Rule 1: $P(s) = 1$; tautologies (must be true)
Rule 2: $P(\phi) = 0$; self-contradictions (must be false)
Rule 3: $0 \leq P(a) \leq 1$; contingencies

Example 2.3: Initial probability assignments New hospital admission procedures should have an impact on hospital income. Roughly stated, they will reduce income r, increase income a, or have no measurable effect on hospital income n. The likelihoods of these impacts are unknown, except that they are between zero and one. It is certain that at least one of these statements will come true and there is no chance that any pair of them or all of them will be true simultaneously. Thus,

$$P(r \text{ or } a \text{ or } n) = P(s) = 1$$

$$P(r \& a \& n) = P(\phi) = P(r \& a) = \text{etc.} = 0$$

$$0 \le P(r) \le 1$$

$$0 \le P(n) \le 1$$

$$0 \le P(a) \le 1$$

The fourth rule in the calculus of probability is called the *disjunctive rule*.* It is, in fact, an axiom. It applies to mutually exclusive statements only. For two statements which cannot be true simultaneously, the likelihood that at least one of them is true is the sum of their individual likelihoods.

Rule 4: $P(a \text{ or } b) = P(a) + P(b)$, when a and b are mutually exclusive

The fifth rule is an extension of the fourth. It is a general disjunctive rule taking into account situations where the two statements are not mutually exclusive, as well as those in which they are. According to the fifth rule, the likelihood that at least one of a and b is true is the likelihood that a is true plus the likelihood that b is true minus the likelihood that both are true simultaneously.

Rule 5: $P(a \text{ or } b) = P(a) + P(b) - P(a \& b)$

Because a and b are not mutually exclusive, they may be true simultaneously. Therefore, the statement $(a \text{ or } b)$ is true whenever a alone is true, whenever b alone is true, and whenever a and b are simultaneously true. Applying the fourth rule, this means that

$$P(a \text{ or } b) = P(a \& \bar{b}) + P(\bar{a} \& b) + P(a \& b)$$

Furthermore, a may be true alone or in combination with b, and b may be true alone or in combination with a. Again applying the fourth rule,

$$P(a) = P(a \& b) + P(a \& \bar{b})$$

and $\qquad\qquad P(b) = P(a \& b) + P(\bar{a} \& b)$

Clearly, the sum $P(a) + P(b)$ overestimates $P(a \text{ or } b)$ by an amount that is equal to $P(a \& b)$. This accounts for the third term in Rule 5.

Example 2.4: Rule 4† Suppose that the likelihoods of a six-sided die landing

---

* The disjunctive rule is also called the *addition rule* because probabilities are added.

† Note that this example alludes to outcomes. This is often more efficient than referring to statements such as, "The die will land with the single-dot side facing up." Since one form is equivalent to the other in a practical sense, convenience and clarity should be one's guides.

with a specified side up are $P(1) = P(2) = \cdots = P(6) = \frac{1}{6}$. The likelihood of its landing with either a 2 or a 3 up is clearly $\frac{2}{6}$, since

$$P(2 \text{ or } 3) = P(2) + P(3) = \frac{1}{6} + \frac{1}{6} = \frac{2}{6}$$

This is so because the outcomes are mutually exclusive.

Example 2.5: Rule 5* Suppose the same die were tossed twice. What would be the likelihood of a 3 occurring on at least one of the tosses? Using subscripts to identify the particular toss, the question in mathematical notation is $P(3_1 \text{ or } 3_2) = ?$ Now there are three mutually exclusive ways to obtain at least one 3 in two tosses. These are: a 3 on the first toss only (3_1 & $\bar{3}_2$); a 3 on the second toss only ($\bar{3}_1$ & 3_2); and 3's on both tosses (3_1 & 3_2). Since the two tosses are independent and the likelihood of a 3 on any toss is $\frac{1}{6}$ while the likelihood of something other than a 3 is $\frac{5}{6}$, the likelihood of at least one 3 is

$$P(3_1 \text{ or } 3_2) = P(3_1 \text{ & } \bar{3}_2) + P(\bar{3}_1 \text{ & } 3_2) + P(3_1 \text{ & } 3_2)$$

$$= \frac{1}{6} \cdot \frac{5}{6} + \frac{5}{6} \cdot \frac{1}{6} + \frac{1}{6} \cdot \frac{1}{6} = \frac{11}{36}$$

Notice, however, that $P(3_1) + P(3_2) = \frac{12}{36}$; the discrepancy arises because this approach counts the outcome (3_1 & 3_2) twice, which becomes obvious when one itemizes all possible outcomes and counts the possibilities.

It is important to know how the likelihoods of a statement, and its denial, are related. Clearly b and $\bar{b}$ are mutually exclusive. Thus $P(b \text{ or } \bar{b}) = P(b) + P(\bar{b})$. Additionally, $(b \text{ or } \bar{b})$ is a tautology, so $P(b \text{ or } \bar{b}) = 1$. Thus, Rule 6 states that for any statement b, the likelihood of $\bar{b}$ is 1 minus the likelihood of b.

Rule 6: $P(\bar{b}) = 1 - P(b)$

The remaining rules depend on the introduction of two additional concepts, namely, independence and conditioning. *Independence*, or, to use the proper term, *stochastic independence*, has a very special meaning in probability. The idea is that statement a is independent of statement b, if knowledge of the truth of b provides no information regarding the truth of a. That is, the likelihood that a is true is, say, $P(a)$, regardless of whether b is true or false. The truth of the statement, "The sky is blue today," is in no way influenced by the truth of the statement, "This ink is wet." Similarly, the outcome of the first toss in Example 2.6 had no bearing on the outcome of the second. The first and second tosses were independent, as evidenced by the fact that the likelihood of any pair of outcomes was always the product of their individual likelihoods; for example, $P(2 \text{ & } 3) = P(2)\,P(3) = \frac{1}{36}$.

---

* Take note that $P(3_1 \text{ or } 3_2)$ is identical to $P(3_2 \text{ or } 3_1)$. That is, the order of tosses is inconsequential. This is an important aspect of conjunctive statements.

Whenever two statements are independent, the likelihood that their conjunction is true is the product of their individual likelihoods. This rule defines stochastic independence;

Rule 7: $P(a \& b) = P(a) P(b)$; if a and b are independent

In tossing a die once, the truth of the statement, "An even-numbered side will land face up," depends very much on the truth of the statement, "Side 4 will land face up." If 4 comes up, the outcome is certainly even. If 4 is not obtained, the outcome may or may not be even. For any toss, there are six equally likely outcomes: 1, 2, 3, 4, 5, 6. Among these, there are three which are "even." Thus $P(\text{even}) = \frac{1}{2}$. However, if one learned that the toss did not or would not produce a 4, there would be only two possible "even" outcomes, namely, 2 and 6. Thus the probability that the outcome will be "even" if it is not a 4, is $\frac{2}{5}$.

The idea of *conditioning* is usually phrased, "Given a, what is the likelihood of b?" and is written, $P(b/a)$. The rule for calculating conditional probabilities is

Rule 8: $P(a/b) = P(a \& b)/P(b)$ or $P(a \& b) = P(a/b) P(b)$

A very special case of this rule holds when a and b are independent. Then, in fact, $P(a \& b) = P(a)P(b)$, so $P(a/b) = P(a)$, which is another way of stating that a is independent of b.

The following observation is extremely important. *Two statements cannot be both independent and mutually exclusive.* If statements a and b were mutually exclusive, the truth of one would prove the falsity of the other. If they were independent, the truth of one would in no way reflect on the truth of the other. More precisely, suppose $P(a) \neq 0$ and $P(b) \neq 0$, to avoid trivialities, and suppose a and b were independent. Then $P(a \& b) = P(a)P(b)$. But $P(a/b) = P(a \& b)/P(b)$. By assumption, then, $P(a/b) = P(a)P(b)/P(b) = P(a)$. Now suppose, instead, that a and b were mutually exclusive. Then $P(a/b) = 0$, which contradicts the previous result, namely, that $P(a/b) = P(a) \neq 0$. This contradiction arose directly from the simultaneous assumptions of independence and mutual exclusivity.

The next rule, called the *multiplicative rule*, defines the procedure for calculating likelihoods for multiple conjunctions.

Rule 9: $P(a \& b \& c \& \cdots \& n \& m) = P(a) P(b/a) P(c/a \& b) \cdots P(m/a \& b \& \cdots \& n)$

This procedure follows from repeated application of the rule for conditional statements. For three statements, the steps are

$$P(a \& b \& c) = P((a \& b) \& c) = P(c/a \& b) P(a \& b)$$

$$= P(c/a \& b) P(b/a) P(a)$$

Further extensions follow suit.

Example 2.6 Application of the rules to data Incidence rates for heart attacks vary among people according to sex and socioeconomic status. Suppose the rates for a randomly selected sample of people are as displayed in the table below.

	Incidences		
	Male	Female	Total
Socioeconomic group 1	3,000	4,000	7,000
Socioeconomic group 2	2,000	3,000	5,000
Total	5,000	7,000	12,000

For a randomly* selected individual from this group, let a be the statement, "The person is female," and let b be the statement, "The person is from group 1." Then,

a. $P(a \text{ or } b \text{ or } \bar{a} \text{ or } \bar{b}) = 1$; tautology
b. $P(a \ \& \ b \ \& \ \bar{a}) = P(\phi) = 0$; self-contradiction
c. $P(a) = 7/12$, $P(b) = 7/12$
d. $P(a \ \& \ b) = 4/12$
e. $P(a/b) = P(a \ \& \ b)/P(b) = 4/7$
f. $P(a \text{ or } b) = P(a) + P(b) - P(a \ \& \ b) = 10/12$
$\qquad = P(a \ \& \ \bar{b}) + P(\bar{a} \ \& \ b) + P(a \ \& \ b)$
$\qquad = 3/12 + 3/12 + 4/12 = 10/12$

The final rule is called *Bayes' rule* or *Bayes' theorem* after Reverend Thomas Bayes, who formulated it in the late eighteenth century. This is the most important rule of all in the context of making decisions involving uncertainty. With Bayes' theorem, one can improve initial probability estimates using new evidence. It is the most interesting rule, from a purely conceptual point of view, as well.

Suppose a physician were diagnosing a patient's illness. The first information available to him would be past history and certain clinical indications observed in his initial examination. He might conclude on the basis of this information that the patient very likely had one of two mutually exclusive diseases, a or b. His probabilistic estimates for the strength of this judgment might be $P(a) = 0.4$ and $P(b) = 0.6$ where, by assumption, $P(a \ \& \ b) = 0$; these are called *prior likelihoods*.

To improve his judgment, he might order a certain laboratory test which could yield positive or negative results and for which he knows the likelihoods

* "Random," as used in this book, means that selection is done in a manner which guarantees that each member of the population has an equal chance of being chosen.

of positive readings when a patient has either disease. These might be $P(+/a) = 0.8$ and $P(+/b) = 0.4$. Suppose the laboratory test indicated positive. Then his revised probability estimates should include the new information as well as the old. In particular, they should depend on both his initial estimates and the evidence offered by the test. Thomas Bayes showed that the revised likelihoods that the patient has either disease are related to the physician's original judgment and the test information, by:

$$P(a/+) = \left[\frac{P(+/a)}{P(+/a)\,P(a) + P(+/b)\,P(b)}\right] \cdot P(a)$$

$$P(b/+) = \left[\frac{P(+/b)}{P(+/a)\,P(a) + P(+/b)\,P(b)}\right] \cdot P(b)$$

Thus the original estimate $P(a) = 0.4$ becomes $P(a/+) \simeq 0.57$, for a positive test result. This is counter to the physician's original judgment, because it supports a diagnosis of a instead of b, which the physician initially favored.

These equations for $P(a/+)$ and $P(b/+)$ are commonly called *posterior likelihoods*, to indicate that they are based not only on the preliminary information but also on new information. The development of these equations is quite straightforward. The rules for conditional probability, Rule 8, and disjunction, Rule 4, are all that are needed. The process is shown here because it is very instructive and demonstrates useful manipulations of the rules.

From the rule for conditional probability, the likelihood that the patient has disease a if the test results are positive is $P(a/+) = P(a\ \&\ +)/P(+)$. Since in this case the test may give positive results for either disease, the likelihood of a positive test result for the patient is the likelihood that the patient has disease a *and* the test result is positive *plus* the likelihood that he or she has disease b *and* the test result is positive, namely, $P(+) = P(+\ \&\ a) + P(+\ \&\ b)$. Now by invoking the conditional probability rule to rewrite each of the terms $P(+\ \&\ a)$ and $P(+\ \&\ b)$ and substituting this form of the expression for $P(+)$, the final form of Bayes' rule for this example is obtained.

The general form of Bayes' rule for any finite number of mutually exclusive statements is formulated as Rule 10.

Rule 10: Bayes' rule. Statements $a_1, a_2, \ldots, a_n$ are mutually exclusive and collectively exhaust all possibilities; that is, $(a_1$ or a_2 or $\cdots$ or $a_n)$ is a tautology. Suppose also that b is a relevant statement, so that there is at least one a_i which can be true simultaneously with b. Then,

$$P(a_i/b) = \frac{P(b/a_i)\,P(a_i)}{P(b/a_1)\,P(a_1) + \cdots + P(b/a_n)\,P(a_n)}$$

for any $i = 1, \ldots, n$.

All 10 rules are summarized in Table 2.3 for convenience.

Table 2.3 Summary of the rules of probability
Inductive reasoning

Rule 1: $P(s) = 1$; s is a tautology (something is true)

Rule 2: $P(\phi) = 0$; ϕ is a self-contradiction (nothing is true)

Rule 3: $0 \leq P(a) \leq 1$; a is any contingent statement (might be true)

Rule 4: $P(a \text{ or } b) = P(a) + P(b)$; if a and b are mutually exclusive

Rule 5: $P(a \text{ or } b) = P(a) + P(b) - P(a \,\&\, b)$; always

Rule 6: $P(\bar{b}) = 1 - P(b)$; any statement, b

Rule 7: $P(a \,\&\, b) = P(a) P(b)$; if a and b are independent

Rule 8: $P(a \,\&\, b) = P(a/b) P(b)$; always

Rule 9: $P(a \,\&\, b \cdots \,\&\, m) \; P(a) P(b/a) P(c/a \,\&\, b) \cdots P(m/a \,\&\, b \,\&\, \cdots)$; always

Rule 10: $P(a_i/b) = \dfrac{P(b/a_i) \, P(a_i)}{P(b/a_1) \, P(a_1) + P(b/a_2) \, P(a_2) + \cdots + P(b/a_n) \, P(a_n)}$;

$a_1, \ldots, a_n$ are mutually exclusive and exhaustive and b is relevent

---

2.4 ILLUSTRATIVE APPLICATIONS

Illustration 1: Conditional probabilities and Bayes' rule Consider this erroneous argument by a radiologist.

When a malignant lesion is present, this x-ray procedure will give a positive indication 80 percent of the time. That is, it will not "see the lesion" 20 percent of the time. Therefore, if on the basis of a negative x-ray we do not proceed to exploratory surgery, there is a 20 percent chance that we are missing a malignancy.

Letting m symbolize, "The lesion is malignant," and $+$ symbolize the statement, "The x-ray indicates the presence of a malignancy," the radiologist's argument translates to:

If

$$P(+/m) = 0.8 \qquad 80\% \text{ are detected}$$

and

$$P(-/m) = 0.2 \qquad 20\% \text{ are not detected}$$

then

$$P(m/-) = 0.2 \qquad 20\% \text{ of the negative results are malignancies}$$

The radiologist has argued that $P(-/m) = P(m/-)$, which is not generally true. Its truth would be accidental. The number $P(-/m)$ is a calibration probability for the procedure. It is determined by starting with a population of persons known to have malignant lesions from biopsy results, applying the x-ray procedure to each and taking the ratio of negative results to the number of trials. On the other hand, $P(m/-)$ cannot be determined experimentally. It must be calculated, using Bayes' rule, from evidence of the presence of a malignancy obtained prior to the x-ray procedure and from the calibration probabilities for the x-ray procedure, and can be very different from $P(-/m)$.

According to Bayes' rule,

$$P(m/-) = \frac{P(-/m)\,P(m)}{P(-/m)\,P(m) + P(-/\bar{m})\,P(\bar{m})}$$

The radiologist reported that $P(-/m) = 0.2$, but did not indicate the likelihood of a negative result when no malignancy is present; $P(-/\bar{m})$. If the other calibration probability $P(-/\bar{m})$ were 0.94, indicating a very selective test, one would find $P(m/-) = 0.176$. This is very close to 0.2, so the radiologist's conclusion would have been nearly correct. However, if $P(-/\bar{m})$ were 0.2, an unselective test, one would find $P(m/-) = 0.5$; this is identical to the prior likelihood, $P(m) = 0.5$, reflecting the fact that the test has no discriminating power [because $P(-/m) = P(-/\bar{m})$]. In this case $P(m/-)$ differs from $P(-/m)$ by a great deal, rendering the radiologist's conclusion wrong.

Do not confuse $P(a/b)$ and $P(b/a)$.

Illustration 2: Judgmental probability* Imagine that in 1976 a United States government official proclaimed, "There is a *very real possibility* of an epidemic of swine flu in the 1976-77 flu season. Therefore, I have asked the Congress to appropriate funds to contract with the pharmaceutical industry for the production and delivery of 200 million doses of a swine flu vaccine. The federal government will distribute the vaccine through standard outlets at no charge. The total cost of this program is expected to be 135 million dollars." How could one judge the reasonableness of this proclamation?

Suppose a number of experts were polled on the likelihood of an epidemic, and their estimates were 5, 25, and 50 percent. Are these estimates and the official statement consistent?

The first problem is to estimate the likelihood strength of the phrase, "very real possibility."† A group of students judged the strength of "possibility" at about 25 percent. The phrase, "very real," they felt increased the strength by about 50 percent. That is, they judged, "very real possibility," to have a likelihood strength of $1.5 \times 25\% = 37.5\%$.

If all the experts were judged to be equally credible, their opinions could be averaged by adding the percentages and dividing by 3, to get 26.7 percent. Thus, the official statement would seem quite inaccurate to the students, under the assumption that the experts were equally credible. On the other hand, if one judged the third expert's opinion to be 3 times as credible as the first, and the

---

* This illustration is based on an article by Philip Boffey in the May 14, 1976 issue of *Science* and a letter by Frederick Mosteller in *Science*, vol. 192.

† Mosteller referenced two pieces of work in this regard: J. Selvidge, "Assigning Probabilities to Rare Events," thesis, Harvard University, 1972; and N. Cliff, *Psychological Review*, vol. 66, 1959, p. 27.

second expert's opinion to be twice as credible as the first, one would obtain a different weighted average, namely,

$$\left(5 \times \frac{1}{6} + 25 \times \frac{2}{6} + 50 \times \frac{3}{6}\right) = 34\%$$

In this case the official statement would seem reasonable to the students.

Clearly, the credibility of the official pronouncement depends very much on what the official means by the phrase, "very real possibility," and on one's judgment regarding the credibility of the individual experts. There is no universal answer, but one can certainly evaluate the statement on a personal basis.

Verbal arguments can sometimes be evaluated quantitatively.

Illustration 3: Reliability Suppose an engineer has designed a production system in which components of an object are produced by separate machines and combined by a third. He has run each machine separately and estimated the likelihood of a breakdown in a workday. His numbers are $P(a) = 0.2$, $P(b) = 0.1$, and $P(c) = 0.2$. He relayed this information to his boss, who quickly observed that $(0.1)(0.2)(0.2) = 0.004$, which he considered the likelihood of a breakdown in the system. Should he move ahead and install the system based on this small number?

He should not. If any one of the machines fails, the system will fail. The likelihood that $(a$ or b or $c)$ is true can be written two ways:

$$P(a \text{ or } b \text{ or } c) = P(a) + P(b) + P(c) - P(a \ \& \ b) - P(a \ \& \ c) - P(b \ \& \ c)$$
$$+ P(a \ \& \ b \ \& \ c) = (0.5) - 0.02 - 0.02 - 0.04 + 0.004 = 0.424$$

$$P(a \text{ or } b \text{ or } c) = 1 - P(\bar{a} \ \& \ \bar{b} \ \& \ \bar{c}) = 1 - (0.9)(0.8)(0.8) = 0.424$$

This is an unreliable system.

If the engineer had two of these systems operating side by side, the likelihood of total production line failure in one workday would be $(0.424)(0.424) \simeq 0.18$, a considerably more reliable system.

Do not confuse conjunction and disjunction.

Illustration 4: Revising initial likelihood estimates with sample data Suppose you are a physician diagnosing a patient's illness. You feel quite confident (95 percent certain) that the patient has disease d. There are two forms of the disease, form a and form b, which can occur individually or coincidentally, and you are uncertain which form your patient has. In your judgment forms a and b are equally likely and, from historical statistics, the likelihood that both forms are present is about 3 percent. The problem is that treatment is risky, because if form a is treated as if it were form b, or vice versa, the patient could suffer serious adverse effects. On the other hand, failure to treat d also has serious implications.

Because of your uncertainty you order a differential diagnostic test for the patient. This test is good at separating cases where both forms a and b are present, from those where only one form is present. There is an 88 percent chance that both forms are present if the test results are positive, and only a 1 percent chance that only one form is present. If the test results were positive, should you feel that your original diagnosis was confirmed? Let us see.

Prior to the test your likelihood estimates were $P(a \text{ or } b) = 0.95$, $P(a \& b) = 0.03$, and $P(a) = P(b)$. But, $P(a \text{ or } b) = P(a) + P(b) - P(a \& b)$. Therefore, $P(a) = P(b) = 0.49$. Furthermore, $P(a \& b/+) = 0.88$ and $P(a \& \bar{b}/+) = P(\bar{a} \& b/+) = 0.01$. Since $P(b/+) = P(a/+) = P(a \& \bar{b}/+) + P(a \& b/+) = 0.89$, $P(a \text{ or } b/+) = 0.89 + 0.89 - 0.88 = 0.9$. Now compare these results.

Prior likelihood		Likelihood, if positive test
0.49	—has a—	0.89
0.49	—has b—	0.89
0.03	—has a and b—	0.88
0.95	—has disease—	0.90

A seemingly strange thing has occurred and it is due to the distinction between deductive and inductive reasoning. It would seem that a positive result should tend to confirm the diagnosis of the disease. A positive result does suggest that both forms are present and does tend to confirm the diagnosis that a is present and that b is present. Yet, a positive test result tends to disconfirm the presence of the disease at all. This occurs because a positive result cannot be interpreted as proof that both forms are present. Therefore, deductive reasoning is not applicable, as shown below.

A positive result occurs only when both forms are present.
The result is positive.
Therefore, both forms are present.
Whenever both forms are present, the disease is present.
Both forms are present.
Therefore, the disease is present.

This form of reasoning would appear to be valid if the test were good enough, however. In this example, if $P(a \& b/+)$ were at least 0.93, the diagnosis would appear to be confirmed by positive test results.

Deductively: a **implies** b **implies** c.
Inductively: Evidence supporting a **may support** b, **yet cast doubt on** c.

2.5 PROBLEMS

2.1 Which of the following pairs of statements are logically equivalent? Which are mutually exclusive? You may use truth tables.

(a) a, $\bar{a}$

(b) $(a \ \& \ \bar{b})$, $\overline{(a \ \& \ b)}$

(c) $(a \text{ or } b) \ \& \ b$, $(a \ \& \ \bar{b})$

(d) $(a \text{ or } a) \ \& \ b$, $a \ \& \ (b \text{ or } b)$

2.2 Identify those pairs of statements in Problem 2.1 which have identical probability values and those which have complementary values. Make the best statement you can about the relative probabilities in the remaining pairs.

2.3 Construct truth tables for each of the following and identify their logically equivalent or mutually exclusive propositions in the tables:

(a) $(a \text{ or } b) \ \& \ (\bar{a} \text{ or } \bar{b})$

(b) $(a \ \& \ b) \text{ or } (\bar{a} \ \& \ b)$

(c) $\overline{(a \ \& \ b)}$

(d) $\overline{(a \text{ or } b)}$

(e) $a \text{ or } b \text{ or } c$

(f) $(a \text{ or } \bar{b}) \ \& \ c$

(g) $a \ \& \ b \ \& \ c$

(h) $\overline{(a \text{ or } b \text{ or } c)}$

(i) $[\overline{(a \ \& \ b) \text{ or } c}]$

(j) $a \ \& \ b \text{ or } c$

2.4 Using truth tables find which of the following are tautologies, self-contradictions, and contingent statements:

(a) $\bar{a} \text{ or } a$

(b) $(a \text{ or } b) \text{ or } (c \text{ or } d) \text{ or } (a \text{ or } d)$

(c) $(a \text{ or } \bar{b}) \text{ or } \overline{(a \text{ or } \bar{b})}$

(d) $\bar{a} \text{ or } a \text{ or } b$

2.5 Show that $[\overline{(a \text{ or } b) \ \& \ c}] \sim [(\bar{a} \ \& \ \bar{b}) \text{ or } \bar{c}]$.

2.6 Let a be independent of b.

(a) Show that b is independent of a.

(b) Show that $\bar{a}$ and b are independent.

(c) Argue for or against the independence of $\bar{a}$ and $\bar{b}$.

2.7 Given that $P(a) \neq 0$, $P(b) \neq 0$, $P \neq 0$, and $P(a \ \& \ b) = P(a) \ P(b)$, $P(a \ \& \ c) = P(a) \ P(c)$, $P(b \ \& \ c) = P(b) \ P(c)$, determine, for each of the following entities, whether it must be true, might be true, or cannot be true:

(a) $P(a \ \& \ b \ \& \ c) = P(a) \ P(b) \ P(c)$

(b) $P(b/a) = P(b/c)$

(c) $P[(a \ \& \ b)/c] = P(a/c) \ P(b/c)$

(d) $P(a \text{ or } b \text{ or } c) < P(a) + P(b) + P(c)$

2.8 Find two argumentative articles in newspapers, newsmagazines, semiprofessional or trade magazines, or publications such as *Science*, *Scientific American*, and *American Scientist*. Analyze the consistency and strength of the arguments, using the rules of inductive reasoning as they were developed in this chapter.

2.9 The color of the Martian surface could be red, with a probability of 0.6, or brown, with a probability of 0.7, and neither, with a probability of 0.25.

(a) What is the probability that it is purely red or purely brown?

(b) What is the probability that it is brown given that it is not red?

2.10 Show that the frequencies for true and false in a truth table do not represent probabilities of true and false. Consider explicitly:

(a) a, b mutually exclusive

(b) a, b exhaustive

2.11 The Society for the Environment is considering suing the Department of the Interior. The directors are trying to establish the odds of winning or losing. Years of experience have shown that the presiding judge is the prime concern.

There are only two judges who could preside over this case: J_1, the most federally biased judge, and J_2, the fairer judge. The society could field one of three lawyers: L_1, their sharpest one, L_2, their not-so-sharp lawyer, and L_3, their least-sharp lawyer.

After careful consideration the directors have made plausible assumptions: Approach L_1 only if J_1 presides; if J_1 presides, there is a $\frac{1}{3}$ chance that they can get L_1 on the case. Because of the judge selection system, there is an even (50–50) chance that J_1 or J_2 will preside, regardless of what the society does. If L_1 does not take the case, L_2 will take the case with a likelihood of $\frac{3}{4}$, no matter what else occurs. The probability that the society will lose if J_1 presides is $\frac{1}{2}$ if L_1 takes the case, $\frac{2}{3}$ if L_2 takes the case, and $\frac{3}{4}$ if L_3 takes the case. The probabilities that the society will win given that J_2 presides are, respectively, 1, $\frac{1}{2}$, and $\frac{1}{3}$ if L_1, L_2, or L_3 takes the case.

(a) What is the likelihood (probability) that J_2 will preside?
(b) What is the likelihood that the society will lose?
(c) If L_2 does not fight the case, what is the likelihood that the society will lose?
(d) If J_2 and L_2 face off, what is the probability (likelihood) that the society will win?

2.12 A politician took a windshield survey of Volkswagens on the highway during her trip from St. Louis to Kansas City along Route 70. She counted 100 Volkswagens going east and only 10 going west. Explain.

2.13 The table below summarizes epidemiological data from a study on a particular disease. Assume it represents the reality of the situation for all people. The categories are mutually exclusive.

(a) Check these data for independence in both dimensions.
(b) Calculate the conditional probabilities: $P(d/a)$; $P(d/s)$; $P(d/n)$; $P(d/b)$.
(c) What is the probability that a person will contract the disease?
(d) What is the probability that a person's parent will contract the disease?

Family incidence data for disease d

	Individual	Total
Parent only, a	25	525
Sibling only, s	30	380
Neither, n	20	720
Both, b	10	130
Total	85	1775

2.14 In environmental health and social policy problems one would like to modify certain factors and observe the consequences under controlled conditions. This is called a *prospective study*. More often than not one must work with retrospective data. Bayes' rule permits one to extract similar conclusions in either case.

For example, let f = causal factor and d = disease. In a prospective study one starts with a disease-free population, introduces a portion of them to the causal factor, and observes the rates of disease among the two groups. If the causal factor is powerful, the prospective likelihood ratio

$$lp = \frac{P(d/f)\,P(\bar{d}/\bar{f})}{P(\bar{d}/f)\,P(d/\bar{f})}$$

should be large. In a retrospective study, one starts with populations of diseased and nondiseased individuals, some of whom have been affected by the causal factor. To evaluate the strength of the relation between d and f, one may calculate a retrospective likelihood ratio

$$lr = \frac{P(f/d)\,P(\bar{f}/\bar{d})}{P(\bar{f}/d)\,P(f/\bar{d})}$$

which also reflects the strength of the relation between d and f. Use Bayes' rule to demonstrate that

$$lr = lp$$

2.15 All children with dyslexia have difficulty reading. John's son, like 20 percent of all nondyslexic children in his age group, has difficulty reading. Experts estimate that about 20 percent of all children in the same age group suffer from dyslexia. What are the chances that John's son suffers from dyslexia?

2.6 SUPPLEMENTAL READING

Ayer, Alfred J.: *Language Truth and Logic*, Dover, New York, 1952.

Bayes, Thomas.: "An Essay toward Solving a Problem in the Doctrine of Chance," *Philosophical Trans. of the Royal Society*, 1973. Reprinted in *Biometrica*, vol. 45, 1958.

Fine, Terrance L.: *Theories of Probability*, Academic Press, New York, 1973.

Good, I. J.: "How to Estimate Probabilities," *Journal of the Institute of Mathematics Applications*, no. 2, 1966, pp. 364–383.

Kahane, Howard: *Logic and Contemporary Rhetoric*, Wadsworth, Belmont, Calif., 1971.

Machol, Robert E.: "Principles of Operations Research—The Titanic Coincidence," *Interfaces*, vol. 5, no. 3, May 1975, pp. 53–54.

Salmon, Wesley C.: "Confirmation," *Scientific American*, vol. 228, May 1973, pp. 75–83.

Skyrms, Brian: *Choice and Chance*, Dickenson Publishing Company, Inc., Belmont, Calif, 1966.

Tribe, Laurence H.: "Trial by Mathematics: Precision and Ritual in the Legal Process," *Harvard Law Review*, vol. 84, no. 6, April 1971, pp. 1329–1393.

THREE

STATISTICAL DESCRIPTION

Gathering and summarizing statistical data are common activities in the public sector. Resource allocation decisions, project evaluations, traffic studies, health studies, and other kinds of studies too numerous to name here are based on the summarization of statistical data and the projection of trends. The purposes of this chapter are to explain the basic concepts and numerical measures used in these activities and to lay the foundation for developing more powerful tools of analysis in subsequent chapters.

The topics discussed in this chapter are quite elementary and easily grasped. A few, such as the arithmetic mean and the standard deviation, border on being common knowledge. However, the elementary nature of this chapter in no way diminishes its importance in policy analysis and public decision making; in fact, the data interpretation techniques presented herein may well be the most widely applied of all quantitative methods in real policy studies.

3.1 AVERAGES

It is convenient, informative, and comfortable to describe things "on the average." It is convenient because it is a means of reducing volumes of data into a single measure describing some appropriate characteristic of a group or population. For example, one talks about the average income of physicians, the average weight of a football team, the average or expected outcome of a surgical procedure, or the average temperature during a month or a year. The concept is informative because it provides a kind of homogeneous measure of some characteristic of a population. It is comfortable because it allows one to generalize and, especially in an argument, permits one to proceed without accounting explicitly for uncertainty. Furthermore, people are most comfortable with deterministic reasoning, and the concept of "average" seems to many to meet that need.

The concept of average must be applied with care, however. Not only is uncertainty masked by the use of "on the average," but different situations require the use of different concepts of average, and some situations require the use of more than one concept of average. Four different measures of "average" are: *arithmetic mean, geometric mean, harmonic mean,* and *mode.* A fifth measure, the *median,* is often referred to as an average even though it is not. All five measures are discussed in this section.

Arithmetic Mean ($\bar{x}$)

The arithmetic mean is the statistical average most often employed. It is, in fact, the fundamental concept which leads to the definition of "expected value" in probability (see Chapter 4). The arithmetic mean is frequently referred to as the weighted average or simply the average. Where a variable x has been measured six times under similar conditions and found to have values $x_1, \ldots, x_6$, the average value is the number $\bar{x}$ where

$$\bar{x} = \frac{x_1 + \cdots + x_6}{6}$$

If the values of the variable were found to be 1, 2, 3, 4, 5, and 6, the arithmetic mean would be

$$x = \frac{x_1 + x_2 + x_3 + x_4 + x_5 + x_6}{6} = \frac{21}{6}$$

Example 3.1: Average cost A hospital administrator is bargaining with Blue Cross about the per diem charge to be covered. The administrator has received from his accounting office a table of actual per diem costs to the hospital over the past 10 weeks. These are shown below.

Week	Per diem costs, $	Week	Per diem costs, $
1	123.00	6	135.00
2	131.00	7	117.00
3	120.00	8	121.00
4	111.00	9	119.00
5	140.00	10	125.00

The administrator feels that these are representative of average, weekly per diem costs today and will probably remain so until the next labor contract is negotiated, 1 year hence. He has decided to bargain for a reimbursement rate equal to the average of these values, plus 10 percent to cover inflation. What is he going to bargain for?

$$\bar{x} = \frac{(123 + 131 + 120 + 111 + 140 + 135 + 117 + 121 + 119 + 125)}{10}$$

$$= \frac{1237}{10} = 123.70$$

Rate $= 123.70 + 0.1(123.70) = 136.07$

Geometric Mean ($\bar{x}_g$)

This version of the mean or average is appropriate when the data exhibit a geometric growth pattern, as in interest compounding in a savings account or population growth where the value in the next period of time is proportional to the current value and the rate of change r is fixed. In such situations the starting value, say, C_0, grows (or declines) according to the following pattern:

Year	Value	
0		C_0
1	$C_1 = C_0 + rC_0$	$= C_0(1 + r)$
2	$C_2 = C_0(1 + r) + rC_0(1 + r)$	$= C_0(1 + r)^2$
$\vdots$	$\vdots$	$\vdots$
n	$C_n = \qquad \cdots$	$= C_0(1 + r)^n$

The objective in calculating an average for any data is to obtain a representative value. There is no single representative value for data exhibiting a geometric pattern. In this case one must estimate a value which fits the pattern. For example, a representative value between the zeroth and second year is the value for the first year. One can calculate this value from the other two values as follows:

$$C_1 = \sqrt{C_0 \times C_2} \quad \text{or} \quad C_0(1 + r) = \sqrt{C_0 \times C_0(1 + r)^2}$$

Similarily, one can calculate the value for the second year from values for the zeroth, first, third, and fourth years by taking the fourth root of the product of all four values, as follows:

$$C_2 = C_0(1 + r)^2 = \sqrt[4]{C_0 \times C_0(1 + r) \times C_0(1 + r)^3 \times C_0(1 + r)^4}$$

Notice, too, that here the C_2 value could be calculated from taking the square root of $C_0 \times C_4$, as follows:

$$C_2 = \sqrt[2]{C_0 \times C_0(1 + r)^4}$$

This works because the pattern is truly geometric, with a growth rate of r. However, real data are almost never so nice. Although the pattern may be geometric, real data exhibit sufficient variability that estimation is greatly improved by taking the product of many, as opposed to a few, data points in the averaging process.

Example 3.2: Accelerating costs National health care expenditures for 10 years are given below. There is a pattern to the data, as shown in the diagram. The values double approximately every 2 years. The magnitude of these costs is growing at a geometric rate of 35 to 40 percent per year. To calculate the average cost one must be careful to account for this characteristic.

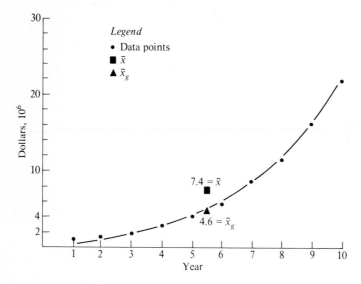

The proper method for averaging these numbers is to calculate the tenth root of the product of the numbers. That is,

$$\bar{x}_g = \sqrt[10]{(1)(1.4) \cdots (21.9)} = 4.72$$

A rough estimate for this number is

$$\bar{x}_g = \sqrt[2]{1\,(21.9)} = 4.68$$

This estimate is fairly good, because the data are nearly geometric with relatively little randomness. On the other hand, the arithmetic mean is $x = 74.09/10 \simeq 7.4$, which is quite different.

Harmonic Mean ($\bar{x}_h$)

The harmonic mean is appropriate for averaging ratios or rates for which the denominator is variable.

Suppose you wished to calculate the average speed of a car, on four legs of a trip. Suppose the speeds on the different legs were

$$x_1 = 4 \qquad x_2 = 5 \qquad x_3 = 8 \qquad x_4 = 10$$

Using the arithmetic mean the average speed would appear to be

$$\bar{x} = \frac{4 + 5 + 8 + 10}{4} = 6.75$$

However, speed is defined as distance divided by time. If the distances are d, $2d$, $4d$, and d, respectively, the times are d/x_1, $2d/x_3$, $4d/x_3$, and d/x_4. Thus the average speed in this case, the harmonic mean, is

$$\bar{x}_h = \frac{8d}{(d/x_1) + (2d/x_2) + (4d/x_3) + (d/x_4)} = 6.4$$

Notice that the harmonic mean is simply the ratio of the total distance $8d$, to the total time. That is,

$$\bar{x}_h = \frac{8d}{[t_1 + t_2 + t_3 + t_4]}$$

Because one usually has the data to calculate this ratio directly, it represents the most useful form of the harmonic mean.

Median (M)

In some cases it is desirable to know the value of a variable x, call it x_0, which partitions data evenly, with values larger than x_0 in one group and values smaller than x_0 in the other group. This concept is particularly valuable in discussions of income distribution, as the following example shows.

> **Example 3.3: Average income** Suppose there were 10 people in a community and they had incomes of 1, 2, 3, 4, 5, 6, 7, 8, 9, and 10 thousand dollars, respectively. Their (arithmetic) average income would be $\bar{x} = 5.5$. On the other hand, their median income would be somewhere between 5 and 6 (possibly 5.5) thousand dollars. Now suppose Mr. Ten, whose income last year was 10,000 dollars, closed a big deal and increased his income by 100 percent this year. Then the town average would be increased to $\bar{x} = 6.5$, indicating an improvement in average income. Yet this gain by Mr. Ten concentrated the total income of the community still further in his hands, which is socially, at least, no improvement. The median reflects this view since it remains unchanged (at about 5500 dollars).

Mode (m)

The mode is simply the most frequently occurring value in a set of data. If no value is repeated, there is no single number representing the mode, although there may be a modal interval in the sense that after dividing the range of the variable into many intervals of equal length, one finds that the data tend to cluster more in one interval. Modal points and modal intervals need not be unique; there may be multiple modes in a set of data.

As an example, if the incomes for 10 people were 1, 2, 3, 4, 5, 6, 7, 8, 9, and 10 thousand dollars, respectively, there would be no modal income for the group. On the other hand, if they were 5, 3, 5, 7, 4, 3, 3, 5, 4, and 5 thousand dollars, the modal income for the group would be 5000 dollars.

3.2 VARIABILITY

Averages gloss over variability and thereby provide a false sense of certainty. Decisions based entirely on averages are only partly informed. Decision makers also need some insight into statistical uncertainty, or variability. Two useful concepts of variability are *range* and *standard deviation*.

Range, or Spread

Range, or spread, is simply the difference between the extreme values in a data set. If $x_1 = 40$, $x_2 = 1$, $x_3 = 55$, $x_4 = 54$, and $x_5 = 50$, the spread is 54 units; from 1 to 55. Note also that the average is 40. These two bits of information provide some insight into how the data are distributed, even though the manager who needs the information may never see the total distribution.

The range is misleading, however, because only one of the values is outside the interval 40 to 55. The value $x_2 = 1$ is so far from the other values that it could represent an erroneous data point, which should be discarded. A clearer impression of how the data are distributed is obtained from a statistical measure of the dispersion of the data around the mean value; it is called the standard deviation.

Standard Deviation (S)*

The standard deviation (SD) for a group of data is defined as the square root of the weighted average of the squares of the deviations of data from the mean $(\bar{x})$. The standard deviation has a desirable property; it weights each deviation by the relative frequency with which it occurs. Thus, single data points which are far from the mean have very little influence on the standard deviation.

---

* The average deviation from the mean is always zero, and for that reason is of no value.

$$\text{Average deviation} = \frac{(x_1 - \bar{x}) + \cdots + (x_n - \bar{x})}{n} = \frac{(x_1 + \cdots + x_n)}{n} - \bar{x} = \bar{x} - \bar{x} = 0$$

The standard deviation S is calculated slightly differently for large and small groups of data. For large groups (over 50 data points),

$$S = \sqrt{\frac{(x_1 - \bar{x})^2 + \cdots + (x_n - \bar{x})^2}{n}} \qquad \text{SD for } n > 50$$

And by convention,

$$S = \sqrt{\frac{(x_1 - \bar{x})^2 + \cdots + (x_n - \bar{x})^2}{n - 1}} \qquad \text{SD for } n < 50$$

when the number of data points is small.* The squared standard deviation S^2 is called the *sample variance*.

Example 3.4: Mean, standard deviation, and range The table below summarizes 10 separate cost estimates for a job. The third and fourth columns indicate deviations and squared deviations of each estimate from the mean.

Estimate number	Cost x, $ (thousands)	Deviation, $x - \bar{x}$	Squared deviation, $(x - \bar{x})^2$
1	10	3	9
2	6	−1	1
3	4	−3	9
4	9	2	4
5	6	−1	1

From these values the mean, standard deviation, and range are, respectively, $\bar{x} = 35/5 = 7$, $S = \sqrt{24/4} \simeq 2.45$, and range $= 10 - 4 = 6$.

3.3 SKEWNESS, OR ASYMMETRY

Very often data tend to cluster at one extreme of the range. This is a characteristic of salaries among professional groups. The extent of the clustering is measured by the *skewness coefficient*, which is defined by

$$g = \frac{(x_1 - \bar{x})^3 + \cdots + (x_n - \bar{x})^3 \dagger}{nS^3}$$

---

* Actually, changing the division to $n - 1$ for small samples is more than a convention. The choice reflects the preference of statisticians for statistical measures which are called *unbiased*. Readers interested in this technicality may refer to any of the statistics texts listed in Section 3.7.

† Here too, for fewer than 50 data points divide by $n - 1$ instead of n.

When the clustering is to the left, the sum in the numerator will tend to be positive, because the relatively few terms far to the right will produce large positive deviations $(x_i - \bar{x})$ while most other deviations on either side of the mean will cancel. Therefore g is positive for clustering to the left. When the clustering is to the right, g is negative.

Example 3.5: Professional salaries Two professions for which annual income is known to be quite skewed are medicine and law. In each profession a very few have enormous incomes while most have incomes clustered around a much smaller value. Hypothetical salaries for six physicians and six lawyers are given in the table below. For each group, the mean, median, standard deviation, and range are reported in thousands of dollars. The skewness coefficients are also reported.

	Physicians				Lawyers			
	Salary x	Deviation $x - \bar{x}$	Squared Deviation $(x - \bar{x})^2$	Cubed Deviation $(x - \bar{x})^3$	Salary y	Deviation $y - \bar{y}$	Squared Deviation $(y - \bar{y})^2$	Cubed Deviation $(y - \bar{y})^3$
	5	-1	1	-1	3	-2	4	-8
	6	0	0	0	4	-1	1	-1
	7	$+1$	1	$+1$	3	-2	4	-8
	8	$+2$	4	$+8$	6	$+1$	1	$+1$
	9	$+3$	9	$+27$	4	-1	1	-1
	1	-5	25	-125	10	5	25	125
Totals	36	0	40	-90	30	0	36	$+106$

$\bar{x} = 6$
$S_x^2 \simeq 8; S = 2.83$
$g_x \simeq -0.68$
Range $= 8$
Median $\simeq 6.5$

$\bar{y} \simeq 5$
$S_y^2 \simeq 7.2; S = 2.68$
$g_y \simeq +0.97$
Range $= 7$
Median $=$ not well defined

With only the summary statistics reported, there is enough information to quickly observe that incomes for physicians tend to be large (g_x is negative), with relatively few earning less than the average, while incomes for lawyers tend to cluster around a smaller value (g_y is positive) and relatively few earn more than the average.

In other professions, such as engineering, one would find that income is more uniformly distributed within the profession; g would be close to zero.

3.4 HISTOGRAMS

Histograms are graphic displays of the relative frequencies with which different values appear in statistical data. They are most useful when many data points are available. Histograms give immediate impressions of how the data are skewed and of the locations of the mean, median, and mode. Many managers are more comfortable interpreting a histogram than they are with only the statistical summaries. However, summary statistics should always be reported on the same page with the histogram; managers need the numbers, too.

Example 3.6: Summarizing data Suppose one were comparing the cost of hospital care for a particular disease in two regions of the country. Data from 10 hospitals in each region might be tabulated as below. Entries represent percent of cases reported for each region which fell into each cost range.

Relative frequencies of reported costs, %

	Cost intervals, $			
	51–150	151–250	251–350	351–450
Region A	10	50	40	0
Region B	0	20	30	50

Histograms for each region are shown below. Note that each bar spans one cost interval and has an area equal to the corresponding relative frequency. Note that area is height times length. Thus height is area divided by length. In this example length is always 100, so height is area, in percent, divided by 100. The total area is 100 percent, as it must be.

The summary statistics for each region are shown on the histograms. The mean, variance, and skewness were *estimated* using the relative frequencies for each interval. The midvalue of each interval is taken as the cost value representing the interval.

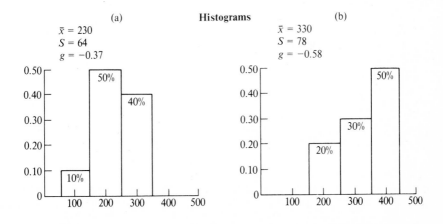

(a)   Histograms   (b)

$\bar{x} = 230$
$S = 64$
$g = -0.37$

$\bar{x} = 330$
$S = 78$
$g = -0.58$

3.5 TRENDS

The changing state of society and, in some ways, the world is chronicled in statistical series collected by public and private agencies. The government of the United States* operates the most extensive statistical system in the world.* It collects, analyzes, and distributes time series data on United States and foreign agriculture, commerce, environment, finance, health, human population, natural resources, science and technology, transportation, and social welfare. These data provide a historical context for monitoring progress, identifying weaknesses and problem areas, evaluating the impacts of new public programs and other intervening factors, and even for making short-term projections of things to come. They are among the most valuable and most used of all statistical data. Public managers could hardly do without them.

Management and policy formulation are forward-looking activities. Consequently, managers are forever trying to anticipate the future. Since the only data available pertain to the present and the past, managers look to them for clues for the future. Of course, forecasting is a very risky business. Long-term forecasts are almost certain to be wrong. One can never anticipate those important conditions which, although they had no bearing on prior situations represented by available data, will strongly influence the future. Typical examples of inaccurate forecasts include predictions in the 1940s that the United States population would be declining in the 1950s; the 1955 forecast by the then president of Chrysler Corporation, Lester Colbert, that automobiles would operate on electronic guidance systems by 1975; and a glorious prediction in 1955 by an agricultural consultant to the Rockefeller Foundation that in 20 years the human race would triumph over nature and reenter the land of Eden. Short-term forecasting, on the other hand, is not so risky. The distinction lies in the fact that trends ordinarily change gradually. The elementary analytical methods discussed in this section are sometimes useful for short-term projections, but must never be relied upon for extended forecasts.

Extrapolation of Trends

In its simplest form, *extrapolation* simply means to continue or extend a trend line into the future. Extrapolating lines with a straightedge may be appropriate in some cases, while fitting temporal data to probability models (see Chapter 4) or calculating geometric averages may be preferable in others. One might even use elementary algebraic procedures to project not only trend lines but cyclic patterns in data.

> **Example 3.7: Geometric trend** Rising costs of health care in the United States exhibit a geometric growth pattern. Actual data are plotted below. Observe that costs have nearly doubled during each 2-year period since

---

* The statistical system of the United States is described in Appendix 1.

1966. To project costs for 1976 and 1977 one could sketch a curve through the data provided and simply extend it for 2 years. French curves available in drawing-equipment supply stores are excellent for such tasks. Or one could merely double 1975 costs as an approximation for 1977 costs. This would be consistent with the trend. Then the 1976 value could be estimated as the geometric average of the 1975 and 1977 values or could be approximated more crudely by linear extrapolation. As a third option one could calculate the annual growth rate and use it to project costs for future years. The (annual) growth rate r may be found by noting that for a starting cost of C_0, the cost after one year will have grown to $C_0(1 + r)$. This in turn will grow to $C_0(1 + r)^2$ by the end of the second year. Since costs are doubling every 2 years, this expression must have a value of $2C_0$. Thus,

$$2C_0 = C_0(1 + r)^2$$

One can solve this expression for r to get

$$r = \sqrt{2} - 1 = 0.414$$

Projections into the future can now be made as follows, assuming that the geometric growth pattern will persist:

$$1976 \text{ costs} = 1975 \text{ costs } (1 + r)$$

$$1977 \text{ costs} = 1975 \text{ costs } (1 + r)^2$$

Etc.

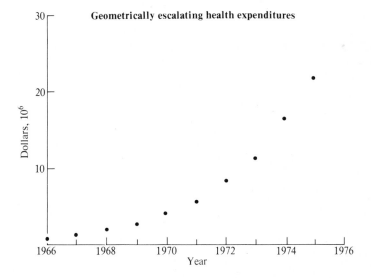

Geometrically escalating health expenditures

Rarely do time series data exhibit such smooth trends as those in Example 3.7. A common pattern is one of fluctuation and cycles around a trend line. Production, for example, declines sharply during labor strikes. It also fluctuates seasonally and may even have a longer-term cyclic pattern reflecting more general economic conditions. Managers must be able to distinguish trends and cyclic patterns from irregular fluctuations if they are to make effective use of time series data. How else could they tell whether a statistic represented a normal situation, a situation in need of official attention, or a consequence of a new policy or program? Statisticians have developed mathematical procedures for separating trends, seasonal variations, other cyclic variations, and irregular fluctuations in time series data.

Consider, for example, the agricultural production data in Table 3.1. When these data are plotted (see Figure 3.1), both a trend and a cyclic pattern become apparent. One cannot extrapolate these data to project production for 1975 without accounting for the cyclic pattern. The first cycle appears to have begun in 1952 or 1953, although these data points are missing. It peaked in 1955 and concluded in 1957. The second cycle began in 1955 and ended in 1962. The third ran from 1962 to 1965, the fourth from 1965 to 1970, and the last cycle began in 1970.

Cycle periods appear to be 5 or 6 years, with troughs following peaks within 3 years. From this pattern one might suspect that production for 1975 would exceed production in 1974. Furthermore, in all but one cycle, production in the first year exceeded production in the peak year of the previous cycle. Consequently, there is strong statistical evidence that production in 1975 should recover to at least 103 units.

The trend line for these data also provides some insight into the level of production to expect for 1975. Cycle averages are also plotted in Figure 3.1. The respective standard deviations are approximately 1.4, 1.9, 1.2, 1.8, and 3.3. Thus, barring major unrepresentative conditions, one would expect production in 1975 to be in the range from 103 to 107 units and very likely less than 105 units.

These are but a few observations which can be made regarding these data. They merely indicate the power of elementary logic and a few statistical observa-

Table 3.1 Agricultural Production, 1954–74

	Production			Production	
Year	Total	Per capita	Year	Total	Per capita
1954	77	97	1965	104	101
1955	82	99	1966	110	98
1956	85	97	1967	113	102
1957	85	95	1968	117	103
1958	91	99	1969	116	103
1959	92	100	1970	119	101
1960	95	101	1971	123	108
1961	95	100	1972	123	108
1962	98	99	1973	123	107
1963	99	101	1974	120	101
1964	104	100	1975	—	—

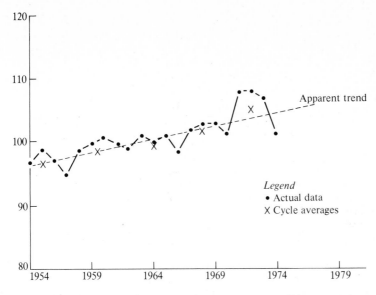

Figure 3.1 Per capita agricultural production from 1954 to 1974.

tions. Wise analysts, in their quest for a reasonable projection for 1975, would pursue these and additional statistical evaluations such as moving averages, discussed below.* But even more importantly, they would go beyond the data to seek reasons for the patterns they observed, with the hope of being able to match conditions in prior years with emerging conditions in 1974–1975 and thereby develop a rational basis for projecting beyond 1974.

Moving Averages

Time series data can be smoothed by calculating moving averages and using these as indicators of trends. For example, the agricultural production data in Table 3.1 exhibit 5-year cycles. Calculating the average for the first 5 years yields a number midway between the maximum and minimum values for the cycle, namely, 97.4. Now replacing the 1974 level with the 1959 level and recalculating the average yields a slightly higher value, 98. Note that both values fall nearly on the trend line indicated in Figure 3.1. Although moving averages do fluctuate, they move slowly by the nature of averaging. Continuing in this manner one can calculate moving averages for all years except the first and last two, as shown in Table 3.2. The trend line so obtained is a basis for short-term projections which might be required.

---

* Other, more sophisticated statistical techniques might be applied with the aid of a computer. One technique, regression over time, is discussed briefly in Chapter 7. More sophisticated techniques are beyond the scope of this book.

Table 3.2 Five-year moving averages, agricultural production data

Year	Per capita production	Year	Per capita production
1954		1965	100.4
1955		1966	100.8
1956	97.4	1967	101.4
1957	98.0	1968	101.4
1958	98.4	1969	103.4
1959	99.0	1970	104.6
1960	99.8	1971	105.4
1961	100.2	1972	105
1962	100.2	1973	
1963	100.2	1974	
1964	99.8		

3.6 PROBLEMS

3.1 The city has commissioned a housing survey of 500 households. The contractor has been told to report the findings statistically and to explain them in narrative form. The survey was carefully done and the data were reported as follows.

Relative frequencies by number of rooms

	Number of rooms									
	1	2	3	4	5	6	7	8	9	10
Percent	0	1	1	10	15	30	20	10	10	3

(a) Compute the mean, mode, standard deviation, and skewness coefficient. Explain your results.

(b) Plot a histogram.

3.2 Over the past 10 years average income in the United States has been approximately as shown.

Year	1968	1969	1970	1971	1972	1973	1974	1975	1976	1977
Income, $	5000	5200	5800	6550	7400	8000	8700	9650	10,500	11,800

(a) Calculate the average increase in income over the 10-year period.

(b) If the value for 1971 were not given, how might it be estimated? (You will have to think about this one.)

3.3 Income data are given below. Calculate the mean, median, mode, standard deviation, and skewness coefficient. Explain your results.

Income range, $	Distribution, %
0–1,999	6.0
2,000–3,999	15.0
4,000–5,999	22.0
6,000–7,999	22.0
8,000–9,999	14.0
10,000–14,999	15.0
15,000–24,999	5.0
25,000 and over	1.0

3.4 The city is in the process of evaluating alternative road surfaces prior to launching a major road resurfacing program. The principal issues are the coefficient of friction and the durability of the surface. The coefficient of friction can vary between zero and unity, and the repair-free lifetime of the surfaces ranges from 5 to 8 years.

(a) Discuss the meaning and pitfalls of average values of these characteristics for decision purposes.

(b) Define some joint statements representing possible characteristics of each road surface.

3.5 The director of public works is keeping a record of the rate at which garbage is collected. Over the previous 5 weeks rates for collection on a particular route have been recorded. The director estimated, respectively, 5, 7, 4, and 9 blocks per hour. Which average is appropriate? Calculate the average rate. Discuss the usefulness of this number to the director.

3.6 The Big Push towboat company operates on the Mississippi River. Moving upriver, operating speeds between five locks and dams are recorded as 5, 8, 9, and 3 land miles per hour. The distances between locks are 10, 8, 13, and 5 miles. When the captain logs in the average speed, what should be recorded?

3.7 A busy executive has a blood pressure problem. He stops in a drugstore downtown when he gets an opportunity and takes his own blood pressure at the new **BP Cuff** self-operated machine. The machine reports the ratio of diastolic to systolic pressures. Discuss the meaning of average blood pressure readings.

3.8 The National Center for Health Statistics published growth data for children from birth to 18 years of age in November 1977.* The following data on height were included in the report. Plot the

* Vital and Health Statistics, ser. 11, no. 165, DHEW Publications no. (PHS) 78-1650.

Observed percentiles of height (stature)
Centimeters

Age, yrs	Sample, n	$\bar{x}$	s	5	10	25	50	75	90	95
						Percentile				
Male										
7	304	121.6	5.1	112.9	114.9	118.1	121.4	125.3	128.5	130.2
8	295	127.8	5.5	118.6	120.4	124.2	128.0	131.3	135.1	137.0
9	312	132.5	6.6	122.6	124.9	128.3	132.4	136.7	141.1	143.5
10	300	134.4	6.9	126.5	128.8	132.9	138.1	141.9	145.1	147.3
11	281	143.0	6.7	132.7	134.7	138.8	143.3	147.3	151.4	153.6
Female										
7	316	120.8	5.6	111.4	113.4	117.2	120.8	124.7	127.5	129.4
8	312	126.3	5.6	117.3	118.9	122.3	126.1	130.3	133.1	135.2
9	294	132.5	6.2	122.4	124.7	127.9	132.6	136.7	140.3	142.7
10	307	138.5	6.8	128.2	130.8	134.8	137.8	142.3	147.6	149.7
11	271	144.1	7.5	131.8	134.7	139.8	144.4	149.1	153.4	155.5

curves and label them completely with the summary statistics. Use graph paper. Label the vertical-axis percentage and the horizontal-axis height. Note that if the 5th percentile is 112.9 centimeters and the 10th percentile is 114.9 centimeters, 5 percent of the sample population measures 112.9 centimeters or less while 10 percent measures 114.9 centimeters or less.

3.9 The curves for weight versus age on page 51 were reproduced from the report reference in Problem 3.8. Use these curves to determine the median, modal, and (arithmetic) average weight for children ages 2, 10, and 18 years. The median may be observed directly from these curves.

To calculate the mode, first convert the data into histograms for each of the ages. The area of each bar on the histogram is the change in percentage from one percentile to the next. For example, for 15-year-old girls the weight change between the 75th and 90th percentiles is about 10 kilograms. Therefore the area of the bar above the weight interval 60 to 70 kilograms must be 15 percent. Since the area of a rectangle is length times height, $15 = 10 \times h$, $h = 1.5$ for this rectangle. The mode is within the interval for which the change in percentage or area per unit change in weight is largest. For 15-year-olds, the rate of change is greatest for the interval from 51 to 57 kilograms, wherein the percentage changes by 25 points; rate $= \frac{25}{6} \simeq 4.2$.

The arithmetic average may be estimated from the histogram by using the midpoint of each bar as the weight value and summing the products of height times the midvalue of weight.

3.10 A food item which cost 10 cents in 1945, cost 19 cents in 1955, 40 cents in 1965, and nearly 80 cents in 1975. If the trend continues, how much would you expect it to cost in 1980? Find the answer in two ways. One method may be graphic extrapolation. The other must be numerical and depend on the character of the trend.

3.11 (*Here is a problem for the more advanced student.*) In Problem 3.10, item cost was doubling about every 10 years. The characteristic of constant doubling intervals is peculiar to geometric (or exponential) growth. As an alternative to the technique used (described in the text) when compounding proceeds continuously, the growth rate in percent may be approximated from the relation, growth rate = 0.7/doubling time. Use this relation to calculate the annual growth rate for costs in Problem 3.10. (Note that the exponential growth model is cost at time t/initial cost $= e^{rt}$, where r is the annual growth rate and e is the number 2.71828. If the cost at time t is twice the initial cost, t is called the *doubling time*. Since the natural log of e^{rt} is just rt and the log of 2 is almost 0.7, the relation above follows from taking the log of both sides of the growth model.)

3.12 Calculate the mean and variance for the data given.

Decade	Average lifetime from birth	Group SD	Decade	Average lifetime from birth	Group SD
1900	57	10	1940	65	5
1910	63	9	1950	68	7
1920	52	12	1960	70	4
1930	65	8	1970	76	5

3.13 Show algebraically that the following equations for calculating the sample variance are equivalent. Check this fact using the data from Problem 3.12.

$$S^2 = \frac{1}{n} \sum (x_i - \bar{x})^2$$

$$S^2 = \frac{1}{n} \sum x_i^2 - \bar{x}^2$$

3.14 Refer to any appropriate publication of the U.S. Bureau of the Census or to "Marriage and Fertility in the Developed Countries," by Charles F. Westoff, *Scientific American*, December 1978.

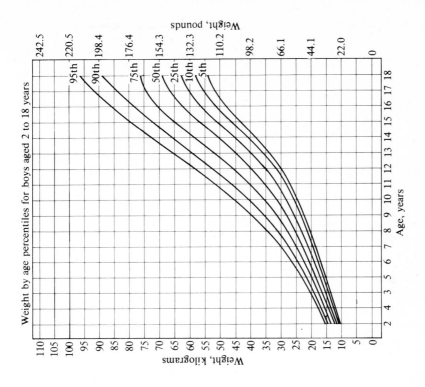

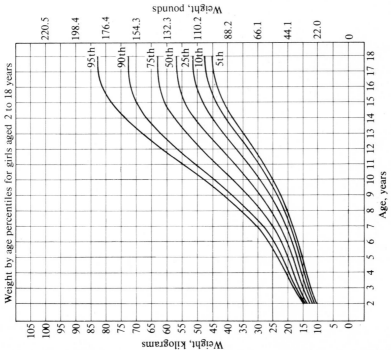

Write a one-page description of the projection technique employed by the Bureau of the Census to project population for the nation.

3.15 Data for this and the following six problems may be found in the *Statistical Abstract for the United States*, U.S. Department of Commerce, Bureau of the Census, U.S. Printing Office, Washington D.C. 1978, which is available in practically all libraries. Compare trends in family income, health care expenditures, and life expectancy. Explain your observations from the perspective of one who is concerned about the health of the nation.

3.16 Contrast the numbers of physicians and other health care professionals per capita with trends in health care expenditure and health insurance benefits paid.

3.17 Explain changing enrollment levels in institutions in terms of median family income and the age distribution of the population. What other factors influence enrollment and how important are they in relation to income and age?

3.18 Explain the procedures employed by the U.S. Bureau of Labor Statistics to calculate the consumer price index. What does this index tell us? How has it changed over time? Refer to the U.S. Bureau of labor Statistics, *Monthly Labor Review*.

3.19 Compare federal research and development expenditure trends for physical sciences and life sciences.

3.20 Compare research and development expenditures by the federal government with expenditures by private industry and foundations.

3.21 Contrast trends in agricultural production with farm population and number of farms. Explain the apparent contradiction in trends.

3.7 SUPPLEMENTAL READING

Benjamin, Jack R., and Cornell, C. Allin: Probability, Statistics and Decision for Civil Engineers, McGraw-Hill Book Company, New York, 1970, chaps. 1 and 2.

Chisholm, Roger K., and Gilbert R. Whitaker, Jr.: *Forecasting Methods*, Richard D. Irwin, Inc., Homewood, Ill., 1971.

Dwass, Meyer: *First Steps in Probability*, McGraw-Hill Book Company, New York, 1967, chaps. 2, 3, and 6.

Hays, William L., and Robert L. Winkler: *Statistics: Probability, Inference, and Decision*, Holt, Rinehart and Winston, Inc., New York, 1971, chap. 6.

Hoel, Paul G.: *Elementary Statistics*, 4th ed., John Wiley & Sons, Inc., New York, 1971, chap. 2.

Tukey, John W.: *Exploratory Data Analysis*, Addison-Wesley Publishing Company, Inc., Reading, Mass., 1977.

TWO

STATISTICAL DATA ANALYSIS

Data analysis actually begins with the basic summarization methods of Chapter 3. At one time more sophisticated methods were rare components of policy studies. However, the sophistication of policy analysts and decision makers has grown in leaps and bounds over the last two decades, until today very sophisticated policy studies are commonplace.

Bear in mind that great technical sophistication is no guarantee of more valuable results. Some analysts revel in technique. They tend to employ a sledgehammer when a tack hammer would do perfectly well. Others employ potentially powerful techniques when in fact, the critical assumptions upon which the techniques are based may have been violated in the data collection process. Thus while some results appear to be compelling, they are not; in fact, they may be misleading. One pressure for excessive analysis is the ease with which anyone with rudimentary training in the use of computers can load data and invoke sophisticated statistical programs to crunch the numbers and print out results. There is a feeling of power in this process even though the participants may have little comprehension of the limitations of the methods employed or the calculational process. As a manager or analyst you should be wary of these practices. When sophisticated analysis is technically justifiable and promises to yield useful information beyond what can be extracted by more elementary methods, by all means use it. But think first and remember—simplicity can be elegant.

The material in Chapters 4 through 7 introduces the more important advanced methods, without using an overwhelming amount of mathematics. The purpose is to introduce concepts and develop an appreciation for their applicability. The idea of probability modeling is introduced in Chapter 4 and the most commonly used models are explained. In Chapter 5 important statistical procedures which are useful for comparing sample data and drawing inferences from them are illustrated. Chapter 6 demonstrates systematic procedures for integrating statistical and nonstatistical data to estimate the likelihoods of decision consequences. Chapter 7 develops the concept of least-squares curve fitting (or regression), which is perhaps the single most frequently and widely employed analytical technique. Chapter 7 also includes an elementary approach to trend analysis and forecasting.

Again, mathematics is downplayed in this part of the book as elsewhere. However, one cannot avoid algebraic manipulation. Neither can one avoid learning a new symbolism. Just keep in mind that the objectives are to acquire concepts and develop an appreciation for applications. Do not allow the symbols to divert your attention. Memorize them quickly and move on to important matters. The symbols serve as shorthand representations for concepts and should be read as one would read words.

FOUR

PROBABILITY DISTRIBUTIONS

It is a fact that statistical data display patterns of symmetry, as depicted by histograms. Mathematicians have spent hundreds of years developing mathematical formulations of many of these patterns. These formulations are frequently called *models* or *distributions*. Statisticians, economists, engineers, and other professionals have identified numerous physical situations which may be described by these models. These accomplishments have paid enormous dividends in the practical world. With the aid of probability models analysts have developed powerful methods of evaluating the effectiveness of drugs, products, and technologies. Efficient means of evaluating public projects and public services have also been devised. And, as we shall see in Chapter 5, techniques for comparing the statistical properties of data from different sources depend on the existence of such models.

The basic purpose of this chapter is to explain the important probability models, which find application in subsequent chapters. These are the Bernoulli, binomial, geometric, Pascal, Poisson, exponential, normal (or Gauss), chi-square, Student's-*t*, *F*, and beta models. To the extent possible, examples of physical situations represented by each model are described. Also, statistical properties of the models are presented.

Your objectives should be to

- Become generally comfortable with the concept of probability models and the processes by which such models are generated.
- Be sure to understand which models are applicable to which situations. Summary Table 4.4 on page 76 will help you in this regard.
- Master the technique of calculating probability values from distributions; probability tables are provided in Appendix 2.

• Try to have a mental image of the shapes of the curves representing the models and commit to memory the expressions for the mean and variance of each model. (Although it is always possible to look up the technical specifics, it is very helpful to store this information in your mind.)

The only way you will master the material in this chapter, and subsequent chapters for that matter, is to work with the concepts. You will be unsuccessful if you attempt to memorize the material, but you will be pleasantly surprised at your progress if you work hard at doing problems thoughtfully. Discussing the text with others is also helpful.

The rules of probability developed in Chapter 2 are precisely the rules which are invoked in this chapter. Not one new rule is developed, although the nomenclature is modified slightly for convenience and for consistency with convention. Since probability models are written in terms of numerical as opposed to verbal statements, the equivalence of these two forms is first established and the idea of random variables is introduced. Next, the concept of histogram is extended to the notion of a probability distribution for a random variable and the concepts of average and variance are redefined and extended in this context. Then the probability models are illustrated.

4.1 STATEMENTS, OUTCOMES, AND RANDOM VARIABLES

Replacing verbal statements with numerical statements in appropriate situations is a notational convenience which facilitates thinking and bookkeeping. Although one could always write appropriate verbal statements about these cases, it is often more convenient to write numerical statements corresponding to possible outcomes and make all probability assignments relevant to these. Just as with verbal statements, numerical statements are either tautologies, self-contradictions, or contingency statements. Collections of numerical statements can be mutually exclusive or independent. Complex statements are formed from simple statements using the logical connectives: "and," "or," and "not." And, finally, the rules of probability apply to numerical statements precisely as they apply to verbal statements.

The only important distinctive feature to be reckoned with is that numerical statements must be made in terms of what is called a *random variable*. For example, an important consequence of a public program might be the number of new jobs created. The random variable corresponds to the concept of jobs created. It may assume values ranging from zero to some finite number. One can break this numerical range into discrete intervals for decision purposes and consider probability statements of the type $P(0 < X \leq 600)$, $P(600 < X \leq 1200)$, etc., or one might be concerned with $P(X \geq 2500)$. A random variable is conceptually distinct from the values it may take. Sometimes one must distinguish between the variable and its values, so uppercase symbols such as X are reserved for the variable while lowercase symbols such as x are reserved for specific values.

Example 4.1: Random variable for the coin problem Consider tossing a coin having one side marked h and the other marked $\bar{h}$. There are two mutually exclusive statements about how the coin will land. Taken together, these are exhaustive. They are

$h \sim$ the coin will land h-side up
$\bar{h} \sim$ the coin will land $\bar{h}$-side up

The probability statements for h and $\bar{h}$ are, of course, $P(h)$ and $P(\bar{h})$, where $P(h \text{ or } \bar{h}) = P(h) + P(\bar{h}) = 1$ and $P(h \ \& \ \bar{h}) = 0$.

One can just as well conceive of a random variable which takes the value 1 when h is true and the value 0 when $\bar{h}$ is true. That is,

$$X(h) = 1; \ X(\bar{h}) = 0$$

Then the probability statements for the possible outcomes are $P(X = 1)$ and $P(X = 0)$. That is,

$$P(X = 1) = P(h)$$
$$P(X = 0) = P(\bar{h})$$

Example 4.2: Random variables for sequential coin-tossing–type situations A nurse gives the same medication daily to a patient. One might be interested in the likelihood that the nurse will make precisely one mistake in three successive days. Using h to represent proper medication, the complex statement relating to the question is $[(h \ \& \ h \ \& \ \bar{h}) \text{ or } (h \ \& \ \bar{h} \ \& \ h) \text{ or } (\bar{h} \ \& \ h \ \& \ h)]$. The probability of only one mistake is

$$P(h \ \& \ h \ \& \ \bar{h}) + P(h \ \& \ \bar{h} \ \& \ h) + P(\bar{h} \ \& \ h \ \& \ h)$$

or $\qquad P(0, 0, 1) + P(0, 1, 0) + P(1, 0, 0)$

There are only eight possible outcomes, namely,

$$(0, 0, 0) \quad (0, 1, 0) \quad (1, 1, 0) \quad (0, 1, 1)$$
$$(0, 0, 1) \quad (1, 0, 0) \quad (1, 0, 1) \quad (1, 1, 1)$$

Assume all are equally likely; the probability of only one mistake must be

$$\frac{1}{8} + \frac{1}{8} + \frac{1}{8} = \frac{3}{8}$$

One can think of a random variable as an assignment function. It assigns numerical values to all possible elementary outcomes of an action or experiment, namely, to the mutually exclusive and exhaustive set of relevant state-

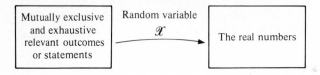

Figure 4.1 Random variables assign numbers to outcomes.

ments. The idea is conveyed in Figure 4.1. The concept of a random variable is essential in statistical analyses of data. While reading this chapter try to identify the random variable implied in each discussion.

4.2 DISTRIBUTION FUNCTIONS

There are two ways of describing distributions of random variables. If a variable takes only integer values, each value will have a specific likelihood assigned to it. This array of likelihood values is sometimes called a *probability mass function* (PMF). The nomenclature derives from the physical concept of body mass. If a random variable is continuous, meaning it can assume any numerical value between two limits, one cannot assign specific likelihoods to each possible value. In this case, the mass function is not appropriate and an analogous function, called a *probability density function* (PDF), substitutes for it. A PDF measures the likelihood that the variable will assume a value within some range just as the histogram does. In either case, one can also describe a distribution in terms of a function called a *cumulative distribution function* (CDF). It represents an accumulation of the elemental likelihood values from the lowest value a variable can assume to any larger value. The CDFs are obtained by summing values of mass functions or integrating density functions.* Mass functions and density functions can be obtained from CDFs by the reverse operations.

The mass function is simply a listing of the probability values for each mutually exclusive numerical statement. These probability values can be written as either $P(X = x)$ or $p(x)$. The CDF simply accumulates these values, as in calculating the probability for a disjunctive statement with mutually exclusive components. For example, if X can assume only the values -1, 0, and 1, with the probabilities $p(-1) = P(X = -1) = 0.1$, $p(0) = P(X = 0) = 0.5$, and $p(1) = P(X = 1) = 0.4$, the PMF and CDF are

Mass function (PMF)   Cumulative function (CDF)

$$p(x) = P(X = x) = \begin{cases} 0.1; \text{ if } x = -1 \\ 0.5; \text{ if } x = 0 \\ 0.4; \text{ if } x = 1 \end{cases} \qquad P(X \le x) = \begin{cases} 0; \text{ if } x < -1 \\ 0.1; \text{ if } -1 \le x < 0 \\ 0.6; \text{ if } 0 \le x < 1 \\ 1.0; \text{ if } 1 < x \end{cases}$$

---

* Integration is simply the summation of infinitesimally small areas. This is all we need to know about this concept, because we never employ it in this text.

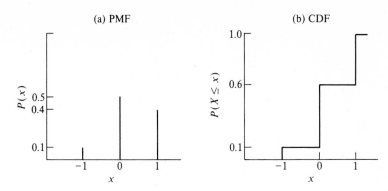

Figure 4.2 Distribution for X.

The PMF and CDF are shown diagrammatically in Figure 4.2. Note that the magnitudes of the jumps or changes in CDF are exactly the PMF values. We observed the same feature regarding histograms and percentile plots in Chapter 3.

Continuous random variables can take on any value between two limits, e.g., between 0 and 1. There are infinitely many values in any such interval, so, if they were each to have some nonzero probability value, as with a PMF, the total probability would add to infinity. Since probability values for the set of relevent, mutually exclusive, exhaustive statements must add to exactly one, according to the rules of probability, the likelihood that a continuous variable takes any specific value must be zero. This is the reason why the concept of PMF is not appropriate for continuous variables.

The PDF is the appropriate and analogous concept for continuous variables. It is usually written $f_x(x)$. It does not specify a probability value directly. Instead, the area beneath the PDF curve between two points measures the likelihood that the variable will assume a value in that interval. The idea is illustrated in Figure 4.3.

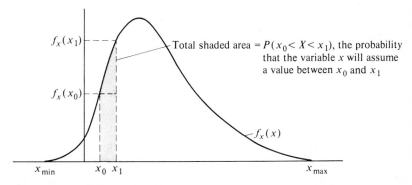

Figure 4.3 The PDF concept.

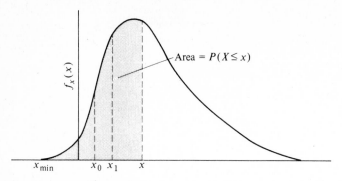

Figure 4.4 Integration of PDF.

To calculate the area between x_{min} and any larger value x, one simply sums many such small areas. The idea is illustrated in Figure 4.4.

As a numerical example, consider the triangular PDF shown in Figure 4.5. Obviously the area under the curve is equal to unity since the area of a right triangle is one-half the base times the height. The sloping line represents the PDF, and since it starts at zero and rises to two, the equation for it is $f_x(x) = 2x$. Thus, the density and cumulative distribution functions are

PDF	CDF

$$f_x(x) = 2x; \; 0 \leq x \leq 1 \qquad P(X \leq x) = \text{Area under the curve between 0 and } x$$

$$\text{or}$$

$$P(X \leq 1/4) = 1/4$$

$$P(X < 1) = 1/4 + 1/4 + 1/2 = 1$$

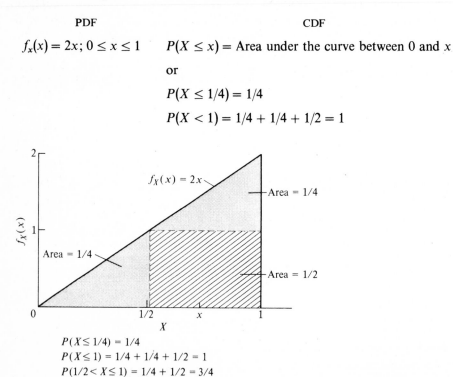

$$P(X \leq 1/4) = 1/4$$
$$P(X \leq 1) = 1/4 + 1/4 + 1/2 = 1$$
$$P(1/2 < X \leq 1) = 1/4 + 1/2 = 3/4$$

Figure 4.5 Integration of a triangular PDF.

Table 4.1 Distribution functions

X is the random variable

Discrete variables
    Values: $X = x_1, x_2, \ldots, x_n$
      PMF: $p(x_1), p(x_2), \ldots, p(x_n)$
      CDF: $P(X \leq x) = p(x_1) + p(x_2)$ $x_2 \leq x < x_3$
             $p(x_1) + \cdots + p(x_n) = 1$ $x \geq x_n$
Continuous variables
    Values: $x_{min} \leq X \leq x_{max}$
      PDF: $f_x(x)$
      CDF: $P(X \leq x) =$ Area under the curve $f_x(x)$ between x_{min} and x

---

Table 4.1 summarizes the concepts introduced in this section.

4.3 EXPECTED VALUE AND VARIANCE

Expected value is an extension of the concept of arithmetic mean. The arithmetic mean for the data in Table 4.2 is

$$\bar{X} = \frac{25 + 10 + 15 + 15 + 10 + 12 + 15 + 12 + 15 + 15}{10} = 14.4 \text{ min}$$

This can be written more conveniently by noting that the value 25 appears as only one of the 10 values (one-tenth the values), 10 appears twice, 15 appears five times, and 12 appears twice. Now simply combine terms to get a more efficient expression for the mean.

$$\text{Mean value} = \left(\frac{1}{10}\right)25 + \left(\frac{2}{10}\right)10 + \left(\frac{5}{10}\right)15 + \left(\frac{2}{10}\right)12 = 14.4$$

The values in parentheses represent the relative frequencies of occurrence. If the sample were representative of the true distribution for the variable X, the relative frequencies would represent the likelihoods that the variable would assume the corresponding values x. That is, the PMF for X would be

$$p(25) = \frac{1}{10} \qquad p(15) = \frac{5}{10}$$

$$p(10) = \frac{2}{10} \qquad p(12) = \frac{2}{10}$$

Table 4.2 Waiting time at a bus stop

Day	1	2	3	4	5	6	7	8	9	10
Minutes	25	10	15	15	10	12	15	12	15	15

In this case, if one were anticipating taking a random sample, one would expect to obtain a set of values having a mean value of

$$\text{Mean value} = p(25) \times 25 + p(10) \times 10 + p(15) \times 15 + p(12) \times 12 = 14.4$$

This would be the expected value of the probability distribution for X, or, more clearly, the *expected value of X*.

In general, a discrete random variable which can take on the (exhaustive and mutually exclusive) values $x_1, x_2, \ldots, x_n$ with probabilities $p(x_1), p(x_2), \ldots, p(x_n)$ has an expected value of

$$E(X) = x_1 p(x_1) + x_2 p(x_2) + \cdots + x_n p(x_n) \qquad (\mu_x)*$$

The *variance* of X is simply the expected value of the squared deviations of the x_i values from the mean. This concept is an extension of the idea of (squared) standard deviation. Referring again to the data in Table 4.2, the square of the standard deviation is

$$S^2 = (25 - 14.4)^2 \frac{1}{10} + (10 - 14.4)^2 \frac{2}{10} + (15 - 14.4)^2 \frac{5}{10}$$

$$+ (12 - 14.4)^2 \frac{2}{10} = 16.44$$

If these observed frequencies of occurrence represented the true PMF for the variable, the variance would be 16.44. The general formulation is

$$\text{var}(X) = p(x_1)(x_1 - \mu_x)^2 + \cdots + p(x_n)(x_n - \mu_x)^2 \qquad (\sigma_x^2)†$$

Accurate calculations of the mean and variance for continuous distributions are accomplished by integration.‡ However, these can be approximated from graphic representations of either the PDF or the CDF as follows. The true mean

---

* The symbol μ_x, pronounced "mu sub ex," is a shorthand notation for $E(x)$.
† The symbol σ_x^2, pronounced "sigma sub ex squared," is a shorthand notation for var (X).
‡ Again, for practical purposes integration means adding together very small (infinitesimal) areas to compile the total area between two points under a PDF.

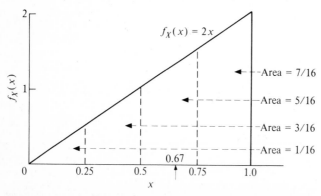

Figure 4.6 Approximating $E(x)$ and var (X) for a triangular distribution.

and variance for the density function $f(x) = 2x$, shown in Figure 4.6, are actually $E(X) = 0.67$ and var $(X) = 0.056$.

Using the midvalue in each interval as the representative value of x for that interval, the mean can be approximated by*

$$\hat{E}(X) = \frac{1}{8}\frac{1}{16} + \frac{3}{8}\frac{3}{16} + \frac{5}{8}\frac{5}{16} + \frac{7}{8}\frac{7}{16}$$

$$\simeq 0.66 \qquad \text{about } 1.5\% \text{ error}$$

Similarly, the variance is approximately

$$\hat{\text{var}}\,(X) = \left(\frac{1}{8} - 0.66\right)^2 \frac{1}{16} + \left(\frac{3}{8} - 0.66\right)^2 \frac{3}{16}$$

$$+ \left(\frac{5}{8} - 0.66\right)^2 \frac{5}{16} + \left(\frac{7}{8} - 0.66\right)^2 \frac{7}{16}$$

$$\simeq 0.049 \qquad \text{about } 12.5\% \text{ error}$$

The large error in the estimate of the variance occurred primarily because of the crudely estimated values of X used in the calculation. Dividing each interval in half would substantially improve the estimate.

To estimate the mean and variance for more complex distributions such as those shown in Figure 4.7, proceed as follows: Divide the x axis into small, equal-size intervals; the smaller the interval, the more accurate will be the estimates. Use the center points in each interval as the representative value x in that interval. The shaded rectangle indicated for the PDF in Figure 4.7 approximates the probability that X will take a value in that interval; for calculational purposes it is treated as the probability that X will take the midvalue in that interval. Thus, for any density function for which n intervals have been specified:

$$\hat{E}(X) = x_1 p(x_1) + x_2 p(x_2) + \cdots + x_n p(x_n) \qquad = \hat{\mu}_x$$

$$\hat{\text{var}}\,(x) \simeq (x_1 - \hat{\mu}_x)^2 p(x_1) + (x_2 - \hat{\mu}_x)^2 p(x_2) + \cdots + (x_n - \hat{\mu}_x)^2 p(x_n) \qquad = \hat{\sigma}_x^2$$

* The caret (ˆ) is used here and elsewhere in this book to indicate that only an approximation to the quantity symbolized is being calculated.

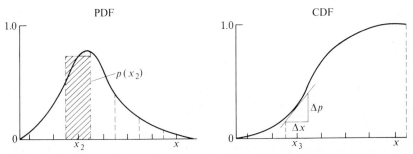

Figure 4.7 Approximating $E(x)$ and var (X) for arbitrary continuous distributions.

The procedure for estimating the mean and variance from a CDF for a continuous distribution (see Figure 4.7) is a bit different. The change in the height of the CDF between two points Δp is the probability that x will take a value in that interval. For calculational purposes treat Δp for an interval as the likelihood that x will take on the midvalue of the interval. Thus for CDFs for which n intervals have been specified:

$$\hat{E}(X) = x_1 \Delta p(x_1) + x_2 \Delta p(x_2) + \cdots + x_n \Delta p(x_n) \qquad = \hat{\mu}_n$$

$$\hat{var}(x) = (x_1 - \hat{\mu}_x)^2 \Delta p(x_1) + \cdots + (x_n - \hat{\mu}_x)^2 \Delta p(x_n) \qquad = \hat{\sigma}_x$$

There are two extremely important facts to keep in mind about $E(X)$ and var (X). It often turns out that one wishes to know the mean and variance of a random variable which is itself a sum, or difference, of random variables. If X_1 and X_2 are random variables, then $Y = X_1 + X_2$ is also a random variable and

• *The expected value of a sum of random variables is simply the sum of the respective expected values, always!*

$$E(Y) = E(X_1) + E(X_2) \qquad \text{if } Y = X_1 + X_2$$

• *The variance for a sum of random variables is the sum of the respective variances, only if the respective variables are independent!*

$$\text{var}(Y) = \text{var}(X_1) + \text{var}(X_2) \qquad \text{if } X_1, X_2 \text{ are independent}$$

The truth of these assertions is easy to demonstrate by example. The two sets of data given in Table 4.3 are sufficient. The row totals are probability values for X_1, while the column totals are probability values for X_2.* The cell

---

* These are commonly referred to as marginal probability values, because they appear in the margins of the table.

Table 4.3 Probability distribution for $Y = X_1 + X_2$

(a) X_1, X_2 independent				(b) X_1, X_2 dependent			
	X_2				X_2		
X_1	0	1	Totals	X_1	0	1	Totals
0	0.12	0.08	0.20	0	0.14	0.06	0.20
1	0.48	0.32	0.80	1	0.32	0.48	0.80
Totals	0.60	0.40	1.00	Totals	0.46	0.54	1.00

$E(X_1) = 0.8$	var $(X_1) = 0.16$	$E(X_1) = 0.8$	var $(X_1) = 0.16$
$E(X_2) = 0.4$	var $(X_2) = 0.24$	$E(X_2) = 0.4$	var $(X_2) = 0.24$
$E(X_1 + X_2) = 1.2$	var $(X_1 + X_2) = 0.40$	$E(X_1 + X_2) = 1.2$	var $(X_1 + X_2) = 0.86$

entries are joint probability values for pairs (X_1, X_2). In Table 4.3(a), for example, the probability that $Y = 1$ is $0.081 + 0.48 = 0.56$. When X_1 and X_2 are independent the cell entries are products of the marginal totals. The cell entries in Table 4.3(b) do not equal products of marginal totals, indicating that X_1 and X_2 cannot be independent. The means and variances calculated from each set of data and summarized in the table clearly demonstrate the points.

4.4 PROBABILITY MODELS

Specific probability models are described in this section. All are used in subsequent chapters. Each model is presented algebraically or graphically and for each the mean and variance are specified.

Bernoulli Model, $B(X/p)$ (Success or Failure in One Trial)

When one is concerned with a single statement and its complement, one is dealing with a *Bernoulli model*. This model was repeatedly employed in Chapter 1. Oddly, many situations in real life are this simple. This model is also the foundation for more complex models. The appropriate random variable is $X = 0$ or 1, and the mass function, mean, and variance are

$$p(x) = \begin{cases} p & \text{if } x = 1 \\ 1 - p & \text{if } x = 0 \end{cases} \qquad \text{Bernoulli}$$

$$E(X) = p \qquad \text{var}(X) = p(1 - p)$$

The model is portrayed graphically in Figure 4.8.

Binomial Model, $B(Y/n, p)$ (Number of Successes in n Trials)

The question, "What is the likelihood that y of the next n *independent* Bernoulli trials will produce positive outcomes?" is answered by the *binomial distribution*.

The general form of the binomial mass function is

$$p(y) = \binom{n}{y} p^y (1 - p)^{n - y} \qquad \text{binomial}$$

$$E(Y) = np \qquad \text{var}(Y) = np(1 - p)$$

The random variable may assume values $0, 1, \ldots, n$, and is a sum of *independent* Bernoulli variables; $Y = X_1 + \cdots + X_n$.*

---

* Since Y is a sum of n independent Bernoulli variables, we know from Section 4.3 that $E(Y) = E(X_1) + \cdots + E(X_n) = np$ and $\text{var}(Y) = \text{var}(X_1) + \cdots + \text{var}(X_n) = np(1 - p)$.

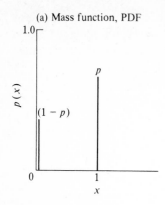
(a) Mass function, PDF

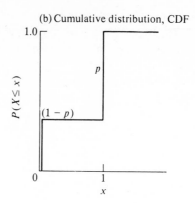
(b) Cumulative distribution, CDF

Figure 4.8 Bernoulli, $B(X/p)$.

The symbol $\binom{n}{y}$ is referred to as "n choose y." It is called a *binomial coefficient*. Its value is the number of unique combinations of y successes in n trials. Its value may be calculated from its definition, which follows:

$$\binom{n}{y} = \frac{n!}{y!\,(n-y)!} \qquad \text{where } n! = n(n-1)\cdots 1$$

As an example, suppose that the statistical likelihood of a female birth were 0.4. That is, 40 percent of all live births in the past were female. A couple planning a family of three children could conclude their family-building with zero, one, two, or three daughters. Letting $X_i = 1$ if the ith child born is female and $X_i = 0$ if the child is male, the random variable $Y = X_1 + X_2 + X_3$, which counts the number of female births, is a sum of independent Bernoulli variables. The mass function for Y and the birth sequences are

Value of X	Birth sequence	PMF
0	(m, m, m)	$p(0) = (0.6)^3 = 0.216$
1	$(m, m, f)\ (m, f, m)\ (f, m, m)$	$p(1) = 3(0.6)^2(0.4) = 0.432$
2	$(m, f, f)\ (f, m, f)\ (f, f, m)$	$p(2) = 3(0.6)(0.4)^2 = 0.288$
3	(f, f, f)	$p(3) = (0.4)^3 = 0.064$

The expected value and variance are $E(Y) = 1.2$ and var $(Y) = 0.72$. The density function and cumulative function are plotted in Figure 4.9.

Geometric Model, $G(N/p)$ (Number of Trials until First Success)

The geometric distribution responds to the question, "What is the likelihood that the next positive outcome in a sequence of independent Bernoulli trials will

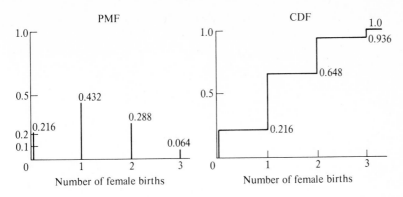

Figure 4.9 Binomial, $B(X/3, 0.4)$.

occur on the nth trial?" The random variable in this case is the number of trials N. It may assume values from unity to infinity, because the next success may occur very far in the future.

The likelihood that the next positive outcome will occur on the tenth trial is the likelihood that the first nine trials are unsuccessful multiplied by the likelihood of a success on the tenth trial, namely, $p(1 - p)^9$. The general form of the geometric mass function is, thus,

$$p(n) = p(1 - p)^{n - 1}$$

geometric

$$E(N) = \frac{1}{p} \qquad \text{var}(N) = \frac{1 - p}{p^2}$$

As an example, suppose an objective of a fire company was to respond to alarms within 5 minutes. If the company were successful 60 percent of the time, the average number of responses from success to success would be 1.67, with a variance of 3.75 alarms, using the geometric distribution. The sequence of successes and failures and the mass function for this situation are

Value of X	Sequence	PMF
1	(1)	(0.6) = 0.6
2	(0, 1)	(0.4)(0.6) = 10.24
3	(0, 0, 1)	(0.4)^2(0.6) = 10.096
4	(0, 0, 0, 1)	(0.4)^3(0.6) = 10.038
⋮		

The mass function and cumulative distribution are plotted in Figure 4.10.

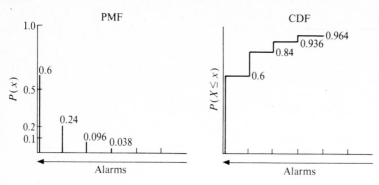

Figure 4.10 Geometric, $G(N/0.6)$.

Pascal Model, $P(N/p, k)$ (Number of Trials until Required Number of Successes)

The number of trials until the kth success in a sequence of independent Bernoulli trials is modeled by the *Pascal* or *negative binomial distribution*. If the kth success is to occur on the nth trial, the first $k - 1$ successes must have occurred in the first $n - 1$ trials. Thus the general form of the Pascal mass function is

$$p(n) = \binom{n-1}{k-1} p^k (1 - p)^{n-k}$$

Pascal

$$E(N) = \frac{k}{p} \qquad \text{var } (N) = \frac{k(1-p)}{p^2}$$

For example, if the likelihood of hitting a target were only 30 percent statistically, the likelihood that it would take only three shots to make three hits would be $(0.3)^3$. Since there are three sequences of four tries to get the third hit, namely, (m, h, h, h), (h, m, h, h), and (h, h, m, h), the likelihood that four shots will be required to get the third hit is $\binom{3}{2}(0.3)^3(0.7)$. The expected number of shots until the third hit is $E(N) = 10$ and the variance is var $(N) = 23.4$.

Poisson Model, $\mathscr{P}(X/\mu)$ (Arrivals per Unit of Time)

All the previously discussed models are classified as *Bernoulli processes*. The important questions were related to the sequence of outcomes and the number of trials. Time was not an important element. In effect, time was incremented only when an event occurred. However, there are numerous situations in which events occur randomly as time progresses. Arrivals of patients at emergency rooms, boats at loading docks, cars at stoplights or toll booths, and airplanes at an airport are but a few examples. In these situations one would often like to know the distribution of occurrences (or arrivals) in time in order to most efficiently design and staff the processing facility. When certain conditions on the occurrence of events can be established, such processes fit the *Poisson model*.

The conditions are

- *Stationarity.* No matter at what time one begins to observe the process, the number of arrivals in a small increment of time Δt is approximately constant and equal to $\mu \Delta t$, where μ is the average rate of arrivals. For this reason the Poisson model is useful in studies of rare events.
- *Singularity.* In an appropriately small time interval, the likelihood of more than one arrival is negligible.
- *Independent time increments.* For distinct time intervals, the number of occurrences in one interval is independent of the number in another.

The random variable for this model counts the number of occurrences (or arrivals) per given time increment. It may assume any integral value from zero to infinity. The Poisson mass function is

$$p(x) = \frac{(\mu t)^x}{x!} e^{-\mu t} \qquad x = 0, 1, \ldots \qquad \text{Poisson}$$

$$E(X) = \mu \qquad \text{var}(X) = \mu$$

The symbol e represents a number which has an approximate value of 2.71828. The letter t represents the number of time units and must be established for each problem.

Exponential Model, $\mathscr{E}(T/\mu)$ (Time between Arrivals)

In traffic studies of all kinds one is often interested in the time until the next arrival. The *exponential distribution* models that situation when the arrivals are Poisson. The probability that the next arrival will occur after t units of time have passed is the Poisson probability that there will be no arrivals in the next t units, namely, $e^{-\mu t}$. Thus, $P(T \le t) = 1 - e^{-\mu t}$, where μ is the average arrival rate; this is actually the CDF for the time between arrivals.

The density function for the exponential distribution is

$$f_t(t) = \mu e^{-\mu t} \qquad t \ge 0$$

exponential

$$E(T) = \frac{1}{\mu} \qquad \text{var}(T) = 1/\mu^2$$

The following scenario illustrates a use of the Poisson and exponential models. The numbers are fictitious and do not remotely reflect reality.

The U.S. Army Corps of Engineers has commissioned a study of the efficiency of Locks and Dam No. 26 on the Mississippi River. The contractors collected data on the arrivals of tows and the rate of servicing them. They found that tows arrive individually and at a rate of about five per day on the average. Only one tow can be processed at a time, and it takes a little less than a day to process five tows when things go smoothly. On the other hand, the number of arrivals on any day is apparently independent of the number which arrives on any other day. Similarly, the number processed in any day seems to be independent of the number processed in any other day.

From these findings, the contractors concluded that the distribution for arriving tows is approximately Poisson, with a mean of five per day. They also concluded that the process rate is about six per day, and the number processed is also Poisson.

The waiting time between arrivals is therefore exponentially distributed, with a mean of $E(T) = \frac{1}{5} = 0.2$ and a variance of var $(T) = 0.04$. The time required to process a barge is also distributed exponentially. The average processing time per barge is $E(T) = \frac{1}{6} = 0.167$, and the variance in processing time is var $(T) = 0.028$.

The contractors will eventually use this information to estimate delay times for tows at the current locks and dam system and to develop computer simulation models to study the potential impacts on delays and the economics of alternative scheduling and operating strategies.

The Poisson and exponential models are the workhorses of systems reliability studies, and they formed the original mathematical basis for modeling service systems in which lines (or queues) tend to form.

The remaining models in this section are less easily explained, because they were derived using mathematics beyond the level of this book. The discussion here is aimed at providing some insight into the kinds of situations they model and to interrelate them, to the extent possible, to facilitate their use in Chapter 5. As one employs them in Chapter 5, one's level of intuition naturally improves.

Normal Model, $N(X/\mu, \sigma)$ (Sums of Variables)

Sums of random variables tend to take on a bell-shaped distribution as the number of variables added becomes large, so long as the variables are relatively independent and each variable has a relatively small effect on the size of the sum. The distributions of the individual variables are not a major consideration so long as the number of variables included is relatively large. This is one of the so-called *central limit theorems.*

Thus, if $X_1, \ldots, X_n$ are independent random variables and if n is fairly large, both the following sums are approximately normal:

$$Y_1 = X_1 + \cdots + X_n$$

$$Y_2 = \frac{1}{n} X_1 + \cdots + X_n$$

The means and variances of the respective normal variables are:

$$E(Y_1) = E(X_1) + \cdots + E(X_n) \qquad \text{var } (Y_1) = \text{var } (X_1) + \cdots + \text{var } (X_n)$$

$$E(Y_2) = \frac{1}{n}[E(X_1) + \cdots + E(X_n)] \qquad \text{var } (Y_2) = \frac{1}{n^2}[\text{var } (X_1) + \cdots + \text{var } (X_n)]$$

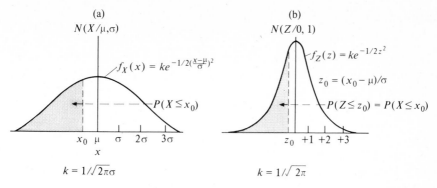

Figure 4.11 Normal density functions.

The density function for a normally distributed random variable is an exponential function of the variable, its mean value, and its variance, as shown:*

$$f_x(x) = ke^{-\frac{1}{2}[(x-\mu)/\sigma]^2}$$ normal

$$E(X) = \mu \qquad \text{var}(X) = \sigma^2$$

A normally distributed random variable X with a mean μ and a variance σ may be converted to a standardized normal variable Z having a mean value of zero and a variance of unity, by writing

$$Z = \frac{X - \mu}{\sigma}$$

Tabulations of the *normal distribution* are given in terms of standardized normal random variables.

Graphic representations for normal distributions are shown in Figure 4.11. Notice that the standardization has the effect of centering the distribution around the origin and squeezing most of the area into a small range of plus or minus three units. For normal random variables, not for others, it is always true that 67 percent of the area is contained within plus or minus one standard deviation $\pm\sigma$ from the mean, about 95 percent is within two standard deviations $\pm 2\sigma$, and all but about 1 percent, i.e., 99 percent, is within three standard deviations $\pm 3\sigma$; these are handy numbers to have memorized.

The normal distribution is the workhorse in statistical data analysis. It is also the foundation for the chi-square, Student's-t and F distributions. Without it, little of what is called statistical inference would exist today.

* The number k is equal to $1/\sigma\sqrt{2\pi}$, where $\pi \simeq 2.414$.

Chi-square Model, $\chi^2(X/n)$ (Sum of Squares of n Independent Standard Normal Variables)

The square of a standard normal random variable has a $\chi^2(X/1)$ distribution. The sum of, say, n independent standard normal variables has a $\chi^2(X/n)$ distribution. Thus, if $Y_1, \ldots, Y_{10}$ are independent normal random variables, with means $\mu_1, \ldots, \mu_{10}$ and variances $\sigma_1^2, \ldots, \sigma_{10}^2$, the following sum is $\chi^2(X/10)$;

$$X = \left(\frac{Y_1 - \mu_1}{\sigma_1}\right)^2 + \cdots + \left(\frac{Y_{10} - \mu_{10}}{\sigma_{10}}\right)^2 \qquad \text{chi-square}$$

The mean and variance for the *chi-square distribution* (sum of n squared normal variables) are

$$E(X) = n \qquad \text{var } (X) = 2n$$

The parameter n is called the number of *degrees of freedom*, for reasons explained below.

The approximate shapes of the chi-square density function for different numbers of variables included in the sum are shown in Figure 4.12.

Notice that when two chi-square variables are added, the sum is also a chi-square variable. That is,

$$\chi^2(X/n) + \chi^2(X/m) = \chi^2[X/(n + m)]$$

because the addition simply increases the number of squared, standardized normal variables in the sum.

The chi-square distribution is especially useful in statistical tests on sample proportions (see Section 5.3) and for accuracy assessments for sample variances (see Section 5.1.).

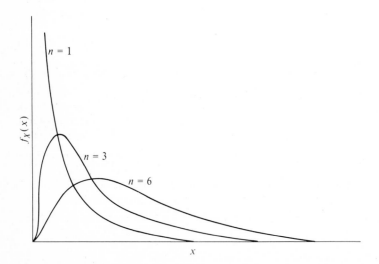

Figure 4.12 Chi-square density functions.

Sample variances are calculated as

$$S^2 = \frac{(y_1 - \bar{y})^2}{n} + \cdots + \frac{(y_n - \bar{y})^2}{n}$$

So, when a sample is taken from a normal distribution,

$$\frac{nS^2}{\sigma^2} = \frac{(y_1 - \bar{y})^2}{\sigma^2} + \cdots + \frac{(y_n - \bar{y})^2}{\sigma^2}$$

is a χ^2 variable. The parameter for this variable is not n but $n - 1$. There are two ways to explain this fact. The easiest explanation is simply that $\bar{y}$ is used for standardizing instead of μ. Since the sum of deviations of sample values around the sample average is always zero, only $n - 1$ of the deviations may be chosen freely; the remaining deviation must be the difference between zero and the sum of those. Consequently, only $n - 1$ of the terms in the chi-square summation are truly random variables. Thus the parameter must be $n - 1$, the number of degrees of freedom. The other explanation follows from the fact that

$$\frac{(y - \bar{y})}{\sigma} = \frac{(y - \mu)}{\sigma} + \frac{(\mu - \bar{y})}{\sigma}$$

which implies that a sum of terms such as

$$\frac{(y - \mu)^2}{\sigma^2}$$

can be written as a sum of terms such as

$$\left[\frac{(y - \bar{y})^2}{\sigma^2} + \frac{(\bar{y} - \mu)^2}{\sigma^2} \right]$$

Now $(y - \mu)^2/\sigma^2$ is $\chi^2(Y/n)$ and, because $\bar{y}$ is normal $N(\bar{Y}/\mu, \sigma\bar{y})$, $(\bar{y} - \mu)^2/\sigma^2$ is $\chi^2(\bar{Y}/1)$, and it must be true that $(y - \bar{y})^2/\sigma^2$ is $\chi^2[Y/(n - 1)]$.

Student's-*t* Model, $S(T/n)$

This distribution was discovered by W. S. Gosset* in 1908. He was concerned with comparing means calculated from small samples with a theoretical value. It is symmetric about the origin like the normal distribution. In statistical tests where the value of the variance for the sample distribution is unknown, the *Student's-t distribution* may be used in place of the normal distribution, as demonstrated in Chapter 5.

---

* Gosset was at the time employed by an Irish brewery which forbade publication of research by employees. He published under the pseudonym, "Student."

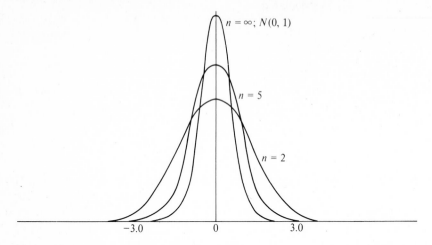

Figure 4.13 Student's-*t* density functions.

The ratio of a standard normal variable to the square root of a chi-square variable divided by the parameter n (degree of freedom) is a Student's-t variable. Symbolically,

$$T = \frac{N(z/0,\ 1)}{\sqrt{\chi^2(X/n)/n}} \qquad \text{Student's-}t$$

$$E(T) = 0 \qquad \text{and} \qquad \text{var } (T) = n/(n-2)$$

The Student's-t distribution is compared to the standard normal distribution in Figure 4.13. For all values of n it is more spread out than the standard normal distribution, because its variance is greater than unity.

F **Model,** $F(X/n_1,\ n_2)$ **(Ratio of Independent Chi-square Variables)**

The ratio of two independent chi-square variables, each divided by its parameter, is an F *variable*. Symbolically,

$$X_{n_1,\ n_2} = \frac{\chi_1^2/n_1}{\chi_2^2/n_2} \qquad\qquad F$$

The mean and variance for F distributions are

$$E(X) = \frac{n_2}{(n_2 - 2)} \qquad \text{and} \qquad \text{var } (X) = \frac{2n_2^2(n_1 + n_2 - 2)}{(n_2 - 4)(n_2 - 2)^2 n_1}$$

Because chi-square variables may not be negative, the F distribution may not be negative either. The F distribution begins at zero and carries to infinity, as indicated in Figure 4.14. Because it is a two-parameter distribution it can be used to approximate distributions of data from many kinds of experiments.

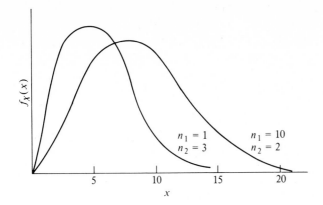

Figure 4.14 F density functions.

The major use of the F distribution in statistical analysis is in deciding whether the sample variances for two sets of data are different. It is used because sample variances are proportional to chi-square random variables. Thus the ratio of two-sample variances is proportional to an F variable. For example;

$$n S_1^2/\sigma_1^2 \quad \text{is} \quad \chi^2[Y/(n-1)]$$

and

$$m S_2^2/\sigma_2^2 \quad \text{is} \quad \chi^2[Y/(m-1)]$$

so

$$\frac{[n S_1^2/\sigma_1^2]/(n-1)}{[m S_2^2/\sigma_2^2]/(m-1)} \quad \text{is} \quad F_{(n-1),\,(m-1)}$$

Beta Model, $\mathscr{B}(X/r, t)$

The *beta distribution* is one of the most flexible of all distributions. Almost any set of statistical data can be approximated by a beta distribution with properly selected parameters r and t. It is an essential distribution in bayesian decision analysis (see Chapter 6) in a large part because of its flexibility, but also because it provides a natural distribution from which to estimate the probability of success p treated as a random variable from data consisting of t trials and r successes. Various shapes the distribution can assume are shown in Figure 4.15. The algebraic form of the density function is

$$f_X(x) = K x^{r-1}(1-x)^{t-r-1} \qquad 0 \le x \le 1$$

$$E(x) = \frac{r}{t} \qquad\qquad\qquad\qquad \text{beta}$$

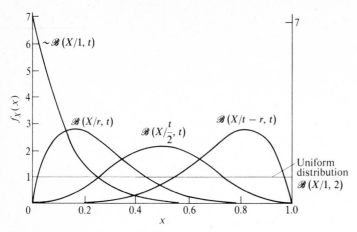

Figure 4.15 Beta density functions.

$$\sigma_x^2 = \frac{r(t - r)}{t^2(t + 1)}$$

Notice that if $r = 1$ and $t = 2$, the beta distribution is simply the *uniform distribution*, namely, $p(x) = 1$ for all x. For the uniform distribution, values in intervals of equal length are equally likely. In bayesian decision analysis the uniform distribution describes the situation where the decision maker has no idea of the likely outcomes of his or her actions except that they are within some bounds (see Chapter 6).

Table 4.4 Summary of probability models

Bernoulli, $B(X/p)$	Success or failure in one trial.
Binomial, $B(Y/n, p)$	Number of successes among n independent Bernoulli trials.
Geometric, $G(N/p)$	Number of trials until next success in series of independent Bernoulli trials.
Pascal, $P(N/p, k)$	Number of trials until k successes in series of independent Bernoulli trials.
Poisson, $\mathscr{P}(X/\mu)$	Number of arrivals per unit of chronological time. Models rare events. Especially useful in traffic studies and study of rare-disease occurrence.
Exponential, $\mathscr{E}(T/\mu)$	Time between Poisson arrivals. Especially useful in traffic studies and study of rare-disease occurrence.
Normal, $N(X/\mu, \sigma)$	Distribution for large sums of independent random variables, so long as none dominates the sum. Averages tend to be normally distributed.
Chi-square, $\chi(X/n)$	For sums of n squared independent normal variables. Estimates of variance for data generated by normal distribution have a χ^2 distribution.

Table 4.4—*Continued*

Student's-t, $S(T/n)$	For variables formed from ratio of a standard normal variable and the square root of a chi-square variable (divided by its parameter). Used in place of the normal distribution in statistical tests when sample sizes are small and σ^2 is unknown.
$F(X/n_1, n_2)$	For ratios of independent chi-square variables (each divided by its respective parameter value). Used in statistical tests comparing estimates of variances from separate sets of data.
Beta, $\mathcal{B}(X/r, t)$	Very flexible model. Especially useful in decision situations in which qualitative information must be converted to combine it with statistical data. Because it has two parameters, r and t, it can be used to model a wide range of statistical patterns. Used primarily in Chapter 6. Provides distribution for possible values of p, treated as random variable when t Bernoulli trials lead to r successes.

4.5 PROBABILITY TABLES

Tables for binomial, normal, chi-square, Student's-t, F, and beta models are included in Appendix 1. These are the most frequently used tabulations. Entire books of probability tables are available from commercial publishers and from government agencies for anyone who needs greater scope. Each tabulation in the appendix includes notes on its use.

4.6 PROBLEMS

4.1 Plot frequency diagrams and histograms for binomial distributions with:

(*a*) $p = 0.3$, $n = 10$ (*d*) $p = 0.5$, $n = 10$
(*b*) $p = 0.3$, $n = 15$ (*e*) $p = 0.5$, $n = 15$
(*c*) $p = 0.3$, $n = 20$ (*f*) $p = 0.5$, $n = 20$

4.2 Use the results of Problem 4.1 to generate PDF and CDF functions for each distribution.

4.3 Standardize the binomial random variables X_1 and X_2 having means $np = 6.0$ and $np = 10.0$, where $n = 20$. That is, create standard random variables Z_1 and Z_2 from the relation

$$Z = \frac{X - np}{\sqrt{np(1 - p)}}$$

(*a*) What are the mean and variance of these standardized variables?

(*b*) Calculate the probabilities that each of these variables will take values within each unit interval starting from $x = -0.5$ and ending at $x = 20.5$.

(*c*) Use the normal tables to calculate the same probabilities for Z.

(*d*) Construct a table comparing (*b*) and (*c*). Which of the two binomial distributions is best approximated by the normal distribution?

This problem demonstrates the power of the central limit theorem, which is discussed in Chapter 5. The point is that for large independent samples, sums of the sample values have a distribution which is nearly normal; binomial variables are sums of independent Bernoulli variables.

4.4 Suppose the likelihood of developing a skin cancer from a chest x-ray were constant and equal to one in one thousand.

(a) Which probability model would be appropriate to estimate the expected number of x-rays until the first lesion is triggered?

(b) What is the expected number of x-rays until the first lesion is triggered? What is the variance?

(c) If a whole population were being screened by x-ray procedures, how many people could be expected to pass through the screen until the first lesion was triggered?

4.5 If a person with tuberculosis (TB) is given a chest x-ray, the probability of detection is 0.95. If a person without TB is x-ray tested, she or he will be incorrectly diagnosed as having TB, with a probability of 0.002. Suppose 0.1 percent of adult residents of the city have TB. If a randomly selected person is diagnosed as having TB on the basis of a chest x-ray, what is the probability that he or she actually has TB?

4.6 A machine is known to fail at an average rate of three times per week. These failures occur at random, and it seems that at most one failure occurs in a small time interval. Furthermore, the likelihood of a failure in a small time interval Δt is about $3\Delta t$. Time increments are independent.

(a) What is the most reasonable distribution for X where X is the number of failures?

(b) What is the distribution for time between failures and what is the average time between failures?

4.7 A maintenance worker can repair a machine in about 2 hours on the average. The time taken to repair is a random variable. Provided that the maintenance worker never has to wait to start work on another machine, the rate at which machines are repaired is about one repair every 2 hours. The maintenance worker cannot repair more than one machine in a small increment of time, and it is reasonable to assume that time increments are independent.

(a) What distribution is reasonable for the time taken to repair? What is the expected time taken to repair?

(b) What distribution is reasonable for the repair rate (number of repairs per hour)? What is the variance for the repair rate?

(c) Suppose there were two maintenance workers and an infinite supply of broken machines. Suppose the average performance for the second maintenance worker, who is independent of the first, is one machine per hour. What is the average repair rate?

(d) For each of the above two cases what is the probability that more than four machines per hour can be repaired?

4.8 Suppose you found a population of people, ages 1 to 65 years, for which there were just as many people in each age interval of the same length; this holds as the interval lengths become arbitrarily small. What is the PDF for this population? What is the CDF?

(a) Write these answers algebraically and sketch them.

(b) Calculate the mean and variance and show them on your sketches.

4.9 A doctor claims to have a procedure which is very valuable in that it gives prospective parents significant control over the sex of their future offspring. The medical profession is skeptical and wants data on the success of the procedure in 12 random cases. If the procedure fails more than three times, the profession will publicly denounce it. Under normal circumstances nature provides for 50 percent boys and 50 percent girls. What is the likelihood that the procedure will be denounced?

4.10 Barges at Locks and Dam No. 27 arrive at a rate of 10 per day. Processing time is about one per hour. The locks operate for only 10 hours per day.

(a) What is the average waiting time for barges (in hours)?

(b) What is the probability that two barges will arrive nearly simultaneously?

4.11 In St. Louis the number of severe tornadoes per year varies independently from year to year. Historical records indicate that the number per year over the last 100 years has a distribution about as follows:

Number	Relative frequency
0	0.37
1	0.36
2	0.19
3	0.06
4	0.02

(a) Which probability model would be useful for answering questions about the expected number of severe tornadoes per year? Calculate its mean and variance.

(b) Compare the relative frequencies predicted by the model to those given in the table.

4.12 Suppose $X_1, X_2, X_3, \ldots, X_n$ are variables whose values you can set. You theorize that some variable Y is related to the X's by

$$Y = a_0 + a_1 X_1 + a_2 X_2 + \cdots + a_n X_n$$

You do some experiments to determine the coefficients $a_0, \ldots, a_n$ and these provide you with probability distributions instead of specific values for the a's. This implies that Y is a random variable. From the probability distributions you calculate the expected values for the a's, namely, μ_0, $\mu_1, \ldots, \mu_n$. What is the expected value of Y?

4.13 Let X, Y be random variables. Let X take values 1, 2, 3 and Y take values 2, 4, 6, with probabilities of $\frac{1}{3}, \frac{1}{3}, \frac{1}{3}$ and $\frac{2}{6}, \frac{3}{6}, \frac{1}{6}$, respectively. Suppose that $p(x, y)$ has values

$$p(1, 2) = \frac{1}{9} \qquad p(2, 2) = \frac{1}{9} \qquad p(3, 2) = \frac{1}{9}$$

$$p(1, 4) = \frac{1}{6} \qquad p(2, 4) = \frac{1}{12} \qquad p(3, 4) = \frac{1}{4}$$

$$p(1, 6) = \frac{1}{18} \qquad p(2, 6) = \frac{1}{36} \qquad p(3, 6) = \frac{1}{12}$$

Are X, Y independent? Why?

4.14 Two discrete variables X and Y are defined to be stochastically independent if $P(X = x \ \& \ Y = y) = P(X = x) P(Y = y)$ for every possible value of X and every possible value of Y. An equivalent definition is $P(X = x/Y = y) = P(X = x)$, of course.

Suppose X represents extended years of life and Y represents annual cost of treatment, and the possible values of X and Y are $X = 0, 1, 2$ and $Y = 50, 75, 100$.

(a) If the data below represent joint occurrences of pairs of X and Y values in an experimental study, can you conclude that the two variables were independent? You will first need to calculate $P(X = x)$ and $P(Y = y)$ for each value of X and Y. For example, $P(X = x) = P(X = x \ \& \ Y = 50) + P(X = x \ \& \ Y = 75) + P(X = x \ \& \ Y = 100)$. Assume that the data are truly representative so that frequencies can be interpreted as probabilities.

Statistical data (X, Y)

(0, 50)	(1, 50)	(0, 75)
(0, 100)	(1, 75)	(2, 100)
(0, 75)	(1, 100)	(1, 50)
(2, 100)	(2, 50)	(1, 100)
(2, 100)	(1, 75)	(2, 75)
(1, 75)	(0, 100)	(0, 100)

X = extended years of life.
Y = treatment costs.

(b) Calculate all the conditional probabilities that a patient's life will be extended by a given amount if certain costs are incurred. Do so first by counting directly. Then do it by calculations of the type $P(X = x/Y = y) = P(X = x \ \& \ Y = y)/P(Y = y)$.

4.15 Repair costs per episode have been recorded as follows:

$ 75	$125	$225	$400
100	75	100	50
25	275	50	25
1750	25	75	150
150	100	50	175

On the basis of the data above,
    (a) Find the expected value of the cost to repair.
    (b) Find the probability that a given repair will cost less than 200 dollars.
    (c) Graph the data with "Cost to repair" as the horizontal axis, and "Probability that repair cost will not exceed \$_____" as the vertical axis.

4.16 The average height of a person at full growth is related to numerous factors including height of parents, nutrition, and so forth. The standard equation for predicting height is

$$H = a_0 + a_1 X_1 - a_2 X_2 + \cdots + a_n X_n$$

Distributions for the x's are not known, but they are assumed to be independent and none dominates. The means and variance are μ_1 through μ_n and σ_1^2 through σ_n^2, respectively. The a's are constants.
    (a) What is the distribution of H?
    (b) What is the mean for H?
    (c) What is the variance for H?

4.17 Diversity is essential for stability of biological systems. A biologist has observed that when the number of species falls to 100, systems become unstable. At a particular site there are 110 species. There is a small chance ($p = 0.01$) that another species will die out in a given year and not be replaced. There is practically no chance that more than one species will die out in any one year.
    (a) What is the expected time until the system becomes unstable?
    (b) What is the variance?
    (c) What is the probability that instability will occur within 5 years; 10 years?

4.18 Name two models from which waiting time can be calculated. Distinguish the two concepts in a sentence or two. Describe a situation to which each concept is applicable; be specific but brief.

4.7 SUPPLEMENTAL READING

Benjamin, Jack R., and C. Allin Cornell: *Probability, Statistics and Decision for Civil Engineers*, McGraw-Hill Book Company, New York, 1970, chaps. 1 and 2.

Chapman, Douglas G., and Ronald A. Schaufele: *Elementary Probability Models and Statistical Inference*, Ginn-Blaisdell, Waltham, Mass., 1970.

Drake, Alvin W.: *Fundamentals of Applied Probability Theory*, McGraw-Hill Book Company, New York, 1967.

Dwass, Meyer: *First Steps in Probability*, McGraw-Hill Book Company, New York, 1967, chaps. 2, 3, and 6.

Hays, William L., and Robert L. Winkler: *Statistics: Probability, Inference, and Decision*, Holt, Rinehart and Winston, Inc., New York, 1971, chap. 6.

Hoel, Paul G.: *Elementary Statistics*, 3d ed., John Wiley & Sons, Inc., New York, 1971, chap. 2.

Mosteller, Frederick, Robert E. K. Rourke, and George B. Thomas, Jr., *Probability with Statistical Applications*, Addison-Wesley Publishing Company, Inc., Reading, Mass., 1961.

FIVE

STATISTICAL COMPARISONS

For the sake of economy, statistical studies are often based on sample data. The underlying assumption is that when sample data are characteristic of the larger population from which they are drawn, sample proportions, means, variances, and other statistics calculated from the data are reasonably good estimates of the corresponding parameters of the true distribution. The questions which must be answered, of course, are, "How accurate are the estimates?" and, "What conclusions can be drawn from them?"

To draw reasonable conclusions from sample data one must know how the data were gathered and how well they represent the population from which they were drawn. Obviously, morbidity data from asbestos workers cannot be expected to reflect morbidity rates for white-collar workers in Montana. Asbestos workers are prone to lung cancer and emphysema because they inhale asbestos particles all day. White-collar workers in Montana live in a very different environment. Similarly, a sample consisting of 85 percent Catholics and 15 percent Protestants could hardly be used to accurately estimate the church attendance rates for the entire United States population. A representative sample would have to consist of individuals from all church-going groups in proportion to their numbers in the population.

The secret to sampling is *random selection*. If a sample is taken in a manner which guarantees that every element in the entire population has an equal chance of being included, the sample is defined as random. The most efficient form of random sampling is *stratified random sampling*. Stratification minimizes sample sizes required for statistical accuracy. The population is first classified according to parameters which reflect its character, such as religion, socioeconomic status, tree density, soil type, or geographical region. The random samples are drawn from each stratum in proportion to the fraction of the total population it represents and are pooled to form the complete sample.

The statistical methods discussed in this chapter are effective tools for analyzing randomly obtained samples and inferring population characteristics from the results. Medical researchers use them to evaluate new drugs, surgical procedures, and diagnostic procedures. Engineers use them to evaluate products or system designs. Physical scientists use them to evaluate the characteristics of materials. Economists and market analysts use them to evaluate consumer behavior or preferences. Pollsters use them to evaluate voter attitudes. Agricultural scientists use them to develop new hybrid plants, cultivation procedures, animal husbandry methods, and the like. City managers use them to evaluate public services.

The accuracy of a statistical estimate based on a random sample can be calculated in terms of the likelihood that the true value differs from the estimate by less than a specified amount. This likelihood is called the *confidence level* for the estimate. Accuracy is the topic of Section 5.1.

The methodology of statistical comparisons and conclusions based on random samples is called *hypothesis testing*. This is the topic of Sections 5.2 through 5.4. Statistical accuracy and statistical comparisons procedures are essential tools for comparing qualities of products, efficacy of drugs, effectiveness of human services delivery procedures, nutritional value of foods, and so forth. For example, the 1977 to 1978 saccharin ban decision was supported by evaluations of the types discussed in Sections 5.2 and 5.3.

In this and subsequent chapters there is no need to distinguish between variables and their values by using uppercase and lowercase symbols. That convention is hereby abandoned.

5.1 CONFIDENCE IN OR ACCURACY OF STATISTICAL ESTIMATES

Estimates of proportions, means, and variances made from sample data almost certainly will be inaccurate. They may be very close to the true parameter values for the population, but it would be pure chance if they were exact. In fact, if the sampling were repeated in precisely the same way, the odds are overwhelming that the second estimate would be different from the first estimate and would also differ from the true value. On the other hand, the average of the two estimates would likely be more accurate than either of the two individual estimates. If the sampling were repeated a great many times, the estimates would tend to cluster around the true value. The larger the sample size, the tighter the cluster. In other words, sample-based parameter estimates, or statistics, as they are called, are actually random variables with probability distributions which tend to peak around the true value of the parameter being estimated. Because of this phenomenon larger samples provide more accurate estimates, and averages

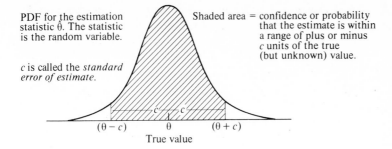

PDF for the estimation statistic $\hat{\theta}$. The statistic is the random variable.

c is called the *standard error of estimate.*

Shaded area = confidence or probability that the estimate is within a range of plus or minus c units of the true (but unknown) value.

$(\theta - c)$ θ $(\theta + c)$

True value

Note: The probability that $\hat{\theta}$, calculated from the sample data will fall within the interval, $\theta \pm c$, is

$$P(\theta - c \leq \hat{\theta} \leq \theta + c) = \text{confidence}$$

Algebraic manipulation of the inequalities within this probability statement yields an equivalent statement

$$P(-c \leq \hat{\theta} - \theta \leq c) = \text{confidence}$$

Further manipulation yields a form which highlights the so-called *confidence interval,* $\hat{\theta} \pm c$, around the estimates within which the true value, θ, is located, with the specified probability

$$P(\hat{\theta} - c \leq \theta \leq \hat{\theta} + c) = \text{confidence}$$

Figure 5.1 Accuracy assessment for sample-based estimate $\hat{\theta}$, of a parameter θ.

of repeated estimates are more accurate than individual estimates. The mathematical formulations of these facts are called the *laws of large numbers.*

The idea is suggested by Figure 5.1. Knowing the form of the distribution for the estimator allows one to calculate the *confidence,* or likelihood, that a specified interval centered on the estimated value actually contains the true value of the parameter. One can also determine the *interval size* associated with a specified confidence level or even the *required sample size* for a specified interval and confidence level.

The calculations are straightforward. From theoretical considerations, the form of the distribution and the appropriate table of probabilities for the random variable are established. These vary depending on whether the parameter is a mean, a proportion, or a variance, as well as on the size of the random sample from which the parameter is being estimated. Finally, using elementary algebra,

- The *error interval,* $\pm c$ units on either side of the estimate, is determined from the appropriate probability table for a specified confidence level (probability), sample size, and sample estimate.
- Or, the *confidence level* is determined from the appropriate probability table for a specified interval, sample size, and estimate.

- Or, the required *sample size n* is determined from the appropriate probability table for a specified confidence level (probability), error interval, and sample estimate.

The required algebraic manipulations are specific to the parameter being estimated and to which of the three questions listed above is being answered. They are demonstrated in the examples which follow.

Proportions and Means

A fundamental theorem of probability called the central limit theorem plays an essential role when proportions and means are being estimated and sample sizes are large ($n > 50$). The theorem states that for a relatively large independent random sample $x_1, x_2, \ldots, x_n$ for any variable x regardless of its distribution, the random variable y corresponding to the ratio

$$y = \frac{x_1 + x_2 + \cdots + x_n}{n}$$

is approximately normally distributed; and the mean value of y is the same as the mean value of x, while the variance of y is simply the variance of x divided by the sample size n. That is,

$$f_Y(y) \text{ is normal} \qquad \mu_y = \mu_x \qquad \text{and} \qquad \sigma_y^2 = \frac{\sigma_x^2}{n}$$

Therefore, as shown in Chapter 4, the standardized variable

$$z = \frac{y - \mu_x}{\sigma_x/\sqrt{n}}$$

is a standard normal variable; it is normal, has a mean value equal to zero and a variance equal to unity. The central limit theorem is directly applicable to both proportions and means, which are estimated in the form of y. These important facts are illustrated in Figure 5.2.

The theorem also states that if two independent variables x and y are at least approximately normally distributed, their difference is also approximately normally distributed. The mean of this difference is the difference in the two means, and the variance is the sum of the two variances.

Proportions (large sample size) Since proportions p are estimated by forming a ratio of the number of successes to the total number of independent Bernoulli trials, $\hat{p}$ is approximately normal, with mean and variance $\mu_{\hat{p}}$ and $\sigma_{\hat{p}}^2$, as shown in Figure 5.2. The probability that an interval of magnitude $c = \pm k\sigma_{\hat{p}}$ around the estimate $\hat{p}$ actually includes the true value p can be written

$$P(p - k\sigma_{\hat{p}} \leq \hat{p} \leq p + k\sigma_{\hat{p}}) = \text{confidence}$$

PDF for y
(approximately normal)

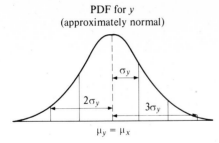

$\mu_y = \mu_x$

General:
Repeated sampling from the distribution x.

$$y = \frac{x_1 + x_2 \ldots + x_n}{n} \; ; \mu_y = \mu_x \, ; \, \sigma_y^2 = \sigma_x^2 / n$$

$$z = (y - \mu_y)/\sigma_y \, ; \, \mu_z = 0 \, ; \, \sigma_z^2 = 1$$

PDF for z

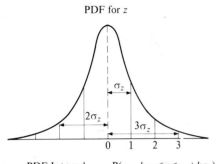

$$0 \quad 1 \quad 2 \quad 3$$

Proportions ($y = \hat{p}$):
Each $x_i = 0$ or 1 because x is Bernoulli, $B(x/p)$.

$$\mu_x = p \, ; \, \sigma_x^2 = p(1 - p)$$

$$\mu_{\hat{p}} = p \, ; \, \sigma_{\hat{p}}^2 = p(1 - p)/n$$

$$z = \frac{(\hat{p} - p)}{\sqrt{p(p - p)/n}} \, ; \, \mu_z = 0 \, ; \, \sigma_z^2 = 1$$

PDF Interval $\mu \pm k\sigma$			$P(\mu_y - k\sigma_y \le y \le \mu_y + k\sigma_y)$ $= P(-k \le z \le k)$	
(k)	(y)	(z)	(Area)	(%)
1	$\mu_y \pm \sigma_y$	1	0.6826	68.26
2	$\mu_y \pm 2\sigma_y$	2	0.9544	95.44
3	$\mu_y \pm 3\sigma_y$	3	0.9974	99.74

Means ($y = \bar{x}$):
Each x_i has the same mean and variance; μ_x, σ_x^2.

$$\mu_{\bar{x}} = \mu_x \, ; \, \sigma_{\bar{x}}^2 = \sigma_x^2 / n$$

$$z = \frac{(\bar{x} - \mu_x)}{\sigma_x / \sqrt{n}} \, ; \, \mu_z = 0 \, ; \, \sigma_z^2 = 1$$

Figure 5.2 Central limit theorem applied to proportions and means.

The corresponding confidence interval from $\hat{p} - k\sigma_{\hat{p}}$ to $\hat{p} + k\sigma_{\hat{p}}$ is written as follows, after substitution for $\sigma_{\hat{p}}$:

$$\hat{p} - k\sqrt{\frac{p(1 - p)}{n}} \le p \le \hat{p} + k\sqrt{\frac{p(1 - p)}{n}}$$

The interval is symmetric because the variable is normally distributed and the normal distribution is symmetric. The product $k\sigma_{\hat{p}} = k\sqrt{p(1 - p)/n}$ is called the *error of estimate*. It is the number c shown in Figure 5.1.

The next two examples demonstrate the appropriate methods for evaluating the accuracy of estimated proportions in the case of large samples, and for determining the sample size required to achieve a prespecified degree of accuracy.

Example 5.1: Confidence in large-sample proportion estimates A drug company tested a new drug for side effects. One hundred patients ($n = 100$) were randomly selected and subjected to identical treatment protocols. Forty percent of the patients experienced certain side effects. The company must file a report with the U.S. Food and Drug Administration, indicating the accuracy of its findings.

1. The company might report the 95 and 99 percent confidence intervals for the proportion of patients in the diseased population who would experience the side effects if treated with the drug according to company recommendations, as the sample group was. Company researchers would calculate these confidence intervals as follows.

Let p symbolize the actual, but unknown, proportion of the diseased population which will suffer the side effects. Then the likelihood that a randomly selected individual will experience such effects is (theoretically) p. Let $\hat{p}$ symbolize the sample estimate of p. According to the central limit theorem, $\hat{p}$ is approximately normal, with a mean value $(\mu_{\hat{p}})$ equal to p and a variance $(\sigma_{\hat{p}}^2)$ equal to $p(1-p)/100$. Thus,

$$z = \frac{\hat{p} - \mu_{\hat{p}}}{\sigma_{\hat{p}}} \qquad \mu_{\hat{p}} = p \qquad \sigma_{\hat{p}} = \sqrt{\frac{p(1-p)}{100}}$$

is a standard normal variable.

The probability that an interval of $\pm k$ standard deviations $(\pm k\sigma_{\hat{p}})$ around $\hat{p}$ will actually contain the true value p can be written conveniently as

$$P(-k\sigma_{\hat{p}} \leq (\hat{p} - p) \leq k\sigma_{\hat{p}}) = \text{confidence}$$

Dividing through by $\sigma_{\hat{p}}$ recasts this statement in terms of the standard normal variable, as follows:

$$P(-k \leq z \leq k) = \text{confidence}$$

The number k is determined from standard normal tables. For a confidence level of 95 percent, the normal tables indicate that k equals 1.96. Consequently, if $\hat{p} = 0.4$, there is a 95 percent chance that the true proportion p for the drug is somewhere within the interval

$$0.4 - 1.96\,\sigma_{\hat{p}} \leq p \leq 0.4 + 1.96\,\sigma_{\hat{p}}$$

Since the sample is large, $\hat{p}$ is probably a good estimate for p, so one may substitute $\hat{p}$ for p in calculating the variance to estimate the numerical limits. Thus $\sigma_{\hat{p}} \simeq \sqrt{0.4(0.6)/100} = 0.049$, and the 95 percent interval is

$$0.304 \leq p \leq 0.496$$

For a confidence level of 99 percent, k has a value of 2.58 (see the normal tables in Appendix 2). The corresponding interval is

$$0.275 \leq p \leq 0.525$$

Thus, the company would report that, according to its study, there is a 99 percent chance that, of all patients treated, some 27.5 to 52.5 percent would experience certain side effects, and a 95 percent chance that the percentage affected would actually be in a smaller interval, between 30.4 and 49.6 percent, provided the drug were administered as recommended.

2. The company could also report accuracy levels for specific intervals. If the interval of plus or minus one standard deviation ($k = 1$) around the estimate were of special interest, and it often is, company researchers would calculate the confidence level from

$$P(-\sigma_{\hat{p}} \le \hat{p} - p \le \sigma_{\hat{p}}) = \text{confidence}$$

or
$$P(-1 \le z \le 1) = \text{confidence}$$

$$\simeq 68\%$$

The 68 percent confidence interval for p is

$$\hat{p} - 0.049 \le p \le \hat{p} + 0.049 \qquad \text{or} \qquad 0.351 \le p \le 0.449$$

Thus, the company would report that it was 68 percent confident that the error in the sample estimate for p was less than or equal to 0.049 units, or equivalently, that there was a 68 percent chance that the true value of p was in the approximate interval from 0.35 to 0.45.

Example 5.2: Choosing a large sample size for proportions Researchers for the drug company of the previous example could have designed their study so that prespecified accuracies would be obtained by testing the drug on a sufficiently large randomly selected patient sample. The required sample size n, to guarantee an error of estimate no greater than, say, 0.03 units ($k\sigma_{\hat{p}} = 0.03$), is easy to calculate.

For a specified minimum confidence level of 95 percent, the probability statement is

$$P\left(-k \le \frac{\hat{p} - p}{\sigma_{\hat{p}}} \le k\right) \ge 0.95$$

The value of k corresponding to 95 percent confidence is $k = 1.96$ from the normal tables. This is the minimum value of k meeting the specified confidence level. The minimum sample size required to meet the specification is the smallest n for which

$$1.96\ \sigma_{\hat{p}} = 0.03$$

To find it, substitute for $\sigma_{\hat{p}}$ and solve for n, as follows:

$$k\sigma_{\hat{p}} = 1.96\sqrt{\frac{p(1-p)}{n}}$$

$$n = \frac{(1.96)^2 p(1-p)}{(0.03)^2} \simeq \frac{(1.96)^2 (0.49)^2}{(0.03)^2} = 1025$$

From these calculations, company researchers would know that they would have to study the effects of the drug on about 1025 randomly selected patients, in order to be 95 percent confident that the true value of p could be within ± 0.03 units (three percentage points) of their estimate; $-0.03 \leq (\hat{p} - p) \leq 0.03$.*

Averages (large sample sizes) Accuracy evaluations for estimated averages proceed in a similar fashion to those for proportions, as demonstrated in the next two examples. Since

$$\bar{x} = \frac{x_1 + x_2 + \cdots + x_n}{n} \qquad \mu_{\bar{x}} = \mu_x \qquad \sigma_{\bar{x}}^2 = \frac{\sigma_x^2}{n}$$

we know from the central limit theorem that $\bar{x}$ is approximately normal. Thus, one works with the standard normal variable

$$z = \frac{(\bar{x} - \mu_x)}{\sigma_x / \sqrt{n}}$$

The probability that an interval of magnitude $c = \pm k\sigma_{\bar{x}}$ around the estimate $\bar{x}$ actually contains the true mean μ_x can be written

$$P(\mu_x - k\sigma_{\bar{x}} \leq \bar{x} \leq \mu_x = k\sigma_{\bar{x}}) = \text{confidence}$$

The corresponding confidence interval is

$$\bar{x} - \frac{k\sigma_x}{\sqrt{n}} \leq \mu_x \leq \bar{x} + \frac{k\sigma_x}{\sqrt{n}}$$

The interval is symmetric due to the symmetry of the normal distribution. The error of estimate is $k\sigma_{\bar{x}} = k\sigma_x / \sqrt{n}$.

Example 5.3: Confidence in large-sample estimates of means The Environmental Protection Agency is studying fuel economy data for new automobiles. A company submitted data for 100 randomly selected cars of a particular model. The average mileage reported was 20 miles per gallon, with a standard deviation of 10 miles per gallon. Agency statisticians have been asked to assess the accuracy of the reported average. The random variable $\bar{x}$ is related to the mileage variables x_i by

$$\bar{x} = \frac{x_1 + \cdots + x_{100}}{100} \qquad x_i \sim i\text{th car}$$

The distribution for $\bar{x}$ is approximately normal, with a mean equal to μ_x, which is unknown, and a variance equal to $\sigma_x^2 / 100$, according to the

* Had the drug company of Example 5.1 done its testing on a small sample of, say, 10 patients, the estimation procedures would have had to be altered to obtain reasonable results. This is because the distribution for the estimate $\hat{p}$ could no longer be assumed normal. Statisticians have developed special techniques for this case. These are technically difficult and are left to more advanced texts.

central limit theorem; the mean μ_x and variance σ_x^2 are the theoretical mean and variance for the auto model as well as for each of the x_i's. Accordingly, to find the 95 percent confidence interval for the estimate $\bar{x}$, the statisticians would use the normal tables to find the minimum k satisfying the relation

$$P(-k \le z \le k) \ge 0.95 \qquad z = \frac{\bar{x} - \mu_x}{\sigma_x/\sqrt{100}}$$

The value of k is 1.96, indicating that there is a 95 percent chance that the model mean μ_x and the estimated mean $\bar{x} = 20$ differ by less than $c = 1.96$ $\sigma_x/\sqrt{100}$ units. Although σ_x is unknown, the sample estimate S_x may be used instead because the sample is large. Since the sample standard deviation is 10, $\sigma_x/\sqrt{100}$ is approximately 1, so the 95 percent interval is

$$20 - 1.96 \le \mu_x \le 20 + 1.96 \qquad \text{or} \qquad 18.04 \le \mu_x \le 21.96$$

To find the confidence level for a specified error of estimate, say, two standard deviations ($k = 2$), the statisticians would simply calculate the probability

$$P(-2 \le z \le 2) = \text{confidence}$$

From the normal tables, the confidence level is about 95.5 percent. The statisticians would therefore report that the data indicated a 95.5 percent certainty that the true, but unknown, average was within approximately 2 miles per gallon of the 20-miles-per-gallon estimate.

Example 5.4: Choosing a large-sample size for means Had the Environmental Protection Agency imposed an accuracy standard of, say, 0.2 units ($k\sigma_{\bar{x}} = 0.2$) at the 95 percent confidence level, a much larger sample than 100 cars would have been required for the mileage study. However, the necessary sample size could not be determined without some prior knowledge of the actual standard deviation σ_x for the fleet. In these kinds of situations, preliminary estimates of the standard deviation, based on smaller samples, and information from engineering studies of previous years are often used to approximate σ_x.

Assuming that $S_x = 10$ is a reasonably good estimate for σ_x, the engineers could estimate the required sample size by finding $k = 1.96$ from the normal tables. Then, solving for n, or $k\sigma_{\bar{x}} = k\sigma_x/\sqrt{n} = 0.2$,

$$n = \frac{(1.96)^2 100}{(0.2)^2} = 9604$$

Averages (small-sample sizes; $n < 50$) For economic reasons small samples are preferable to large samples. The normal distribution is not applicable to small-sample data unless one already knows the variance σ_x, because S_x is no longer a good approximation of σ_x. Instead one uses the Student's-t variable (with parameter $n - 1$), which incorporates S_x in place of σ_x,

$$t = \frac{\bar{x} - \mu_x}{S_x/\sqrt{n}} \qquad \text{approximately Student's-}t$$

It is applicable even for samples as small as $n = 5$, so long as the sample is drawn from a distribution which is approximately normal.* Many statistical studies are done with small samples, so the Student's-t distribution is frequently employed. The accuracy evaluation process is essentially the same as in previous examples, as demonstrated in Example 5.5. The Student's-t distribution is symmetric around the origin, so confidence intervals are symmetric.

Example 5.5: Confidence in small-sample mean estimates The county highway department tested a new road surface in nine locations. One-mile segments of the surface were laid in each location. The number of vehicles crossing the surface was counted electronically, and the lifetime of the surface was measured in terms of the number of vehicles which crossed the surface before the first major flaw was observed.

Traffic engineers calculated the average lifetime $\bar{x}$ to be 100,000 cars, with the sample standard deviation of $S_x = 900$. The 95 percent confidence interval for the estimate of the mean was determined as usual. The value $k = 2.3$ in the relation

$$P\left(-k \le \frac{\bar{x} - \mu_x}{S_x/\sqrt{n}} \le k\right) \ge 0.95$$

was extracted from the tables of the Student's-t distribution with parameter $v = n - 1 = 8.$† The interval was then calculated from the relation

$$\bar{x} - \frac{2.3S_x}{\sqrt{9}} \le \mu_x \le \bar{x} + \frac{2.3S_x}{\sqrt{9}}$$

By substitution, the interval limits were found to be 99,310 and 100,690. The engineers therefore reported 95 percent confidence in the surface lifetime estimate of $100,000 \pm 690$ cars.

By similar calculations they determined that there was only about an 80 percent chance that the true value and the estimate differed by no more than 1.4 standard deviations, or 420 cars.

Sample-size requirements may be calculated using the Student's-t distribution, just as they were calculated using the normal table. However, there is no

---

* As in Chapter 4, the ratio of a standard normal variable to the square root of a chi-square variable divided by its parameter value is a Student's-t variable. The normal variable here is $(\bar{x} - \mu)\sqrt{n}/\sigma_x$. The χ^2 variable is $(n - 1)S_x^2/\sigma_x^2$. The parameter is $(n - 1)$.

† The Greek symbol v, pronounced "nu," is commonly used to symbolize the parameter value of the distribution to avoid confusing it with sample size.

guarantee that the sample size determined will be small. If the required confidence level were 95 percent and the specified estimate of error were two standard deviations, the necessary sample size would be 61, because sample size is calculated as $n = v + 1$, and v, the parameter of the Student's-t distribution, must have a value of 60, as determined from the table.

Variances

Accuracy assessments for the sample variance are also easy to make when the sample is taken from a normal distribution or when sample sizes are large. Then the ratio nS_x^2/σ_x^2 is approximately a chi-square variable with $v = n - 1$;*

$$\frac{nS_x^2}{\sigma_x^2} = \chi_{n-1}^2$$

Therefore, the 95 percent confidence interval can be calculated from the relation

$$P\left(k_2 \leq \frac{nS_x^2}{\sigma_x^2} \leq k_1\right) = P(k_2 \leq \chi_{n-1}^2 \leq k_1) = 0.95$$

where k_2 represents the upper 97.5 percent tail and k_1 represents the upper 2.5 percent tail of the distribution.† This relation may be rewritten as

$$P\left(\frac{nS_x^2}{k_1} \leq \sigma_x^2 \leq \frac{nS_x^2}{k_2}\right) = 0.95$$

which indicates the 95 percent interval limit on σ_x^2. To find the numerical limits on the interval, one refers to the chi-square table for the appropriate parameter value and converts the results as suggested by these equivalent relations. The next example demonstrates the process. The general shape of the χ^2 distribution is illustrated in Figure 5.3 on page 92.

Example 5.6: Confidence in large-sample variance estimates In Example 5.3 the Environmental Protection Agency was evaluating mileage data submitted by an auto manufacturer. The sample size was $n = 100$ and the sample standard deviation, $S_x = 10$ miles per gallon. Agency statisticians could have estimated the 95 percent confidence interval for S_x^2 by noting that $v = (n - 1) = 99$; so the upper 2.5 percent region of this chi-square distribu-

---

* There are methods appropriate to small samples when the normality assumption does not hold, but they are much more advanced and are not discussed in this text. The procedure illustrated here works for large sample sizes even in the absence of normality, because for large n, χ_{n-1}^2 is nearly normal. Hence, although the normal model should be used, the χ^2 model is approximate.

† Refer to Figure 5.3 or to the chi-square table of probabilities, where the shape of the chi-square distribution is illustrated. Notice especially that it is not symmetric. For this reason one must obtain a k value for each tail of the distribution.

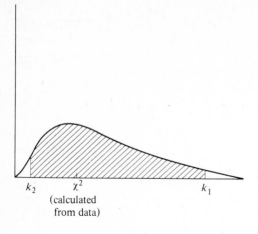

k_2 χ^2 k_1
(calculated
from data)

Figure 5.3 Confidence intervals for the χ^2 PDF.

tion begins at about $k_1 = 128$, while the upper 97.5 percent region begins at about $k_2 = 73$. Thus, the interval bounds on σ_x^2 are approximately

$$\frac{nS_x^2}{k_1} = 78 \qquad \frac{nS_x^2}{k_2} = 137$$

$$78 \le \sigma_x^2 \le 137$$

That is, the engineers could have determined that there was a 95 percent chance that the true value σ_x^2 was actually in the interval from 78 to 137 miles per gallon.

For convenience, the methodological highlights of this section are summarized in Table 5.1.

Table 5.1 Accuracy assessment
Means, proportions, and variances

	Variable	Confidence level
Large-sample sizes (n)		
Proportions	$z = (\hat{p} - p)\sqrt{n}/\sigma_p$	$P(-k \le z \le k)$
Means	$z = (\bar{x} - \mu)\sqrt{n}/\sigma_{\bar{x}}$	$P(-k \le z \le k)$
Variances	$\chi_{n-1}^2 = nS^2/\sigma^2$	$P(k_2 \le \chi^2 \le k_1)$
Small-sample sizes (n)		
Proportions*		
Means	$t = (\bar{x} - \mu)\sqrt{n}/S_x$	$P(-k \le t \le k)$
Variances†	$\chi_{n-1}^2 = (n-1)S^2/\sigma^2$	$P(k_2 \le \chi^2 \le k_1)$

  * See advanced texts, e.g., J. V. Bradley or J. L. Fleiss.
  † This procedure works for small samples only if sampling is from a normal distribution. There are special techniques for other situations which are covered in advanced texts.

5.2 COMPARING TWO ESTIMATES

The statistical method of comparing estimates of proportions, means, and variances is called hypothesis testing. The so-called null hypothesis, which is symbolized H_0, makes the claim that differences observed are not real, but are simply chance occurrences. The alternative hypothesis H_1 may claim nothing more than that the differences are real or it may claim that the differences are either positive or negative.* The object of hypothesis testing is to *reject* or *not reject* the null hypothesis, based on the statistical data.

Two kinds of error are associated with hypothesis testing. One may reject the null hypothesis when it is, in fact, true or one may fail to reject it when it is false. The first type of error is called a *Type I error*, or *significance* by statisticians. The second is called a *Type II error*. The likelihood of making an error of either type depends on the decision rule one employs. Obviously, the likelihood of being wrong is greater when one rejects the null hypothesis on the basis of very small statistical differences, although the larger the sample size, the greater the chance that even small statistical differences are real. The *power* of the test is 1 minus the Type II error level, because it is the likelihood of correctly rejecting H_0 when the alternative is true. Symbolically, the significance and power of a hypothesis test are written

Significance

$$P(\text{reject } H_0/H_0) = \text{Type I error level} \qquad H_0 \text{ is assumed true}$$

Power

$$P(\text{reject } H_0/H_1) = 1 - \text{Type II error level} \qquad H_1 \text{ is assumed true}$$

The two types of error are not independent. Minimizing one type in an evaluation increases the other type. The tests described in this section are designed to make the probability of a Type II error as small as possible for any preselected Type I error; quite naturally, they are called *most-powerful tests*. We have no need to discuss the concept of power further in this text.

Hypothesis tests proceed as follows. Suppose θ is the parameter of interest; it may be a proportion, a mean, a variance, or something else. On the basis of prior information it is known or theorized to have a value of θ_0. Either to test the theory or to see if the value of the parameter has changed for some reason, one collects a random sample of n observations from the population of interest and calculates $\hat{\theta}$ as an estimate of θ. If the estimate $\hat{\theta}$ is sufficiently close to the anticipated value θ_0, one continues to accept the value θ_0 as the true value of the parameter; that is, one does not reject the null hypothesis. If on the other hand $\hat{\theta}$ is substantially different from the anticipated value θ_0, one rejects the notion that $\theta = \theta_0$; one rejects the null hypothesis.

The magnitude of the difference required for a decision to reject the null

---

* Which form H_1 takes must be decided before the data are analyzed, to avoid biased results.

hypothesis is a function of one's willingness to risk making a Type I error. For example, if one specifies an error level of no more than 5 percent, one is saying that

$$P(\text{reject } H_0 / H_0) \leq 5\%$$

This in turn, via the probability distribution for $\hat{\theta}$, determines the magnitude of the required difference to reject H_0 with no more than a 5 percent chance of being wrong. When H_0 is rejected at a specified error level, the test is called *significant* at that level.

Notice the logical connection between hypothesis testing and confidence interval estimation. A 95 percent confidence interval calculated for $\hat{\theta}$ tells one that there is no more than a 5 percent chance that the difference between $\hat{\theta}$ and the true value of θ is outside the interval. If this interval does not contain θ_0, the assumed true value of θ under the null hypothesis, the null hypothesis is rejected at the 5 percent significance level.

Proportions and Means

Here, again, the central limit theorem plays an important role when sample sizes are large. For small-sample sizes, a special technique called the *Fisher's exact method* is appropriate for comparing proportions, whereas the Student's-t distribution is used for comparing means.

For large samples, $\hat{p}$ and $\bar{x}$ are approximately normal. Thus hypothesis testing is accomplished using the standard normal tables. Using $\hat{\theta}$ as a generic symbol for either statistic, one generally proceeds as portrayed in Figure 5.4.

Proportions (large samples) The next two examples demonstrate hypothesis testing for proportions when sample sizes are large. In Example 5.7, a long-standing, experimentally or theoretically determined proportion is taken as the standard against which a recent estimate, based on a modified experimental situation, is tested. In Example 5.8, two proportions estimated from different experimental data are compared.

> **Example 5.7: Comparing a proportion against a standard** A physician claims to have an effective method for preselecting the sex of babies at the time of conception. Since the sex ratio for all live births is about unity (the standard proportion of males is almost 0.5), the physician is claiming that with this method the proportion can be made greater or less than 0.5, at will.
>
> The medical association commissioned a research group at a city hospital to test this claim. First, 100 couples were randomly selected from among 1000 childless couples who wanted a son as their first child. The group was instructed in the method and agreed to follow it meticulously. The sex of the first child born to each couple was recorded and the ratio of males to total births was observed to be 60 percent.

Comparing a sample-based estimate $\hat{\theta}$ to a standard θ_0

($\hat{\theta} = \hat{p}$ or $\hat{\theta} = \bar{x}$; θ_0 is a specific number)

Null* Hypothesis $H_0 : \theta = \theta_0$

Alternative Hypotheses†	Decision rules and type I error level‡

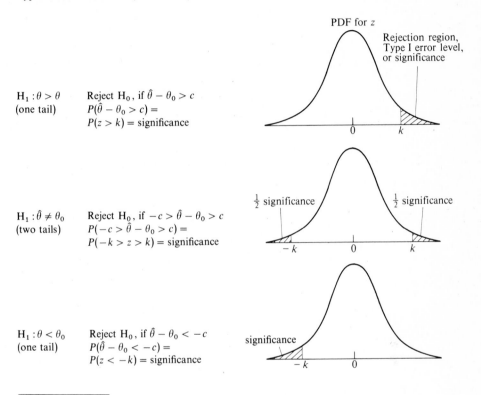

PDF for z

Rejection region, Type I error level, or significance

$H_1 : \theta > \theta$
(one tail)

Reject H_0, if $\hat{\theta} - \theta_0 > c$
$P(\hat{\theta} - \theta_0 > c) =$
$P(z > k) =$ significance

$\frac{1}{2}$ significance $\frac{1}{2}$ significance

$H_1 : \hat{\theta} \neq \theta_0$
(two tails)

Reject H_0, if $-c > \hat{\theta} - \theta_0 > c$
$P(-c > \hat{\theta} - \theta_0 > c) =$
$P(-k > z > k) =$ significance

significance

$H_1 : \theta < \theta_0$
(one tail)

Reject H_0, if $\hat{\theta} - \theta_0 < -c$
$P(\hat{\theta} - \theta_0 < -c) =$
$P(z < -k) =$ significance

---

*Asserts that $\mu_{\hat{\theta}} = \theta = \theta_0$, so $z = (\theta - \theta_0)/\sigma_{\hat{\theta}}$.

† Establishes the decision rule.

‡ That is, $P(\text{reject } H_0/H_0) = P(-c < \hat{\theta} - \theta_0 \leq c) = P(-c/\sigma_{\hat{\theta}} \leq z \leq c/\sigma_{\hat{\theta}}) =$ Type I error. The number, $k = c/\sigma_{\hat{\theta}}$, is found in the normal tables for a specified Type I error level. *The number c is called the critical value of the test.*

Figure 5.4 Comparing a sample-based estimate $\hat{\theta}$ to a standard θ_0 ($\hat{\theta} = \hat{p}$ or $\hat{\theta} = \bar{x}$: θ_0 is a specific number).

The study was designed to test the null hypothesis $p = 0.5$ against the alternative $p > 0.5$. The association did not specify an acceptable Type I error level, so the research group reported its findings for two error levels, namely, a 5 percent chance and a 1 percent chance of mistakenly rejecting the standard in favor of the physician's claim. To determine how much larger than the standard the experimental proportion of males would have to be to justify rejecting the null hypothesis for each level, they proceeded as follows:

Contending hypotheses	Decision rule
$H_0 : p = 0.5$ $H_1 : p > 0.5$	Reject H_0 if $\hat{p} - 0.5 > c$ one-tail test

The likelihood of mistakenly rejecting the null hypothesis with the decision rule must be no greater than the error level specified. Therefore,

$$P(\hat{p} - 0.5 > c) \leq \text{specified error level}$$

Since the sample is large, $n = 100$, p is approximately normally distributed, and $z = (\hat{p} - p)/\sqrt{p(1 - p)/100}$ is approximately a standard normal variable. The previous probability statement can be rewritten in terms of this normal variable by invoking the null hypothesis $(p = 0.5)$; thus for error levels of 5 percent (or 1 percent)

$$P\left(z > \frac{c}{\sqrt{(0.5)(0.5)/100}}\right) \leq 0.05 \quad \text{(or 0.01)}$$

For the error level of 5 percent, the normal table indicates that the smallest value of k for which the error limit is maintained is $k = 1.64$. That is,

$$k = c/\sqrt{(0.5)(0.5)/100} = 1.64 \quad \text{or} \quad c = 1.64\sqrt{(0.5)(0.5)/100}$$

Calculation yields $c = 0.08$. Therefore, H_0 may be rejected at the 5 percent error level only if the experimental proportion $\hat{p}$ is greater than the theoretical value 0.5 by at least 0.08 units. That is, reject H_0 only if $\hat{p} \geq 0.58$.

By the same logic, the hypothesis that the sex ratio is still unity $(p = 0.5)$ may be rejected at 1 percent significance only if $\hat{p} \geq 0.62$.

The researchers reported their findings to the association as follows: The standard proportion of males to total live births is 50 percent. The experimental study, using the physician's sex selection method described in the attached technical report, yielded 60 percent males. The likelihood is less than 5 percent, but greater than 1 percent, that a difference of this magnitude would be observed if the method were not effective. That is, the findings are significant at the 5 percent level but not at the 1 percent level.

Example 5.8: Comparing two experimental proportions Medical scientists recently studied the efficacy of a new drug. Using random-selection procedures they composed a group of 500 people, 100 of which were used as a control group and treated by standard methods. The researchers had no way of determining which 400 were being treated with the new drug; that was determined by a computerized random-selection process to avoid biasing the evaluations. Forty percent of the control group lived beyond the 1-year study period, $\hat{p}_1 = 0.4$. Fifty percent of the experimental group lived that long, $\hat{p}_2 = 0.5$. The question of whether the new drug was more effective than the standard treatment protocol was answered as follows.

The assumption at the beginning of the study was that the new drug would be more effective than standard methods of treatment. Thus:

$$H_0 : p_2 = p_1 \quad \text{or} \quad p_2 - p_1 = 0 \quad \text{reject } H_0 \text{ if } \hat{p}_2 - \hat{p}_1 > c$$

$$H_1 : p_2 > p_1 \quad \text{or} \quad p_2 - p_1 > 0 \quad \text{one-tail test}$$

The two Type I error levels used were 5 percent and 1 percent.

The samples are large and independent, so $\hat{p}_1$ and $\hat{p}_2$ are each approximately normal, as is their difference. The mean and variance for their difference are

$$E(\hat{p}_2 - \hat{p}_1) = E(\hat{p}_2) - E(\hat{p}_1) = p_2 - p_1$$

$$\text{var } (\hat{p}_2 - \hat{p}_1) = \text{var } (\hat{p}_2) - \text{var } (\hat{p}_1) = \frac{p_2(1 - p_2)}{n_2} + \frac{p_1(1 - p)}{n_1}$$

$$r_{.2} = 400 \qquad n_1 = 100$$

Accordingly,

$$z = \frac{(\hat{p}_2 - \hat{p}_1) - (p_2 - p_1)}{\sqrt{\dfrac{p_2(1 - p_2)}{400} + \dfrac{p_1(1 - p_1)}{100}}}$$

In this case $\hat{p}_1$ is the standard for comparison. Furthermore, H_0 asserts that $p_1 = p_2$. Thus, after substitution the likelihood of a Type I error, $P(\text{reject } H_0 / H_0)$, is

$$P\left(z > \frac{c}{\sqrt{(0.4)(0.6)(5/400)}}\right) = 0.05 \qquad \text{(or 0.01)}$$

For the 5 percent error level, the normal table indicates that the smallest value of k possible is $k = 1.64$. That is,

$$\frac{c}{\sqrt{(0.4)(0.6)(5/400)}} = 1.64 \quad \text{or} \quad c = 1.64\sqrt{(0.4)(0.6)(5/400)} = 0.09$$

At the 1 percent error level the smallest value of k is 2.33, or $c = 0.12$.

The researchers reported their findings significant at the 5 percent level, because the difference $\hat{p}_2 - \hat{p}_1 = 0.1$ was larger than the required 0.09 value but not significant at the 1 percent level.

There was some dispute about the form of the alternative hypothesis. Some observers felt that it should have been stated as

$$H_1 : p_2 \neq p_1 \quad \text{or} \quad p_2 - p_1 \neq 0 \quad \text{two-tail test}$$

because there was no a priori reason to suppose that the new drug would be superior to standard treatment procedures. Therefore, the research group tested the significance of their findings for this formulation as well.

In this case, the decision rule is to reject the null hypothesis for absolute differences greater than some number c; that is

$$\text{Reject if} \quad -c < \hat{p}_2 - \hat{p}_1 < c$$

The likelihood of a Type I error was, in this case,

$$P(\text{reject } H_0 / H_0) = P(-c > \hat{p}_2 - \hat{p}_1 > c/H_0) = 0.05 \quad \text{(or 0.01)}$$

which reduced to

$$P\left(\frac{-c}{\sqrt{(0.4)(0.6)(5/400)}} > z > \frac{c}{\sqrt{(0.4)(0.6)(5/400)}}\right) = 0.05 \quad \text{(or 0.01)}$$

For the 5 percent error level, the normal table indicates that $k = \pm 1.96$ for two-tail tests; $2\frac{1}{2}$ percent of the distribution is above 1.96 and $2\frac{1}{2}$ percent is below -1.96, so that 5 percent is outside the range ± 1.96. Therefore $c = 0.107$. At the 1 percent level, the limiting values are $k = 2.58$ or $c = 0.141$ units.

The researchers reported their findings in the context of this new formulation of the alternative hypothesis as not significant at either the 5 percent or 1 percent levels.

Proportions (small samples) When samples are small, hypothesis tests depending on the assumption that $\hat{p}$ is nearly normal are not valid, because the central limit theorem does not hold. Statisticians favor the Fisher's exact method in this situation. The method depends on the true or exact distribution for the problem, not on the normal approximation. The probability model for this test is easy to derive; it is based on binomial sampling. Imagine an extremely large population. Randomly select one sample of size n and another of size m. Assume that the likelihood that an individual in the large population has some particular characteristic of interest is p. The likelihoods that k of the first group and, independently, that l of the other group have the characteristic, respectively, are then

$$\binom{n}{k}p^k(1-p)^{n-k} \quad \text{and} \quad \binom{m}{l}p^l(1-p)^{m-l}$$

Now the total number in the combined sample is $N = n + m$. A total of $K = k + l$ have the characteristic. The likelihood of this total is

$$\binom{N}{K}p^K(1-p)^{N-K} = \binom{n+m}{k+l} = p^{k+l}(1-p)^{n+m-k-l}$$

For these totals N and K, the probability of realizing the specific distributions in the two samples is the conditional probability of k successes in n Bernoulli trials and l successes in m Bernoulli trials, given that the total number of trials is

$N = n + m$ and the total number of successes is $K = k + l$. Because the two samples are independently drawn, this converts to the ratio

$$\frac{\binom{n}{k}\binom{m}{l}}{\binom{N}{K}} \left[\frac{p^{k+l}(1-p)^{(n-k+m-l)}}{p^K(1-p)^{N-K}} \right] = \frac{\binom{n}{k}\binom{m}{l}}{\binom{N}{K}}$$

The use of this hypothesis testing procedure is deferred to Chapter 8, where the 1977–78 saccharin ban decision is discussed.

Means (large samples) The next example demonstrates the procedure for testing the significance of the differences in two sample means. The procedure is basically the same as for proportions. Sample sizes are large, so the normal tables are used. The procedure for testing a sample mean against an experimentally or theoretically established value is not demonstrated; it is identical to the procedure for proportions.

Example 5.9: Comparing two means from large samples The Tooth Paste Corporation was challenged under the truth in advertising law two years ago and conducted some controlled studies of the effectiveness of its product as a cavity-prevention agent. A control group of 100 randomly selected children within a specific age range brushed with a paste of baking soda for 1 year and a matched experimental group brushed with the corporation's product, under an identical protocol. The average number of cavities for the control group was 4 per year, $\bar{x}_1 = 4$. The average for the experimental group was 3.75 cavities, $\bar{x}_2 = 3.75$. The standard deviations were $S_1 = 1.0$ and $S_2 = 2.0$.

Justice Department statisticians interpreted these findings using the following logic:

$$H_0 : \mu_1 - \mu_2 = 0 \qquad \text{reject } H_0 \text{ if } \bar{x}_1 - \bar{x}_2 > c$$

$$H_1 : \mu_1 - \mu_2 > 0 \qquad \text{one-tail test}$$

The idea was to reject the hypothesis that the toothpaste was no better than baking soda if the control group had a sufficiently greater average. Because the samples were large, $\bar{x}_1$ and $\bar{x}_2$ were approximately normal variables, with means μ_1, μ_2 and variances $\sigma_1^2/n_1, \sigma_2^2/n_2$, respectively. Thus

$$z = \frac{(\bar{x}_1 - \bar{x}_2) - (\mu_1 - \mu_2)}{\sqrt{(\sigma_1^2 + \sigma_2^2)/100}} \qquad n_1 = n_2 = 100$$

For a 5 percent maximum acceptable Type I error, the likelihood $P(\text{reject } H_0/H_0) \le 0.5$ reduced to

$$P(z > c/\sqrt{(\sigma_1^2 + \sigma_2^2)100}) \le 0.05$$

The minimum value of k for which this would hold was $k = 1.64$. Therefore,

$$c = 1.64 \sqrt{\frac{\sigma_1^2 + \sigma_2^2}{100}}$$

Although the variances are unknown, they can be reasonably approximated by S_1 and S_2, because the samples are large. Hence, c is approximately equal to

$$c = 1.64 \sqrt{\frac{5}{100}} \simeq 0.37$$

The average numbers of cavities for the two groups differed by only 0.25 units. This is less than the required 0.37 units, so one cannot reject (at the 5 percent level) the hypothesis that baking soda is as effective as the toothpaste tested.

Of course the form of the alternative hypothesis has an impact on the level of significance found. If instead of formulating H_1 to express the belief that one mean is larger than the other, as in the previous example, it is formulated to express the belief that the two means are merely unequal, experimental results which prove significant in a one-tail test will prove significant in a two-tail test only at the cost of higher Type I error levels. The reason for this is that the two-tail test divides the error level between the two extremes of the distribution, thereby requiring greater differences in estimates for the same level of significance.

Means (small sample) The next example demonstrates a two-tail test for small samples. The Student's-t distribution is employed on the assumption that the sampled random variables x_1 and x_2 are approximately normally distributed, are independent, and have the same variance. Statisticians have determined that the random variable for use in this test has a Student's-t distribution. It has the form

$$t = \frac{(\bar{x}_2 - \bar{x}_1) - (\mu_2 - \mu_1)}{\sqrt{(n_2 - 1)S_2^2 + (n_1 - 1)S_1^2}} \sqrt{\frac{n_1 n_2 (n_1 + n_2 - 2)}{n_1 + n_2}}$$

where n_1 and n_2 are the respective sample sizes.* This variable has a parameter value $v = n_1 + n_2 - 2$.

---

* If a small-sample estimate of a mean were being tested against a standard μ, the random variable would be simplified to

$$t = \frac{\bar{x} - \mu_x}{S_x / \sqrt{n}}$$

and have a parameter $v = n - 1$.

Example 5.10: Comparing two means from small samples Error rates for employees working under two different sets of operating conditions were studied. Two groups of 10 employees each were tested. Error rates for the two situations were assumed to be normally distributed and independent. Further, the variances for error rates in both situations were assumed to be equal. The first group averaged five errors per day, $\bar{x}_1 = 5$, and the second group averaged seven errors per day, $\bar{x}_2 = 7$. The sample standard deviations were $S_1 = 2$ and $S_2 = 2$. There was no reason to assume that the mean for either group would necessarily be larger. Thus:

$$H_0 : \mu_1 = \mu_2 \qquad \text{reject } H_0 \text{ if } -c > \bar{x}_2 - \bar{x}_1 > c$$

$$H_1 : \mu_1 \neq \mu_2 \qquad \text{two-tail test}$$

Management was conservative and insisted that the likelihood of mistakenly rejecting the null hypothesis be no greater than one chance in 100. That is, $P(\text{reject } H_0/H_0) \leq 0.01$, or

$$P(-k \geq t \geq k) \leq 0.01$$

The probability value for each tail of the Student's-t distribution was thus 0.005. The parameter was $v = 10 + 10 - 2 = 18$. The minimum value of t which met these conditions was determined from the table to be $k = 2.89$. From the sample data, the actual value of t was calculated to be

$$t = \frac{(\bar{x}_2 - \bar{x}_1)\sqrt{(100 \times 18)/20}}{\sqrt{4(9 + 9)}} = 2.2$$

which was less than k. Hence the difference in error rates was not significant at the 1 percent level.

Means (paired data) If instead of two groups only one group of subjects is involved in a study, first in one situation and then in another, the data will not represent independent random samples. Such data are described as *paired*. To test the difference in averages for the two situations one would proceed as follows: Let n be the number of subjects. Calculate differences in performance, $d_i = x_{i_2} - x_{i_1}$, for each subject in the two situations. Calculate the average and standard deviations for the differences. Thus

$$d_i = x_{i_2} - x_{i_1} \qquad \bar{d} = \frac{1}{n} \sum d_i \qquad \mu_d = \mu_2 - \mu_1 = \mu_d$$

$$S_d^2 = \frac{1}{n-1} \sum (d_i - \bar{d})^2 \qquad t = \frac{\bar{d} - \mu_d}{S_d/\sqrt{n}} \qquad \sigma_d = \frac{\sigma_d}{\sqrt{n}}$$

With these variations, hypothesis testing proceeds as previously discussed. The parameter for the t variable is $v = n - 1$.

Example 5.11: Comparing means for paired data Test scores for students before and after taking a refresher course are shown. The course was given on the basis that it would significantly improve scores.

Grades		Difference	
Before (x)	After (y)	$d = (y - x)$	
50	55	5	$\bar{d} = 3$
75	72	-3	
63	68	5	$S_d^2 = 12$
82	85	3	
77	82	5	

The hypothesis was tested at the 5 percent level.

$$H_0 : \mu_y = \mu_x \qquad \text{reject if } \bar{d} > c$$

$$H_1 : \mu_y > \mu_x$$

$$P(\text{reject } H_0/H_0) = P(t > k) \leq 5\%$$

From the t tables with $v = 4$, $k = 2.13$. Invoking the null hypothesis $\mu_d = 0$, calculate

$$t = \frac{\bar{d}}{S_d/\sqrt{n}} = \frac{3}{0.69} = 4.35$$

which exceeds the value of k. Thus the null hypothesis is rejected at the 5 percent level. (Of course, the sample size was extremely small, so the legitimacy of this conclusion may be open to questions.)

Variances

Variances are compared using the F distribution. Instead of focusing on differences in variances, one phrases the hypothesis in terms of ratios. That way the random variable for the test is the ratio of two chi-square variables divided by their parameter values, assuming the sample variables are independent and normal. Each of the chi-square variables has the form $[(n_i - 1)S_i^2/\sigma_i^2]$. Assuming $S_2^2 > S_1^2$, the F variable for the test is

$$F = \frac{(n_2 - 1)S_2^2/\sigma_2^2(n_2 - 1)}{(n_1 - 1)S_1^2/\sigma_1^2(n_1 - 1)} = \frac{S_2^2/\sigma_2^2}{S_1^2/\sigma_2^2}$$

Under the null hypothesis $\sigma_1^2/\sigma_2^2 = 1$, so the variable reduces to the ratio

$$F = \frac{S_2^2}{S_1^2}$$

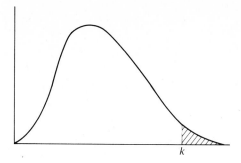

Figure 5.5 One-tail critical region for F.

The parameters for this variable are, respectively, $n_2 - 1$ for the numerator and $n_1 - 1$ for the denominator.

A one-tail critical region for the F distribution is illustrated in Figure 5.5.

Example 5.12: Comparing two sample variances Ballistics engineers tested the muzzle velocity of a particular cartridge design, using two different gunpowder mixes. They had the experimental-production division prepare 10 cartridges loaded with each powder mix and fired them in an instrumented indoor experimental firing range. In addition to comparing average velocities they compared variations in velocity within samples, because ammunition reliability is critical.

The engineers anticipated that the muzzle velocity for the second sample would be more variable, on the basis of previous experience with one of the components in the powder mix. Thus, the hypotheses were taken to be:

$$H_0 : \frac{\sigma_2^2}{\sigma_1^2} = 1 \qquad \text{reject } H_0 \text{ if } S_2^2/S_1^2 > k$$

$$H_1 : \frac{\sigma_2^2}{\sigma_1^2} > 1 \qquad \text{one-tail test}$$

The random variable for the test was

$$F = \frac{S_2^2}{S_1^2} \qquad \begin{matrix} n_2 - 1 = 9 \\ n_1 - 1 = 9 \end{matrix}$$

At the 5 percent Type I error level, $P(\text{reject } H_0 / H_0) = P(S_2 / S_1 > k) \leq 0.05$. From the F table for $v_1 = v_2 = 9$, the value of k was found to be 3.18. Thus the engineers adopted the practical decision rule

$$\text{Reject if} \quad S_2^2/S_1^2 > 3.18$$

Two-tail tests, when the alternative hypothesis is in the form

$$H_1 : \frac{\sigma_2^2}{\sigma_1^2} \neq 1$$

may also be done. However, since the F distribution is not symmetric, the Type I error statement breaks into two pieces, one for each tail of the F distribution. For an error limit of 2 percent these would be:

$$P\left(\frac{S_2^2}{S_1^2} > k_1\right) \le 0.01$$

$$P\left(\frac{S_2^2}{S_1^2} < k_2\right) \le 0.01 \qquad \text{or} \qquad P\left(\frac{S_1^2}{S_2^2} > \frac{1}{k_2}\right) \le 0.01$$

The limit k_2 would be near the origin; for $n_1 = 10$ and $n_2 = 21$, $k_1 = 4.80$ and $k_2 = 0.21$.

The highlights of the methods for comparing two means, proportions, or variances are summarized in Table 5.2 for easy reference.

Table 5.2 Comparing sample estimates

	Variable
Large samples	
Proportions	
Standard, μ	$z = \dfrac{\hat{p} - p}{\sqrt{p(1-p)/n}}$
Two samples	$z = \dfrac{(\hat{p}_2 - \hat{p}_1) - (p_2 - p_1)}{\sqrt{\dfrac{p_2(1-p_2)}{n_2} + \dfrac{p_1(1-p_1)}{n_1}}}$
Means	
Standard, μ	$z = \dfrac{\bar{x} - \mu}{\sigma_x/\sqrt{n}}$
Two samples	$z = \dfrac{(\bar{x}_2 - \bar{x}_1) - (\mu_2 - \mu_1)}{\sqrt{\dfrac{\sigma_1^2}{n_1} + \dfrac{\sigma_2^2}{n_2}}}$
Variance	$F_{n_2-1,\,n_1-1} = \dfrac{S_2^2/\sigma_2^2}{S_1^2/\sigma_1^2}$
Small samples	
Proportions	Fisher's exact method
Means	
Standard, μ	$t_{n-1} = \dfrac{(\bar{x} - \mu)}{S/\sqrt{n}}$
Two samples	$t_{n_1+n_2-2} = \dfrac{(\bar{x}_2 - \bar{x}_1) - (\mu_2 - \mu_1)}{\sqrt{(n_2-1)S_2^2 + (n_1-1)S_1^2}} \sqrt{\dfrac{n_1 n_2(n_1 + n_2 - 2)}{n_1 + n_2}}$
Variances	Same as for large samples if sampling is normal

5.3 COMPARING MANY PROPORTIONS

Accuracy assessments and comparisons of statistical parameters by the methods in Sections 5.1 and 5.2 depend on assumptions about the distributions from which sample data are obtained. It is not always clear that these assumptions are justified from a theoretical perspective. Consequently, it may be desirable to use the sample data to test the validity of such assumptions. Technically, the same problem occurs when engineers, scientists, or operations researchers use sample data to develop probability models for the time distribution of floods by magnitude, the rate of decay of atomic particles, the distribution of plant or animal populations, traffic flow, ballistic accuracy, and the like. In these situations, the objective is to fit the statistical or experimental data to known probability models and test the accuracy or "goodness of fit." How to do this is the first topic discussed in this section.

The second topic of this section concerns data classification. Economists, other social scientists, and educators frequently study multiple characteristics of a population for the purpose of determining whether the characteristics are somehow interdependent or are statistically independent. For example, educators are concerned about the relations of grades to teaching method, age of the teacher, or classroom environment. Social scientists are interested in determining links between health status and income, between income and attitudes toward authority, and between socioeconomic status and political attitudes. The method of testing hypotheses about such interrelationships is basically similar to the methods of testing the consistency of data with assumed probability models.

In both cases, the chi-square distribution is used to assess the degree of consistency of theoretical and experimental proportions. Squared deviations between sample frequencies and hypothesized or theoretical frequencies are divided by the theoretical frequencies and summed to form chi-square variables of the form

$$D = \sum_{i=1}^{m} \frac{(n_i - np_i)^2}{np_i}$$

$$= \sum_{i=1}^{m} \frac{n(\hat{p}_i - p_i)^2}{p_i}$$

where m = number of groupings of the data
n = sample size
p_i = theoretical proportion expected for the ith group
$\hat{p}_i$ = actually observed proportion for the ith group
n_i = number of data points in the ith group

The parameter v for these chi-square variables is $m - 1$ unless the same data are used to estimate model parameters such as μ and σ, in which case the value of v must be reduced by 1 for each parameter estimated.

The decision rule for these tests is of the form

$$\text{Reject } H_0 \text{ if } D > k$$

where k is determined from the chi-square table.

The next example demonstrates the method of testing the goodness of fit of data to a probability model. Although the normal model is used, because it is so important in statistical data analysis, any other model could have been used in its place, without affecting the process.

Example 5.13: Comparing sample data to a model Ballistics engineers measured the muzzle velocity of 40 rounds of an experimental ammunition. This was a small sample and the engineers now wish to compare the average muzzle velocity of the experimental design to the standard design, using the Student's-t variable. Before proceeding, the engineers wish to determine whether the velocity data obtained are consistent with the assumption that velocity is distributed normally, since the Student's-t test may not be used legitimately unless this assumption is true.

The engineers observe that if velocities are normally distributed, the standardized velocity, calculated from the formula

$$\text{Standard velocity} = \frac{v - \mu_v}{\sigma_v}$$

is a standard normal variable and sample velocities are equally likely to fall into each of the ranges indicated in the data table below. To check this null hypothesis ($H_0: p_1 = p_2 = p_3 = p_4 = 0.25$), standardize all velocities, using the relation

$$v_s = \frac{v - \bar{v}}{S_v}$$

where $\bar{v}$ is the average velocity for the 40 rounds, and S_v^2 is the sample variance. Then count the number of occurrences of velocities in each interval and tabulate them as follows:

Interval	Theoretically expected frequency $(40 \times 0.25) = np$	Experimental frequency n_i
0.675 and above	10	6
0–0.675	10	13
−0.675–0	10	15
Below −0.675	10	6

Next, calculate the sum of standardized squared deviations from the relation

$$D = \frac{(n_1 - np_1)^2}{np_1} + \frac{(n_2 - np_2)^2}{np_2} + \frac{(n_3 - np_3)^2}{np_3} + \frac{(n_4 - np_4)^2}{np_4}$$

and obtain the value $D = 6.6$. Since the data are classified into four intervals, and two statistics, namely, $\bar{v}$ and S_v, have been estimated from the data already, consult the chi-square table for $v = 1$ $(n - 1 - 2 = 1)$ and observe that the likelihood of obtaining a value as large as 6.6, provided that velocities are normally distributed, is between 2.5 and 1 percent. Therefore, there is only slightly more than a 1 percent chance of being wrong in rejecting the null hypothesis.* Consequently, the engineers had better resort to large samples to estimate velocity statistics (and perhaps find another probability model).

In the previous example the data were classified in only one dimension. Social scientists often classify data into two dimensions to explore the degree of independence of the dimensions. In this situation there are no probability models from which to determine the theoretical proportions for each cross category. Instead, the null hypothesis is an assertion of independence between the variables of classification, which means that the proportion of the total sample which falls into any cross category (A, B) should be the product of the proportion of the total found in one category, A, and the proportion found in the other category, B. Assuming that the observed proportions in each category represent accurate classifications, they may be treated as probability estimates. Then the number of observations expected in any cross category may be estimated from the relation

$$\text{Cell frequency} = \frac{(\text{number in A})(\text{number in B})}{\text{sample size}}$$

The next example demonstrates the process of assessing the degree of independence of two variables of classification.

Example 5.14: Strength of relationship and contingency tables Medical sociologists are studying attitudes of physicians toward government intervention in the form of standards for health care delivery. They suspect that age is somehow related to attitudes toward these new events and have classified responses from 90 randomly selected physicians in a two-way contingency table, as shown. The numbers in parentheses are the theoretical frequencies calculated from the marginal totals and rounded to the nearest integers; for example, $(31 \times 22)/90 = 8$ appears in the cell of the third row and first column. These numbers characterize the null hypothesis: $(H_0: p_{11} = 7, p_{21} = 7, p_{31} = 8, p_{12} = 10, p_{22} = 11, p_{32} = 11, p_{13} = 12, p_{23} = 12, p_{33} = 13)$, which will be rejected only if the sum of squared deviations is sufficiently large.

---

* Again, the decision rule is

$$\text{Reject } H_0 \text{ if } D > k$$

The value k in this example is 6.6349 for the 1 percent error limit and about 5.0239 for the 2.5 percent error limit.

Age	Strong dislike	Indifference	Strong favor	Total
39	10 (7)	10 (10)	9 (12)	29
40–59	6 (7)	12 (11)	12 (12)	30
60	6 (8)	10 (11)	15 (13)	31
Total	22	32	36	$n = 90$

If the two variables of classification are independent, the sum of standardized squared deviations between theoretical and actual cell frequencies should be small. That is, the chi-square variable D should be small. The actual value of D is

$$D = \frac{(10 - 7)^2}{7} + \frac{(6 - 7)^2}{7} + \frac{(6 - 8)^2}{8} + \frac{(10 - 10)^2}{10} + \frac{(12 - 11)^2}{11}$$

$$+ \frac{(10 - 11)^2}{11} + \frac{(9 - 12)^2}{12} + \frac{(12 - 12)^2}{12} + \frac{(15 - 13)^2}{13} = 3.17$$

The parameter value for this variable is determined from the relation

$$v = \left(\begin{array}{c} \text{One less than} \\ \text{the number of} \\ \text{rows } (r - 1) \end{array} \right) \times \left(\begin{array}{c} \text{One less than} \\ \text{the number of} \\ \text{columns } (c - 1) \end{array} \right)$$

The value of k, from the chi-square table with $v = 2 \times 2$, is about 9.5 at the 5 percent level. Hence $D < k$.

The researchers would report these findings as not significant at the 5 percent level. By this they would mean that the data do not support the assumption of a link between age and attitude toward government intervention, at the 5 percent level.

One final and very important point: The chi-square distribution is only an approximation for the distribution of standardized squared deviations D. Therefore, if too few data points are distributed among the dimensions of classification, the approximate results obtained in the analysis are likely to be erroneous. *The rule to observe is that each cell in a contingency table, or each interval for the theoretical random variable in goodness of fit tests, must contain a frequency at least as large as 5 for the results to be acceptable. Before proceeding one may have to combine cells to meet this requirement.*

5.4 COMPARING MULTIPLE MEANS

If science, engineering, or management were constrained to progress one step at a time, always comparing a variation with the past before testing another variation, progress would be painfully slow. The agricultural researchers would have to vary a single component of a complex feed mix and assess its impact on animal growth before varying any other component. The ballistics engineer would have to vary a single component of a gunpowder and assess its impact on muzzle velocity before varying any other component. A city manager would have to evaluate the impact on the cost of trash collection by varying a single component in the system before varying another. Yet, in each of these cases, optimizing the system for each component separately and using these optimal values as system parameters probably would not result in the best technical or most economical system, because of interactions among the components. Progress has never proceeded in this lockstep fashion; and statisticians have developed a method called the *analysis of variance*, by which the average impacts of simultaneous changes in many components may be assessed. Although the hypothesis being tested is of the form

$$H_0 : \mu_1 = \mu_2 = \cdots = \mu_n$$

the analysis focuses on sample variances, thus the name "analysis of variance."

The next example demonstrates the method for analyzing the average impacts of varying two components simultaneously. The method also applies to situations where only one component is varied but more than two variations are evaluated simultaneously—as, for example, when ballistics engineers load many ammunition samples with basically the same gunpowder except that each sample has a different density of one of the components, and they compare the muzzle velocities of all the samples at once. Quite similar methods exist for comparing the impacts of varying many variables. These are generally available in computerized statistical analysis packages.

Example 5.15: Two-variable analysis of variance The department of education is testing three methods of teaching reading to first-grade students with similar aptitudes. A sample of 18 students is involved in the project. Performance scores for the year are summarized in the table below according to teaching method and which of the three elementary schools in the city the student attends. Scores shown are the average for the two students in each school being taught by the same method, and are rounded to the nearest integer for purposes of illustration.

School	Teaching method 1	2	3	Row average ($\bar{R}$)
1	85	79	93	86
2	77	95	79	83
3	82	90	80	84
Column average ($\bar{C}$)	81	88	84	Overall average $\bar{x} = 84$

The variance for these data may be estimated in four ways, each of which should give essentially the same estimate if there were no real differences due to school or teaching method. Using the notation x_{ij} to represent the score in the ith row and jth column of the table, r to represent the number of rows, and c to represent the number of columns, the general forms for the four estimation equations are:

Assuming that table entries represent independent sample values from the sample distribution

$$S^2 = \frac{1}{rc - 1} \sum_{\substack{\text{rows } (i) \\ \text{columns } (j)}} (x_{ij} - \bar{x})^2 \qquad \text{standard estimate}$$

Assuming actual row means or column means are equal, one could estimate the variance from row averages or column averages as

$$S_R^2 = \frac{c}{r - 1} \sum_{\text{rows } (i)} (\bar{R}_i - \bar{x})^2 \qquad \text{row estimate}$$

$$S_c^2 = \frac{r}{c - 1} \sum_{\text{columns } (j)} (\bar{C}_j - \bar{x})^2 \qquad \text{column estimate}$$

Assuming the true row means and column means and the overall mean are equal, one could estimate the variance from

$$S_I^2 = \frac{1}{(r - 1)(c - 1)} \sum_{\substack{\text{rows } (i) \\ \text{columns } (j)}} (x_{ij} - \bar{R}_i - \bar{C}_j + \bar{x})^2 \qquad \text{interaction estimate}$$

It turns out that

$$(rc - 1)S^2 = (r - 1)S_R^2 + (c - 1)S_c^2 + (r - 1)(c - 1)S_I^2$$

which suggests that S_R^2 measures variations among row averages, S_c^2 measures variations among column averages, and S_I^2 measures the remaining effects (due, perhaps, to interaction between the variables of classification).

Thus, if the school setting itself has no influence on scores, the ratio S_R^2/S^2 should have a value near unity. If the teaching method is not influential, the ratio S_c^2/S^2 should be near unity, and if the two variables have no interactive impact on scores, the ratio S_I^2/S^2 should be near unity. Each of these ratios is an F random variable, since the numerator and denominator of each is an independent χ^2 variable.* Thus:

$$F_R = \frac{S_R^2}{S^2} \qquad v_1 = r - 1 \qquad v_2 = rc - 1 \qquad \text{row}$$

$$F_c = \frac{S_c^2}{S^2} \qquad v_1 = c - 1 \qquad v_2 = rc - 1 \qquad \text{column}$$

$$F_I = \frac{S_I^2}{S^2} \qquad v_1 = (r - 1)(c - 1) \qquad v_2 = rc - 1 \qquad \text{interaction}$$

If any one of these ratios calculated from the data are larger than the corresponding theoretical value from the F table for the specified v_1 and v_2 values, the null hypothesis of equal means may be rejected at the appropriate Type I error level $(P(F_R > k_R / H_0) \leq$ error limit).

From the data provided $F_R = 0.16$, and for $v_1 = 2$, $v_2 = 8$ and a 5 percent error level, the theoretical value of k_R is 4.46. Thus, the school setting is not a significant factor at this level. Also, $F_c = 0.83$; again, the theoretical value is 4.46. Thus, neither is teaching method a significant factor. Finally, $F_I = 1.68$ while for $v_1 = 4$, $v_2 = 8$, the theoretical value of F is 3.84. Consequently, even in interaction the two variables are not significant at the 5 percent level. The calculations are summarized in the table below.

Estimate	Ratio (F)	F Parameters	Theoretical value (5%)	Reject?
$S^2 = 44.8$	—	—	—	—
$S_R^2 = 7.5$	0.16	$v_1 = 2$ $v_2 = 8$	4.46	No
$S_c^2 = 37.5$	0.83	$v_1 = 2$ $v_2 = 8$	4.46	No
$S_I^2 = 75.3$	1.68	$v_1 = 4$ $v_2 = 8$	3.84	No

In the previous example, although there was one observation per cell, each was composed of the average of two observations. Multiple observations per cell are called *replication*. Sufficient replication is essential for controlling for bias which may be introduced from outside an experimental design; in the last example bias could be introduced in the student selection process. Without replication, tests for interaction could be misleading.

In addition, sampling was implicitly assumed to be from a normal distribution. This was the basis of the assertion that the sums of squared deviations were chi-square variables. When the normality assumption is coupled with uniform replication (the same number of data points in each cell), statisticians say the experimental design is *orthogonal*, to indicate that the row, column, and interactive tests are not influenced one by the other.

One can directly evaluate the degree of variability introduced into an orthogonal two-way analysis of variance by sources other than the factors of concern. One simply divides the components of variance into row, column, and interaction terms, as previously, plus a special error term which accounts for variation

---

* Clearly, $\sum_{i,j} (x_{ij} - \bar{x})^2$ is χ^2. Since there are rc table entries, $v = rc - 1$. Equivalent arguments hold for row and column terms. Then since the sum of χ^2 variables is χ^2, the number of degrees of freedom attributable to $\sum_{i,j} (x_{ij} - \bar{R}_i - \bar{C}_j + \bar{x})^2$ is found by solving

$$(rc - 1) = (r - 1) + (c - 1) + v$$

for v. Independence is not so easy to show.

within cells. In particular, using $\bar{I}_{ij}$ to symbolize the average value within a cell and x_{ijk} to represent on individual entry within cell (i, j),

$$S^2 = \frac{1}{rcn - 1} \sum_{\substack{\text{rows } (i) \\ \text{columns } (j) \\ \text{cell entries } (k)}} (x_{ijk} - \bar{x})^2 \qquad \text{standard}$$

$$S_R^2 = \frac{nc}{r - 1} \sum_{\text{rows } (i)} (\bar{R}_i - \bar{x})^2 \qquad \text{row}$$

$$S_C^2 = \frac{nr}{c - 1} \sum_{\text{columns } (j)} (\bar{C}_j - \bar{x})^2 \qquad \text{column}$$

$$S_I^2 = \frac{n}{(r - 1)(c - 1)} \sum_{\substack{\text{rows } (i) \\ \text{columns } (j)}} (\bar{I}_{ij} - \bar{R}_i - \bar{C}_j + \bar{x})^2 \qquad \text{interaction}$$

$$S_E^2 = \frac{1}{rc(n - 1)} \sum_{\substack{\text{rows } (i) \\ \text{columns } (j) \\ \text{cell entries } (k)}} (x_{ijk} - \bar{I}_{ij})^2 \qquad \text{error}$$

Each term is a chi-square random variable. The degrees of freedom of the first three are, respectively: $rcn - 1$, $r - 1$, and $c - 1$. The error term has $rc(n - 1)$ degrees of freedom. This follows because

$$(rcn - 1)S^2 = (r - 1)S_R^2 + (c - 1)S_C^2 + (r - 1)(c - 1)S_I^2 + rc(n - 1)S_E^2$$

is a sum of chi-square variables. Thus the degrees of freedom must add, namely,

$$rcn - 1 = (r - 1) + (c - 1) + (r - 1)(c - 1) + v$$

leads to $v = rc(n - 1)$

Suppose the data for Example 5.15 had been tabulated so as to show the individual student scores. It might have appeared as in Example 5.16 and could have been analyzed as demonstrated to reflect variability from student selection.

Example 5.16: Further analysis of variance The data below are equivalent to those in the previous example, when cell variability is ignored. The variance estimates are:

$$S^2 = 65.7 \qquad S_C^2 = 75.0 \qquad S_E^2 = 45.3$$
$$S_R^2 = 15.0 \qquad S_I^2 = 146.5$$

	Teaching method			
School	1	2	3	Row average ($\bar{R}$)
1	83/87	79/79	90/96	86
2	70/84	90/100	78/80	83
3	80/84	87/93	70/90	84
Column average ($\bar{C}$)	81	88	84	Overall average $\bar{x} = 84$

If the teaching method or the school setting are themselves important factors in test results, the ratios S_C^2/S_E^2 and S_R^2/S_E^2 will be large, implying that these factors are more powerful than variability introduced in the student selection process. If the interaction is relatively important, the ratio S_I^2/S_E^2 will be large. These ratios are:

$$F_C = 1.66 \qquad F_R = 0.33 \qquad F_I = 3.23$$

From the F tables, the critical values are, respectively, $k = 4.26$, 4.26, and 3.63 at the 5 percent level. Thus student selection appears to be the primary factor in grades.

There are many computer programs for performing analysis of variance for multiple classification studies. Among the best are the Statistical Package for Social Sciences (SPSS) and the Statistical Analysis System (SAS) programs. If complex problems are being analyzed, the wisest course of action is to learn how to use these packages. If one or two variables of classification are sufficient in a given situation, a simple eight-function, hand-held calculator with memory is sufficient.

5.5 PROBLEMS

5.1 Describe procedures for obtaining random samples appropriate to the following situations. Use any references you wish, but be specific in your answers.

(a) Forecasting an election when two candidates are running
(b) Determining the average age of an animal population in a particular local habitat
(c) Estimating the strength of machined tools
(d) Evaluating a new surgical procedure
(e) Determining the peak pressure limit on a particular type of container

5.2 Show that:

$$\text{(a)} \qquad\qquad E(\bar{x}) = \mu$$

$$\text{(b)} \qquad\qquad E(\hat{p}) = p$$

where μ is the mean value for the population being sampled and p is the theoretical probability of a "success" for independent Bernoulli trials. Assume that sampling is independent and random.

5.3 Show that:

$$\text{(a)} \qquad\qquad E(\bar{x} - \mu) = 0$$

$$\text{(b)} \qquad\qquad E(\hat{p} - p) = 0$$

when the conditions of Problem 5.2 are satisfied.

5.4 Show that the following are true when sampling is independent and from the same distribution:

$$\text{(a)} \quad P(-k\sigma_{\hat{p}} \le \hat{p} - p \le k\sigma_{\hat{p}}) = P\left(-k \le \frac{\hat{p} - p}{\sqrt{p(1-p)/n}} \le k\right)$$

$$\text{(b)} \quad P(-k\sigma_{\bar{x}} \le \bar{x} - \mu \le k\sigma_{\bar{x}}) = P\left(-k \le \frac{\bar{x} - \mu}{\sigma_x/\sqrt{n}} \le k\right)$$

$$\text{(c)} \quad P\left(k_2 \le \frac{(n-1)S_x^2}{\sigma_x^2} \le k_1\right) = P\left(\frac{(n-1)S_x^2}{k_1} \le \sigma_x^2 \le \frac{(n-1)S_x^2}{k_2}\right)$$

5.5 How large a sample must be taken to determine the proportion of voters who will vote for candidate A over candidate B, if an error of no greater than five percentage points is required at 99 percent confidence?

5.6 The U.S. Army Corps of Engineers is studying barge traffic flow through Locks and Dam No. 26 on the Mississippi River. A consultant has been charged with estimating average traffic flow for each season of the year. How should she sample traffic flow rates for summer months? If she records flow rates on only 60 days during the summer and finds that the average number of barges passing through the locks is 50 daily with a standard deviation of 5, how accurate are her estimates?

5.7 The National Health Department recently reported findings of an evaluation of x-rays as a diagnostic tool for early detection of cancer, compared to the diagnostic protocol utilizing clinical and thermographic detection methods. The competing protocols were independently applied to the same random sample of 200 patients. The x-ray protocol identified 20 of 25 lesions smaller than 2 centimeters in diameter. The other protocol identified only 17 lesions of that size.

(a) How accurate are these estimates?

(b) Was the Department justified in reporting that the x-ray protocol was significantly better than the other protocol?

5.8 In the study described in Problem 5.7 the sample consisted of 500 patients between 18 and 25 years old, 700 patients in the age range from 26 to 40, and 800 patients over 40 years of age. The distribution of small lesions identified by the x-ray protocol was 16 for the younger group, 15 for the middle group, and 20 for the older group. For the other protocol the respective numbers were 15, 16, and 17. Analyze the influence of age on the effectiveness of the two protocols, and contrast your findings for the two protocols.

5.9 The medical research department is planning to study the effectiveness of a chemical extracted from apricot pits as an anticancer drug. Experimental animals with a specific form of cancer will receive injections of the chemical, and the progress of their tumors will be observed for 6 months. If tumor growth ceases or reverses, the injection will be counted as a success. The principal investigators on the project are trying to decide how many experimental animals to include in the study; they are considering using either 75 or 100.

(a) Calculate 90 and 95 percent confidence intervals for both sample sizes, assuming they were to observe cure rates of 10 and 50 percent, respectively, for the control and experimental groups.

(b) Also calculate the 1 percent significant critical region for both sample sizes. That is, find c for $\hat{p}_1 - \hat{p}_2$.

(c) Use the results of part (b) to calculate the power of the test for each sample size.

5.10 Which sample size would you recommend if experimental animals cost 1 dollar apiece; if they cost 1000 dollars apiece? Show your analysis. (Reference is to Problem 5.9.)

5.11 Sociologists are studying the influence of weight at birth and socioeconomic status on intelligence, as measured by a standard IQ test. They randomly selected five individuals from each cross category and tested them. The averages of each set of five scores are shown in the table below.

IQ scores

Weight at birth, pounds	Socioeconomic status		
	Low	Medium	High
Less than 6	95	130	110
6–7	100	120	115
7–8	110	140	170
8 or above	115	110	120

Evaluate the influence of the two factors on IQ scores.

5.12 Conservationists posted signs along a highway and then studied their impact on the highway death rate for small animals. In the year before they posted the caution signs, an average of 10 small animals were hit and killed by cars each week of the summer. This year only 9 were killed per week on the average. These estimates are based on counts at one randomly selected location in each of the 15 weeks during the summer. The respective standard deviations were 2 and 2.5.

    (*a*) How good are the estimates of averages and variances?

    (*b*) Are the averages and variances for the 2 years significantly different?

5.13 There is one fundamental assumption regarding data, which is the foundation for every analytical technique covered in Chapter 5. What is it?

5.14 Often children are pretested and subsequently posttested to evaluate a particular educational intervention. Suppose 10 children were tested under those circumstances. Explain the statistical test you would use to determine whether posttest scores were better than pretest scores on the average. Be sure to mention any assumptions you make and rationalize your selection of method.

5.15 At what point in the analysis do we employ the null hypothesis when conducting a chi-square test of independence? Explain.

5.16 One hundred rats were fed a certain chemical in their diet. Daily dosage was fixed. Data on the lifetime of the rats are given. Are these data consistent with the hypothesis that lifetime is normally distributed? Test the hypothesis.

Lifetime, yrs	Rats
0–0.5	11
0.5–1.0	27
1.0–1.5	30
1.5–2.0	20
2.0–2.5	9
2.5–3.0	3

5.17 A Student's-*t* variable is a ratio of two variables. What are their distributions? Specify the distribution of a squared Student's-*t* variable.

5.18 For large samples, $\bar{x}$ is approximately normal. For small samples from a normal population, $\bar{x}$ is normal. In either case,

$$z = \frac{\bar{x} - \mu}{\sigma/\sqrt{n}}$$

is standard normal. Therefore, under these conditions hypothesis tests on $\bar{x}$ can be done using normal tables. Notice that

$$P(Z \le z) = P(Z^2 \le z^2)$$

Because of this fact you may do your tests using a different probability model. What is it?

5.19 Physicians are being scored on the quality of hospital care they are providing. The scoring technique is considered to be patient and disease independent. The health services researchers conducting the study think that both age and the existence of peer review may be important factors. Consequently, they randomly selected five physicians from each of three age groups in each of two hospitals, one with peer review and the other without. They randomly selected two patient cases for each physician and retrospectively scored him or her on each case. Then they averaged individual and group (cell) scores to obtain the data in the table below. Are the suspected factors significant?

	Age groups, yrs		
	(30–39)	(40–49)	(50–59)
Peer review	80	82	85
No peer review	70	79	82

What other sources of variability might have been evaluated? How would you account for them?

5.20 Economists studying the difference between two prepaid health plans interviewed subscribers to obtain the responses tabulated below. Your job is to test the significance of the differences in proportions indicated.

Subscribers would consult physician for	Plan 1 $N = 926$ (%)	Plan 2 $N = 890$ (%)	Significance level 5%
Rash	28	22	
Tiredness	13	9	

5.21 What sampling and distributional assumptions are fundamental to testing the significance of:

(a) Mean   (d) Many proportions

(b) Variance   (e) Multiple means

(c) Proportion

5.22 We have said very little about the power of a hypothesis test. Recall that power is defined as

$$P(\text{reject } H_0/H_1) = (1 - \text{Type II error level}) = \text{power}$$

The power of a test is calculated *after* you find the critical value c for a specified significance level. Using this value of c and a range of numerical specifications for H_1, one calculates a range of power values. The usual practice is to plot the power curve or the Type II error level (operating characteristic curve). These curves provide insight into the ability of a test to find differences in parameter values when they exist.

For the following data plot the power and operating characteristic curves in the same diagram (use $n = 100$ and then $n = 81$):

$$H_0 : \mu = 10$$

$$H_1 : \mu > 10$$

A significance of 5 percent gives $k = 1.64$. Calculate c for each n. The decision rule is: Reject H_0 when $\bar{x} > c$. Assume $\sigma_x = 10$.

Set

$$z = \frac{\bar{x} - \mu}{10/\sqrt{n}} = \frac{(\bar{x} - \mu)\sqrt{n}}{10}$$

Then

$$P(\bar{x} > c/\mu) = P\left(\frac{(\bar{x} - \mu)\sqrt{n}}{10} > \frac{(c - \mu)\sqrt{n}}{10}\right)$$

which can be found in the normal tables for any value of n. This probability value is the power. Proceed in this manner. Interpret the results.

5.23 Life expectancy is an important issue when deciding the value of health screening programs. A group of physicians has postulated that annual physical examinations prolong life. To test this claim the government followed the health practices of 500 people from birth until death. From this group statisticians compared 75 people who had annual physical examinations at government expense from age 30 onward to a matched sample of 75 who sought medical advice only when physical problems arose. Data on the lifetimes of these individuals are given below.

  (*a*) How accurate are the estimates of expected lifetimes?

  (*b*) What assumptions must you make to answer this question?

Control group	Study group
$\sum x_i = 5250$	$\sum y_i = 5625$
$\sum x_i^2 = 377,500$	$\sum y_i^2 = 428,875$

5.6 SUPPLEMENTAL READING

Benjamin, Jack R., and C. Allin Cornell: *Probability, Statistics and Decision for Civil Engineers.* McGraw-Hill Book Company, New York, 1970, chaps. 1 and 2.

Bradley, James V.: *Distribution Free Statistical Tests*, Prentice-Hall, Inc., Englewood Cliffs, N.J., 1968.

Chapman, Douglas G., and Ronald A. Schaufele: *Elementary Probability Models and Statistical Inference*, Ginn-Blaisdell, 1970.

Cohen, Jacob: *Statistical Power Analysis for the Behavioral Sciences*, Academic Press, Inc., New York, 1969.

Drake, Alvin W.: *Fundamentals of Applied Probability Theory*, McGraw-Hill Book Company, New York, 1967.

Dwass, Meyer. *First Steps in Probability*, McGraw-Hill Book Company, New York, 1967 chaps. 2, 3, and 6.

Fleiss, Joseph, L.: *Statistical Methods for Rates and Proportions*, John Wiley & Sons, Inc., New York, 1973.

Hays, William L., and Robert L. Winkler: *Statistics: Probability, Inference, and Decision*, Holt, Rinehart and Winston, New York, 1971, chap. 6.

Hoel, Paul G.: *Elementary Statistics*, 3d. ed., John Wiley & Sons, Inc., New York, 1971, chap. 2.

Mosteller, Frederick, Robert E. K. Rourke, and George B. Thomas, Jr.: *Probability with Statistical Applications*, Addison-Wesley Publishing Company, Inc., Reading, Mass., 1961.

Nie, Norman H., et al.: *Statistical Package for the Social Sciences*, SPSS, McGraw-Hill, New York, latest edition.

SIX

STATISTICAL MODIFICATION OF PRIOR JUDGMENT

Rarely do important decisions turn on the basis of statistical analyses of sample data. Qualitative and quantitative information exclusive of sampling may be crucial. Unfortunately, the methods of analysis treated in Chapter 5 are appropriate for analyzing sample data only. They cannot be modified to explicitly account for such other information. Information available prior to sampling can be incorporated into statistical analyses only to the extent that it provides a basis for designing the sampling procedure, suggests hypotheses to be investigated, or influences the interpretation of the statistical results. These are ad hoc influences and do not adequately integrate such information with subsequent sample data to provide a balanced perspective for decision making.

Special methods of analysis, called *bayesian methods*, have been developed to systematically integrate sample data with prior information in certain circumstances. One very important feature of bayesian methods is that sample information and prior information are weighed into the analysis according to their respective credibilities. When prior information is fuzzy, absent, or conflicting, the sample data dominate. When sample data are limited or display a large variation, prior information can dominate. Furthermore, when sample data are used to reevaluate prior judgment regarding some parametric value, relatively small, more economic samples produce levels of accuracy (confidence) which could be achieved only with larger, more expensive samples using the methods of Chapter 5.

Bayesian methods for quantifying prior information as probabilistic estimates of likelihoods, proportions, and means, and subsequently modifying these estimates with additional sample information are the topics of this chapter. Bayes' rule, first introduced in Chapter 2 as Rule 10, is again discussed in Section 6.1. Quantification of prior information is the subject of Section 6.2. The specific procedures for using sample data for revising initial estimates of likelihoods, proportions, and means are demonstrated in Sections 6.3 and 6.4.

Although bayesian methods are important tools of data analysis per se, perhaps their greatest value is in the context of decision making under uncertainty. The objective of this chapter is to communicate technique and briefly illustrate some applications. The full power of bayesian methods in decision situations is made clear in Chapters 18 through 21 on formal decision analysis.

6.1 BAYESIAN MODIFICATION OF PRIOR INFORMATION

The process of revising initial likelihood estimates with experimental evidence using Bayes' rule was explained in Chapter 2. Briefly, the objective was to estimate the likelihoods for a set of mutually exclusive and exhaustive statements or outcomes, a_1 and a_2, for example, when a certain set of circumstances prevail. Preliminary estimates $P(a_1)$ and $P(a_2)$ were taken as a starting point. These were based on prior information in the same sense that a physician converts information about a patient's medical history, recent activities, and symptoms into a preliminary diagnosis. They were revised via Bayes' rule on the basis of experimental evidence. The assumptions made in the process were first that the estimates could be revised only indirectly by observing an experimental or test result, and second, that calibration probabilities for the test were known or could be determined empirically or theoretically. The calibration probabilities represented the likelihoods of certain experimental results, say, r or $\bar{r}$, when either a_1 or a_2 was true. The revised likelihood estimates for a_1 and a_2 for such dichotomous experimental results r and $\bar{r}$ were shown to be:

$$P(a_i/r) = \frac{P(r/a_i)P(a_i)}{[P(r/a_1)P(a_1) + P(r/a_2)P(a_2)]}$$

$$P(a_i/\bar{r}) = \frac{P(\bar{r}/a_i)P(a_i)}{[P(\bar{r}/a_1)P(a_1) + P(\bar{r}/a_2)P(a_2)]} \qquad i = 1 \text{ or } 2$$

$P(r/a_i)$ and $P(\bar{r}/a_i)$ represent the calibration probabilities.

In the first illustration in Section 2.4, an x-ray procedure was employed to revise a preliminary diagnosis of a malignant lesion. The prior likelihood for the presence of a lesion was given as $P(m) = 0.5$, and the calibration probabilities for the x-ray procedure were $P(+/m) = 0.8$ and $P(+/\bar{m}) = 0.06$. From Bayes' rule, the revised diagnostic probabilities were given as $P(m/+) = 0.93$ and $P(m/-) = 0.176$. Thus, if the x-ray test indicated a positive result, a physician would revise his or her prior estimate of the likelihood of a malignancy from 50 percent to 93 percent; if it indicated a negative result, a physician would take the likelihood of malignancy to be only 17.6 percent.

Table 6.1 Discrete and continuous forms of Bayes' rule

Discrete

$$P(a_i/r_j) = \frac{P(r_j/a_i)P(a_i)}{\sum_k P(r_j/a_k)P(a_k)}$$

where a_i = one of a set of mutually exclusive and exhaustive verbal or numerical statements
r_j = one of a set of mutually exclusive and exhaustive experimental results
$P(r_j/a_i)$ = a calibration probability

Continuous

$$f(x/r) = \frac{f(r/x)f(x)}{f(r)}$$

where x = continuous random variable, the value of which is being estimated
r = continuous random variable representing possible experimental results
$f(x)$ = preliminary estimate for x, in the form of a density function
$f(r/x)$ = calibration probability density for experimental results r, for a specified value of x
$f(r)$ = probability density function for all possible experimental results

The revision process is essentially the same whether one is dealing with discrete statements and experimental results such as a_1, a_2, r, and $\bar{r}$ as above, or one is dealing with a continuous array of statements and experimental outcomes. One merely uses the discrete form of Bayes' rule in the former case, and the continuous form of Bayes' rule in the latter case. The two forms of Bayes' rule are shown in Table 6.1. Specific probabilities are used in the discrete form, and density functions are used in the continuous form.

Examples of the process when the continuous form of Bayes' rule is appropriate are deferred to Sections 6.3 and 6.4. In those sections very elementary procedures which entail only basic arithmetic are developed for two specific applications. To demonstrate the use of the continuous form of the rule in general would entail introducing numerical integration methods better left to advanced texts. Keep in mind, however, that the process is, in fact, generally applicable—*it is not restricted to the special applications demonstrated in this chapter. It can be used to revise the distribution of any random variable*, including correlation coefficients, introduced in Chapter 7, not only likelihoods, proportions, and means.

6.2 INITIAL ESTIMATES FROM PRIOR INFORMATION

To employ bayesian methods, prior information must be cast in the form of a probability distribution. Empirical studies suggest strongly that the best procedure for converting nonstatistical information into probabilistic estimates is to form hypothetical bets regarding the value of the variable being estimated. One should form bets in terms of interval values for the variable. The idea is to

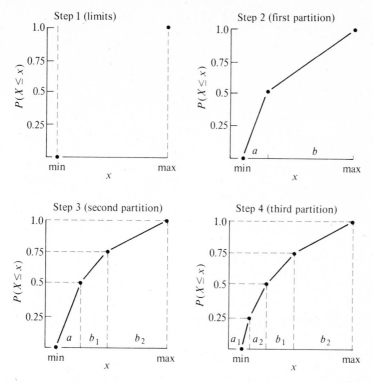

Figure 6.1 Generating a CDF for a variable x.

place odds that the variable has a value within one or another interval, say, from 0 to $\frac{1}{4}$ versus from $\frac{1}{4}$ to 1. One should not bet that the variable has a particular value, such as $p = 0.5$, for two reasons: first, point estimates are usually extremely inaccurate; second, point estimates have an aura of certainty about them which is unjustified.

The objective of this betting process is to generate a CDF for the variable. The CDF can be converted to a PDF for use in the continuous form of Bayes' rule. Also, from the CDF or the PDF one may calculate the expected value μ and variance σ^2. These numbers provide initial estimates for the variable in the form of the expected value μ or to explicitly display uncertainty $\mu + k\sigma$.

The first step in the quantification process is to determine the limiting values for the variable. For proportions the limits are zero and unity ($0 \le p \le 1$), by definition. The same is true of likelihoods, of course. For means and most other variables there are no universal limits and one must use the prior information as a guide. When the limits are not tautologically determined, experience has shown that a good rule of thumb is to set limits which are twice as distant as seems necessary.

Next, subdivide this range (x axis) into two intervals, say, a and b, as in Figure 6.1, and decide whether a bet that the variable assumes values in interval

a is better than a bet on interval b. If one interval seems a better bet than the other, adjust the boundary until there seems to be no advantage to betting on one versus the other. The boundary between these intervals is the median of the distribution, because $P(a) = P(b) = 0.5$. That is, the CDF must have a value of 0.5 at the boundary point between the two intervals. Now partition intervals a and b, each, into two subintervals, say, a_1 and a_2, b_1 and b_2, by exactly the same betting process. This yields four intervals of equal likelihood, namely, $P(a_1) = P(a_2) = P(b_1) = P(b_2) = 0.25$, and provides two more values for the CDF. The point dividing intervals a_1 and a_2 must correspond to the 0.25 level on the CDF, and the point dividing intervals b_1 and b_2 must correspond to the 0.75 level on the CDF. The three points on the x axis corresponding to the 0.25, 0.5, and 0.75 levels are called the 0.25, 0.5, and 0.75 fractiles of the distribution.

This process may be continued indefinitely, but as a practical matter at this point one might just as well sketch the CDF or fit it to a known probability model from which μ and σ^2 may be calculated. One may also convert it into a density function for use in Bayes' rule. This is the third step in the process.

There are many approaches to fitting a CDF to a probability model. One way is to use the χ^2 "goodness of fit" procedure, presented in Chapter 5. A second method, called the *method of fractiles*, is handy when great accuracy is not essential and when fractile tables are available. This is the most useful method for the purposes of this chapter, except when distributions on the mean are concerned, as in Section 6.4. The beta distribution is ideal. It is exceptionally flexible and easily fitted to judgmental CDFs. It has additional advantages, as indicated in Section 6.2.

The "fitting" process is quite simple. Suppose, for example, one had estimated the 0.25, 0.5, and 0.75 fractiles as 0.14, 0.23, and 0.34, respectively. Looking down the respective fractile columns of the beta table, one observes that the beta distribution with $t = 8$ and $r = 2$ provides a relatively good (visual) fit. Unless further perusal were to yield an apparently better fit, one would use this distribution as the prior distribution for the variable. From it one could calculate the mean and variance of the variable using only the parameter values r and t.

Now consider the more complex situation where the fractiles were estimated as 0.66, 0.77, and 0.86. This distribution clusters in the upper range of the variable, between $x = 0.5$ and $x = 1.0$. The tables provided in Appendix 2 include only distributions which cluster in the lower range $x = 0$ to $x = 0.5$, i.e., $r \le t/2$. However, this is no problem, since distributions clustering in the upper range are merely reflections of those clustering in the lower range. That is, the distribution $\mathscr{B}(x/r, t)$ is the mirror reflection of the distribution $\mathscr{B}(x/t - r, t)$, as shown in Figure 6.2. Consequently $\mathscr{B}(x/r, t) = \mathscr{B}(y/t - r, t)$, where $y = 1 - x$. And if $x_{0.25}$, $x_{0.5}$, and $x_{0.75}$ are the 0.25, 0.5, and 0.75 fractiles of $\mathscr{B}(x/r, t)$, then

$$y_{0.25} = 1 - x_{0.75}$$
$$y_{0.5} = 1 - x_{0.5}$$
$$y_{0.75} = 1 - x_{0.25}$$

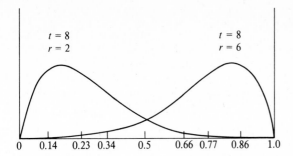

Note: Beta tables in Appendix 2 contain only distributions for which $r \leq t/2$. These are the distributions which peak to the left of $x = 0.5$. Distributions which peak above $x = 0.5$ may be found as indicated in the text.

Figure 6.2 Beta density functions.

Because of these relations, one proceeds to find the required approximate distribution $\mathscr{B}(x/r, t)$ by looking for $\mathscr{B}(y/t - r, t)$ in the beta tables of Appendix 2. Thus:

$$x_{0.25} = 0.66 \qquad y_{0.75} = 0.34$$

$$x_{0.5} = 0.77 \qquad y_{0.5} = 0.23$$

$$x_{0.75} = 0.86 \qquad y_{0.25} = 0.14$$

From the tables the distribution for y is $\mathscr{B}(y/2, 8)$. Hence the distribution for x is $\mathscr{B}(x/6, 8)$ and:

$$\mathscr{B}(y/t - r, t) = \mathscr{B}(y/2, 8) \qquad \mathscr{B}(x/r, t) = \mathscr{B}(x/6, 8)$$

$$\mu_y = \frac{2}{8} = 0.25 \qquad \mu_x = \frac{6}{8} = 0.75$$

$$\sigma_y^2 = \frac{(2)(6)}{(64)(9)} = 0.21 \qquad \sigma_x^2 = \frac{(6)(2)}{(64)(9)} = 0.21$$

Certainly this process of quantifying prior information in the form of a probability distribution is imperfect. But it turns out to be quite useful. Perhaps the most important concern one should have is that the true value of the variable lie somewhere toward the middle of the distribution. If it does not, the distribution will be of limited value, in that larger samples will be needed to revise it to give a reasonably accurate indication of the true value of the variable than otherwise would have been needed. (Even worse, if the true value is completely outside the range of the prior distribution, no amount of sampling can modify the distribution so as to encompass it.) Yet, studies done by Dr. Howard Raiffa and others indicate that most people tend to generate very narrowly constrained judgmental prior distributions. Dr. Amos Tversky and others explored the reasons for this and found that people anchor their judgment around events they can readily call to mind. They generate prior distributions which are tightly concentrated around what seems to them to be the most likely value of the variable. As a result their prior distributions sometimes do not adequately reflect their uncertainty and may not include the true value of the variable in mid-distribution.

To overcome this tendency, Raiffa suggests generating a prior distribution to the best of one's ability and then discounting the implied knowledge by about one-half. This helps to ensure that the prior distribution is sufficiently spread out (diffuse) to include the true value of the variable within its midrange. The procedure of discounting prior distributions for proportions or likelihoods is particularly easy. If the prior distribution fits a beta distribution with parameters $t = 20$ and $r = 12$, for example, one imagines that it actually represents a revised uniform (no information: $r = 1$, $t = 2$) distribution incorporating experience and observation equivalent to a sample of 18 in which there were 11 successes. To discount the prior distribution, divide the implied sample results $n = 18$, $k = 11$ by 2 to get 9 and 5.5. Then instead of accepting the prior distribution as is, substitute a discounted (more diffuse) prior distribution with $t = 9 + 2 = 11$ and $r = 5 + 1 = 6$. The next example illustrates the technique.

Example 6.1: Fitting and discounting a prior distribution Fit the following data to a beta distribution, determine the implied level of sample information, discount this prior distribution as suggested above and compare the means and variances for the original prior distribution and its discounted version: $x_{0.5} = 0.4$, $x_{0.25} = 0.3$, $x_{0.75} = 0.5$.

The beta distribution with $r = 5$ and $t = 12$ is a pretty good fit. After discounting, one has $r = 3$, $t = 7$. The means for these prior distributions are quite close, but the variances differ by quite a bit:

$\mathscr{B}(x/5, 12)$	$\mathscr{B}(x/3, 7)$	change, %
$\mu = 0.42$	$\mu = 0.43$	2.4
$\sigma^2 = 0.019$	$\sigma^2 = 0.046$	142

6.3 REVISING PROPORTIONS AND LIKELIHOODS

The procedure for revising proportions and likelihoods is based on three facts:

- First, prior distributions for these parameters can always adequately be approximated by a beta distribution with appropriate parameters t and r. Even in the situation where nothing more can be said about the value of a parameter p except that it is bounded between zero and unity, there is an appropriate beta distribution. The uniform distribution, a beta distribution with $t = 2$ and $r = 1$, characterizes this state of ignorance in which prior information provides no insight into the value of p, other than its limits.

- Second, one can always devise a Bernoulli sampling situation from which a statistical estimate of p can be made by forming a ratio of the number of successes to the number of trials.
- Finally, combining Bernoulli sampling results with a beta prior distribution via Bayes' rule produces a revised distribution which is itself a beta distribution. In fact, the values t' and r' for the revised distribution are simple arithmetic combinations of the sample results and the values t and r from the prior distribution.

Calculating the revised parameters is a three-step procedure. The first step is to fit the CDF for the proportion or likelihood p to a beta distribution. This provides a prior density function of the form:*

$$f(p) = K p^{r-1}(1-p)^{t-r-1} \qquad \text{where } t > r, r > 0, 0 \le p \le 1$$

One could use x to represent the variable, but p is more suggestive here.

The next step is to devise a sampling experiment of independent Bernoulli trials which can be used to estimate p directly by calculating the ratio of successes to trials. The results of the experiment depend on the true value of p. For example, the probability of obtaining k successes in n trials is calculated from the binomial distribution†

$$p(k) = \binom{n}{k} p^k (1-p)^{n-k}$$

Finally, the prior distribution $f(p)$ is modified by adjusting the parameters t and r according to the experimental results as follows:

$$t' = t + n$$

$$r' = r + k$$

The revised density function for p, which includes both prior information and experimental evidence, is‡

$$f(p/k) = K' p^{k+r-1}(1-p)^{t+n-k-r-1}$$

The sampling procedure could as well be designed to make either the geometric or pascal distribution appropriate. That is, one could continue sampling until one obtained one or more sucesses, instead of deciding on an overall sample size at the outset. It makes no difference which of the three sampling

---

* The constant K is included to ensure that the area under the density curve is equal to 1. Its value is dependent on the two parameters t and r, by the relation $K = (t-1)!/(r-1)!(t-r-1)!$.
† The number $\binom{n}{k}$ is similar in form to K, thus $\binom{n}{k} = n!/k!(n-k)!$.
‡ The parameters of the revised distribution are t' and r'. Thus

$$K' = \frac{(t'-1)!}{(r'-1)!(t'-r'-1)!}$$

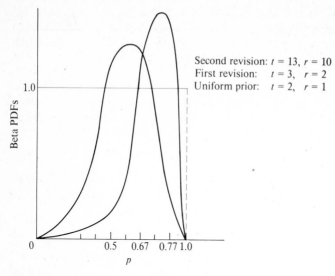

Figure 6.3 Uniform prior and revised beta density functions.

procedures is used. The posterior distribution always has the same algebraic form; its shape is dependent only on the parameters t and r, the total number of trials n, and the number of successes k.

This wonderful property of beta distributions, that they are modified only to the extent that the sampling results are added to the parameters of the prior distribution, makes modification via Bernoulli sampling almost trivial. If the prior distribution is beta with parameters t and r, the revised distribution will be beta with parameters $t + n$ and $r + k$. Revising t and r on the basis of the sample results provides all the necessary information about the revised distribution. From these two numbers, the mean and standard deviation for the revised estimate of p may be calculated immediately. In addition, confidence intervals for p may be calculated directly from beta tables.

As a simple numerical illustration of the process, consider the situation in which the prior distribution was approximated by a beta distribution with parameters $t = 2$, $r = 1$. This is the uniform distribution over the interval zero to unity, as shown in Figure 6.3. It represents the situation where nothing is known about p prior to the statistical experiment, except that it is bounded by zero and unity. Assume a Bernoulli trial is observed and the result is a "success." The revised distribution for this single experiment is $\mathscr{B}(p/2, 3)$; $\mu_p = \frac{2}{3}$.* Now assume 10 more Bernoulli trials are observed and 8 of them are a "success." These data

---

* Combining results from two samples, $n = 1$; $k = 1$ and $n = 10$; $k = 8$, as done here, is permissible only when the sampling situations are identical. Otherwise the results will be senseless.

Table 6.2 Prior, revised, and sample estimates for p

Bernoulli sample $n = (1 + 10)$, $k = (1 + 8)$

Prior estimate	Revised estimate	Combined sample estimate
$(r = 1; t = 2)$	$(r' = 10; t' = 13)$	
$\mu_p = 0.500$	$\mu'_p = 0.769$	$\hat{p} = 0.818$
$\sigma_p^2 = 0.083$	$\sigma_p'^2 = 0.013$	

Bernoulli sample $n = 101$, $k = 81$

Prior estimate	Revised estimate	Combined sample estimate
$(r = 1; t = 2)$	$(r' = 82; t' = 103)$	
$\mu_p = 0.500$	$\mu'_p = 0.796$	$\hat{p} = 0.802$
$\sigma_p^2 = 0.083$	$\sigma_p'^2 = 0.0016$	

revise the distribution further, to $\mathscr{B}(p/10, 13)$; $\mu'_p = \frac{10}{13}$, as indicated in Table 6.2. Notice how the revised distribution concentrates around a value of p which is to the right of $p = 0.5$. Notice, too, that the expected value ($\mu'_p = 0.769$) calculated from the distribution for $r = 11$, $t = 13$ is not equal to the direct statistical estimate of p from the combined sample ($\hat{p} \simeq 0.82$). As the sample size increases, the revised distribution becomes increasingly concentrated and the estimates μ'_p and $\hat{p}$ become more nearly equal with values very near the mode of the distribution. For example, for a sample size of 101 yielding 81 successes, the respective estimates would be $\hat{p} \simeq 0.802$ and $\mu'_p \simeq 0.796$, as shown in Table 6.2.

The fact that the direct statistical method of estimating p and the bayesian method do not produce identical results, even when prior information is assumed to provide no insight into the value of p, is unsettling. Apparently, the uniform distribution does not really represent a truly informationless situation in the context of the bayesian method of estimation. Unfortunately, there is as yet no way out of this difficulty except to suggest that either estimation procedure is probably acceptable in the case of useless or no prior information. However, whenever prior information provides some insight into the value of p, it should be integrated with sample information via Bayes' rule, as described in this section.

Example 6.1 illustrates fitting a judgmental CDF to a beta distribution.

Example 6.2: Discrete probability estimates An analyst is faced with estimating the likelihoods for two mutually exclusive scenarios S_1 and S_2, which he believes characterize the possible decision-making environments of the next year. He is working with numerous experts with critical knowledge bearing

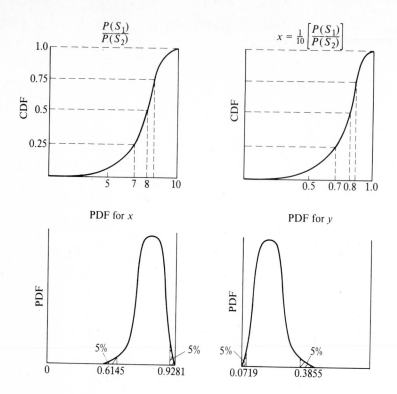

PDF for x

PDF for y

on the situation in an attempt to estimate $P(S_1)$ and $P(S_2)$ as a basis for formulating policy recommendations for his superiors. Since $P(S_1) + P(S_2) = 1$, they need only estimate one of the probabilities or the ratio $P(S_1)/P(S_2)$. They have decided to do the latter, because they feel they have more information on the relative likelihood than on the individual likelihoods.

After much discussion and study the analyst and his team of experts generated the prior CDF shown in the diagram. They are nearly 100 percent confident that the true value of the ratio $P(S_1)/P(S_2)$ is not beyond the range shown. To fit this judgmental CDF to a beta distribution they first divide the fractile numbers 7, 8, and 8.5 by 10 to convert to a distribution for $x = \frac{1}{10}[P(S_1)/P(S_2)]$ over the unit interval from 0 to 1. They observe that the steepest portion of the CDF curve is beyond the midpoint on the x axis, which means that this distribution peaks to the right of the midpoint. To find a good approximation of the distribution for x in the beta tables they proceed with the knowledge that $\mathscr{B}(x/r, t) = \mathscr{B}(y/t - r, t)$, and

Actual fractiles	Converted fractiles
$x_{0.25} = 0.70$	$y_{0.75} = 0.30$
$x_{0.5} = 0.80$	$y_{0.5} = 0.20$
$x_{0.75} = 0.85$	$y_{0.25} = 0.15$

A reasonably good approximate distribution for y is found in the tables with $t = 13$ and $r = 3$. Hence the equivalent distribution, for x has the parameters $t = 13$ and $r = 13 - 3 = 10$. Since $P(S_1)/P(S_2) = 10x$ the mean and variance of the ratio are

$$\text{Expected odds ratio} = (10) \frac{r}{t} = 7.7$$

$$\text{Variance of odds ratio} = (100) \frac{r(t - r)}{t^2(t + 1)} = 1.0$$

A 90 percent confidence interval for $P(S_1)/P(S_2)$ is obtained by finding the 5 percent tails of the distribution for x. Again one must use the beta tables, so one first finds the 5 percent tails of $y = 1 - x$ for $t = 13$, $r = 3$. These are 0.0719 for the lower tail and 0.3855 for the upper tail. Hence for x they are 0.6145 and 0.9281, respectively. Since $P(S_1)/P(S_2) = 10x$, the 90 percent confidence interval for it is 6.145 to 9.281. The intervals for x and y are shown in the diagrams.

Interval estimates of $P(S_1)$ and $P(S_2)$ may be calculated using the estimated mean and variance. For example,

$$(\mu - 2\sigma) \leq \frac{P(S_1)}{P(S_2)} \leq (\mu + 2\sigma)$$

$$(\mu - 2\sigma)P(S_2) \leq P(S_1) \leq (\mu + 2\sigma)P(S_2)$$

Since $\mu = 7.7$ and $\sigma = 1.0$, upper and lower bounds on $P(S_1)$ are

$$P(S_1) = 8.7P(S_2) \qquad \text{upper bound}$$

$$P(S_1) = 6.7P(S_2) \qquad \text{lower bound}$$

Since $P(S_2) = 1 - P(S_1)$, substitution yields

$$P(S_1) = \frac{8.7}{9.7} = 0.897 \qquad \text{upper bound}$$

$$P(S_1) = \frac{6.7}{7.7} = 0.87 \qquad \text{lower bound}$$

Again invoking the condition $P(S_1) + P(S_2) = 1$, one finds the corresponding lower and upper bounds on $P(S_2)$, namely,

$$P(S_2) = \frac{1}{9.7} = 0.103 \qquad \text{lower bound}$$

$$P(S_1) = \frac{1}{7.7} = 0.13 \qquad \text{upper bound}$$

Similar manipulations can be applied to the limits of the 90 percent confidence interval on $P(S_1)/P(S_2)$, as follows:

$$6.145 \leq \frac{P(S_1)}{P(S_2)} \leq 9.281$$

Upper bounds	Lower bounds

$$P(S_1) = \frac{9.281}{10.281} \approx 0.90 \qquad P(S_1) = \frac{6.145}{7.145} \approx 0.86$$

$$P(S_2) = \frac{1}{7.145} \approx 0.14 \qquad P(S_2) = \frac{1}{10.281} \approx 0.10$$

6.4 REVISIONS FOR MEANS

Within the bounds of certain assumptions, the procedure for revising prior distributions for a mean value is as elementary as the procedure for proportions. The required assumptions are that the prior distribution for μ be normal and that the statistical evidence be in the form of an independent random sample from a normal distribution, with known variance. If these assumptions hold at least approximately, the revised distribution is also normal and its parameters (mean and variance) are algebraic combinations of the means and variances of the prior distribution and the sampling distribution, the sample average $\bar{x}$, and the sample size n.

If the prior distribution for the mean μ were fitted to a normal distribution with a mean value m' and standard deviation σ', and a random sample $x_1, \ldots, x_n$, taken independently from the appropriate distribution with unknown mean μ and known standard deviation σ, the revised distribution would be normal, with mean and variance as follows:

- Prior distribution for μ: $f(\mu/m'; \sigma')$
- Sampling distribution with known variance: $f(x/\mu, \sigma)$
- Sample average: $\bar{x}$
- Revised distribution: $f(\mu/m'', \sigma'')$

- $m'' = \left(\dfrac{m'}{\sigma'^2} + \dfrac{n\bar{x}}{\sigma^2} \right) \left(\dfrac{\sigma'^2 \sigma^2}{\sigma^2 + n\sigma'^2} \right)$

- $\sigma''^2 = \dfrac{\sigma'^2 \sigma^2}{\sigma^2 + n\sigma'^2}$

To illustrate how the prior distribution is modified by this process, assume that the normal approximation to a judgmentally determined prior distribution has a mean of 12 and a variance of 1. Suppose an independent random sample of size n were taken from a normal population with a variance of 9, and that the average value for the sample were 11. Then for various sample sizes the mean and variance for the revised distribution would be as shown in Table 6.3.

Notice how sample estimates dominate when sample sizes are large. It is also true that for a large variance σ^2, more weight is given to the prior information relative to the sample information and uncertainty is decreased little from

Table 6.3 Revision of a normal prior distribution

Prior distribution		Sample	
Mean = 12	(m')	Statistical average = 11	$(\bar{x})$
Variance = 1	(σ'^2)	Assumed variance = 9	(σ^2)
Same size		Revised mean	Revised variance
n		m''	σ''^2
100		11.08	0.08
10		11.47	0.47

the prior to the revised distribution unless the sample size is increased to compensate.

The impact of sample information on the prior distribution is illustrated graphically in Figure 6.4. Notice that the revised distribution is always a compromise between the prior and sample distribution, indicating that the mean and variance of the revised distribution have values which lie between those of the prior and sample distributions.

The following example illustrates the use of the bayesian procedure to assess the accuracy of an estimate of the mean.

Example 6.3: Confidence of estimation for means A public agency is studying time overruns on its contracts. Previous studies suggest that the average is about 11.5 months with a standard deviation of one month. Thus overruns vary from 9.5 months to about 13.5 months, depending on many factors. Officials have made a judgmental assessment of current overruns in the form of a prior distribution with an average of 11 months and a standard deviation of about $\frac{3}{4}$.* Using a normal distribution as an approximation for this

---

* Officials could have ignored their knowledge about the current situation and used the old statistics as the parameters of their prior distribution. However, they would have lost information in the process.

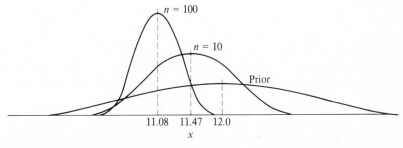

Figure 6.4 Prior and revised PDF.

judgment, they conclude that there is probably about a 95 percent chance that the true value of the mean is within the interval from 9.53 to 12.47 months, since 95 percent of the area of a normal distribution is contained within $\pm$ 1.96 standard deviations of the mean.

For economy reasons they monitored a random sample of only 10 current contracts. The average overrun was 12 months. The sample variance of one month was consistent with previous data, so the researchers assumed that the sample was drawn from a normal population with a variance of unity. They calculated the revised mean and standard deviation as:

$$m'' = 11.85 \text{ months}$$

$$\sigma'' = 0.92 \text{ months}$$

With these new estimates they calculated a revised 95 percent accuracy interval, using the familiar fact that

$$P(-1.96\sigma'' \le \mu - 11.85 \le 1.96\sigma'') = 0.95$$

Thus their revised 95 percent confidence interval was from 10.05 to 13.65 months.

They felt confident enough in their estimate to use the value 11.85 months as the current average overrun period. Notice that their final estimate was a combination of their original estimate and the sample estimate.

Had the researchers of Example 6.2 depended exclusively upon the sample data to make their estimates, they would have assessed the accuracy of the sample mean and variance by the methods of Chapter 5, using the Student's-t distribution for the mean and the chi-square distribution for the variance. From the Student's-t with $v = 9$ they would have found the 95 percent confidence interval for the mean to be approximately $(5 \le \mu \le 19)$, which is a great deal larger than the 95 percent confidence interval calculated from the revised distribution. The reason for this difference is that the methods of Chapter 5 ignore all prior information. Had the sample been very large, the sample data would have dominated the prior information and the confidence intervals would have been practically identical.

A detailed comparison of the bayesian methods of this chapter with the classical methods of Chapter 5 in a real policy situation is undertaken in Chapter 8.

6.5 PROBLEMS

6.1 You are on the verge of making an investment. You have estimated payoffs for each investment contingent on some set of government decisions a_1, a_2, and a_3, which are mutually exclusive and exhaustive. You are at a loss to estimate the likelihoods that these events will occur, so you have assigned $\frac{1}{3}$ to each. You intend to make your investment choice after you read the next economic report of the President, because it will indicate whether a certain economic state b has been achieved. This information will allow you to select conditional likelihoods for a_1, a_2, and a_3 because you have determined the likelihoods that b can be achieved for each decision: $P(b/a_1) = 0.4$, $P(b/a_2) = 0.5$, and $P(b/a_3) = 0.8$.

- Calculate the conditional probabilities

$$P(a_i/b) \text{ for } i = 1, 2, \text{ and } 3$$

- If b has not been achieved, which of the decisions has most likely been selected? What are the likelihoods?

6.2 The captain of the Cajun Queen towboat knows by experience that there is only a 20 percent chance that she will not have to wait for another tow to clear Locks and Dam No. 2 before she can begin locking through. She also knows that the chances are about 80 percent that she will not have to wait at any of the other locks. What is the likelihood that she will have to wait at Locks and Dam No. 2 if she was not delayed at the previous lock? Assume that she has been delayed at neither lock on about 10 percent of her trips.

6.3 Assume you are intent on estimating the proportion of all young Americans born in the year 1960 who will marry by the age of 16. Based on similar studies done about 20 years ago and information you have regarding changing norms of behavior, you have generated a prior distribution for p which is closely approximated by a beta distribution with $t = 10$ and $r = 5$. Assume you interviewed a random sample of this cohort and found that 10 percent of the group interviewed had married by age 16.

For sample sizes 5 and 500,
(a) Calculate the parameters t and r of the revised distributions for p.
(b) Calculate the mean, mode, and variance of the revised distributions.
(c) Calculate the direct statistical estimates $\hat{p}$ and compare to $\bar{p}$ from part (b).

6.4 The director of the department of social services has a tentative objective of screening 90 percent of the eligible welfare children next year and treating all with "serious health defects," under the EPSDT/Medicaid program. The problem is that contracts to health providers for treating disorders must be budgeted now and their is no statistically valid way to determine the number of professionals needed, since this is the first year that more than a token number of eligible children will have been screened.

The kinds of information known are the following:

- About 90 percent of those screened in the past have needed treatment for severe defects.
- A recent study, by B Corporation, in eight states, concluded that during the early years of the EPSDT program in states where people were discouraged from using the services, the proportion of children with serious defects started high but decreased sharply (by about 50 percent) as officials responded to the federal mandate to implement the program fully.
- The department had previously discouraged people from using the program and was not quite prepared to serve the entire population.
- The EPSDT program is not organized here in quite the same way as it is in any of the other states. In the other states either private clinics or public health clinics did the screening; here, children's hospitals and public health clinics will be doing the screening.

The director and his staff have done a great deal of investigating and thinking and have finally managed to make interval estimates for the proportion p which have serious defects. Their estimates are

$$P(0 < p \le 0.25) = 0.4 \qquad P(0.25 < p \le 0.5) = 0.3 \qquad P(0.5 < p \le 0.75) = 0.2$$

$$P(0.75 < p \le 1) = 0.1$$

The director plans to use this information, the census estimates of the number of eligibles in the state, and regional data on the average cost of correcting serious defects, to estimate his budget needs. He will argue for enough money to treat the expected number of cases of serious defects, plus enough to account for one standard deviation for p. That is, he will request $(\mu_p + \sigma_p)NC$ dollars, where C is the average cost per patient and N is the number of eligibles. If he cannot sell this estimate, the director will ask for $NC(\mu_p - \sigma_p)$ dollars and run the program on a pilot basis to better estimate p before requesting a budget supplement.

(a) Plot the prior distribution for p implied in this discussion.

(b) Fit the distribution to a beta distribution and calculate μ_p and σ_p.

(c) Assume a random sample of 100 were taken from the pilot program and 20 percent had serious defects. Reestimate μ_p and σ_p. Discuss your results.

6.5 Calculate the 95 percent confidence interval for a beta prior distribution with parameters $t = 3$, $r = 1$, for the revised beta distribution with parameters $t = 15$, $r = 5$, and for the direct statistical estimate $\hat{p} = \frac{1}{3}$.

6.6 On this problem you *must* work alone. Each response will be uniquely yours. There are no correct answers. Your grade will be determined by whether you honestly and thoroughly answered all the questions. If you have completed the exercise in that fashion, you will receive a maximum score; if you fail to respond reasonably to any question, you will receive partial credit.

Your basic task is to generate your judgmental probability distribution for the following items:

1. The population of the United States as of today
2. The population of Missouri as of today
3. Medicare outlays by the federal government this past year

Before starting please write a brief paragraph expressing your views:

(a) Does it make sense to generate such probability distributions? Why?

(b) In which cases is it more or less reasonable to do this?

For each of items 1 through 3 please:

(a) Write an outline of your argument for choosing the fractile points. This should include indications of relevant information you drew upon.

(b) Indicate keys to your thinking process. For example, did you mull it over and finally choose a point, did you reason methodically from a known base, or did you flit from thought to thought and eventually home in on a point? There is no best method!

(c) Record the length of time you spent considering each question and generating the distribution.

(d) How confident are you that the true value will fall within the middle 70 percent range of your distribution:

- Less than 50 percent
- Between 50 and 59 percent
- Between 60 and 69 percent
- Between 70 and 79 percent
- Better than 80 percent

6.7 You would like to estimate average income for physicians in the Midwest for this year. You have data on the national average for last year from a partial census survey, but you know things have changed since then, and you are aware that national incomes are not representative of Midwest incomes.

You have generated a prior distribution for the average μ, which is approximately normal, with mean $m = \$30,000$ and a standard deviation of 4000 dollars. Using a directory of licensed physicians as a source of names, you decide to write to 100 Midwestern physicians to find out how much they expect to earn this year. Only 18 respond and the average reported is 41,000 dollars. Taking the standard deviation reported in the national survey as approximately accurate for the sampled group, you assume $\sigma = \$5000$ for this group.

(a) Calculate the mean and variance for the revised distribution.

(b) What is the skewness coefficient for both the prior and revised distributions?

(*c*) What is the 95 percent confidence interval for the revised distribution? Contrast this with the 90 percent confidence interval for the prior distribution.

6.8 Describe how you could use a prior distribution for either a proportion or a mean value to formulate a hypothesis to be tested, using the methods of Chapter 5.

6.9 The table below contains incomplete CDFs for age at marriage for Americans born in different years. Table entries are percentage of cohort.

Cohort age in 1973, yrs	Age at marriage						
	12	13	14	15	16	17	18
15	1	4	6				
16	1	2	3	8			
17	–	1	2	5	17		
19	–	–	1	2	7	17	30

(*a*) Fit the data to beta distributions (eyeball). Specify *r* and *t* for each.
(*b*) What are the mean and variance for each?
(*c*) Compare the results.

6.10 The local Professional Standards Review Organization knows that the (national) average length of hospital stay for disease *D* is 10 days with a standard deviation of 3 days. Standard deviations for length of stay can be expected to be uniform nationwide, but average lengths of stay vary significantly from hospital to hospital. The organization has sampled ($n = 24$) the records for hospital H and found that the average length of stay there for *D* is 13 days.

(*a*) Use the national data as the prior distribution for H and use the sample average ($\bar{x} = 13$) to revise it.

(*b*) Use the revised distribution to calculate the probability that the competing hypotheses are true:

$H_0: \mu > 13$ outside the acceptable range

$H_1: \mu \leq 13$

$H_0: 9 \leq \mu \leq 12$ very near the national average

$H_1: \mu > 13$

6.6 SUPPLEMENTAL READING

Behn, Robert D., and James W. Vaupel: "Teaching Analytical Thinking," *Policy Analysis*, vol. 2, no. 4, Fall 1976, p. 663.

Benjamin, Jack R., and C. Allin Cornell: *Probability, Statistics and Decision for Civil Engineers*, McGraw-Hill Book Company, New York, 1970, chap. 5.

Brown, Rex V., Andrew S. Kahr, and Cameron Peterson: *Decision Analysis for the Manager*, Holt, Rinehart and Winston, Inc., New York, 1975, chaps. 30–35.

Good, I. J.: "How to Estimate Probabilities," *Journal of the Institute of Mathematics Applications*, vol. 2, 1966, pp. 364–383.

Hays, William L., and Robert L. Winkler: *Statistics: Probability, Inference, and Decision*, Holt, Rinehart and Winston, Inc., New York, 1971, chap. 8.

Hoel, Paul G.: *Elementary Statistics*, 3d ed., John Wiley & Sons, Inc., New York, 1971, chap. 10.

Kassouf, Sheen: *Normative Decision Making*, Prentice-Hall, Inc., Englewood Cliffs, N.J., 1970, chap. 4.

Moore, P. G., and H. Thomas: *The Anatomy of Decision*, Penguin Books, Inc., Baltimore, 1976, chaps. 7 and 8.

Raiffa, Howard: *Decision Analysis: Introductory Lectures on Choices under Uncertainty*, Addison-Wesley Publishing Company, Inc., Reading, Mass., 1968, chaps. 2 and 5.

Tversky, Amos, and Daniel Kahneman: "Judgment under Uncertainty: Heuristics and Biases," *Science*, vol. 185, September 1974, pp. 1124–1131.

LINEAR RELATIONS AMONG VARIABLES

Unless two variables are stochastically independent, knowing the value of one provides some information regarding the value of the other.* In fact, one can often predict the value of one variable quite accurately from the value of the other. Commonplace examples of this phenomenon include predicting academic achievement for college students from their high school grade point averages, academic achievement from a child's weight at birth, characteristics of offspring from characteristics of parents, sales volume from the price of a product, the strength of materials from the concentrations of embedded impurities, and annual health care costs from the age of the client population.

Regression, or, more precisely, *least-squares curve fitting*, is a statistical procedure for developing equations to predict the value of one variable from the value(s) of one or more others. Today, almost all professions employ regression as a staple analytical tool. Scientists and engineers have used the method of least-squares for more than a century to determine mathematical relationships between physical variables such as temperature and heat, pressure and heat, force and velocity, electromagnetic energy and light intensity, and the like. Economists could hardly do without regression. When an economist refers to a model of consumer preferences, a demand model, or a labor model, for example, the economist is most likely referring to a regression equation expressing the named variable as a function of one or more predictive variables. The least-squares method is applicable whether relations between variables are linear (straight lines) or not. However, linear regression alone is discussed in this chapter, because it is the least cumbersome mathematically.

* Stochastic independence is defined in Rule 7 of Chapter 2.

Clearly, the first step in linear regression is to assess the reasonableness of assuming a linear predictive equation. Unless variables are strongly linearly related, linear predictions of one variable in terms of others are likely to be very inaccurate. The *correlation coefficient*, which measures the degree of linearity between two variables, therefore plays a crucial role in linear regression. The use of the correlation coefficient is the subject of Section 7.1.

The procedure for developing a regression equation for two variables is developed in Section 7.2. Limiting the discussion to this most basic regression problem, called *simple linear regression*, facilitates developing an appreciation of the concept and its limitations using simple two-dimensional diagrams and basic algebra.

More complex, many-variable regression methods are discussed briefly in subsequent sections. The basic ideas of simple linear regression are extended to *multiple linear regression* in Section 7.3. The discussion is brief because the objective is to prepare the reader to formulate multiple-regression problems and interpret the computer-generated results. Finally, a special computerized approach to selecting the most important predictive variables for inclusion in multiple regression is discussed in Section 7.4. This approach, called *stepwise multiple regression*, is perhaps the most rational approach to multiple regression. Both forms of multiple regression may be accomplished using standard computer programs such as the Statistical Package for Social Sciences (SPSS) and the Statistical Analysis System (SAS).

The reader should concentrate on understanding the concept of regression and its appropriateness to certain situations. Calculators and computers are available to crunch the numbers; the analyst must do the thinking and interpret the results.

7.1 LINEAR CORRELATION

Educators know that academic performance in the first year of college tends to be proportional to the level of performance in high school. Farmers know that crop yields increase, to an extent, in proportion to fertilization intensity. And the

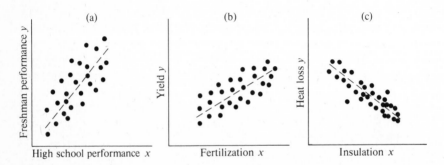

Figure 7.1 Scatter diagrams.

fact that heat loss from dwelling units decreases in proportion to the thickness of ceiling and wall insulation is widely appreciated these days. Data for each of these situations suggest an overall linear relationship between each of these pairs of variables; but they also exhibit a fair amount of variability, as is indicated in the scatter diagrams of Figure 7.1. The strength of the linear relationships between the pairs of variables in these and similar examples is measurable using a statistical measure called the correlation coefficient.

The correlation coefficient is defined as the ratio of the covariability of two variables to the product of their standard deviations.* For a set of data consisting of n pairs of values (x_i, y_i) the correlation coefficient is estimated by the equation

$$r = \frac{\text{cov }(x, y)}{S_x S_y} = \frac{\frac{1}{n} \sum_{i=1}^{n} (x_i - \bar{x})(y_i - \bar{y})}{S_x S_y} = \frac{\frac{1}{n} \sum_{i=1}^{n} x_i y_i - \bar{x}\bar{y}}{S_x S_y}$$

The numerator, called the *covariance* for the variables, written cov (x, y), tells the tale of correlation.† Each term in the covariance is a product of deviations from average values. Therefore, as indicated in Figure 7.2, when the deviations $(x_i - \bar{x})$ and $(y_i - \bar{y})$ are predominantly of the same sign, their products tend to be positive, making the numerator large and positive and indicating an inclining trend line. When the deviations are predominantly of opposite signs, their products tend to be negative, making the numerator large but negative and indicating a declining trend line. However, when the data points are distributed somewhat symmetrically around both means, the products of deviations tend to cancel in the summation, making the numerator small in magnitude and indicating no trend line, as in Figure 7.2c.‡

---

* The theoretical expression for the correlation coefficient is

$$\rho = \frac{E[(x - \mu_x)(y - \mu_y)]}{\sigma_x \sigma_y}$$

The Greek symbol ρ (rho) is used to represent the theoretical value of the coefficient, whereas r is used to indicate a statistical estimate of ρ.

† The covariance arises naturally when one calculates the variance for the sum of two variables. Theoretically,

$$\text{var }(x + y) = E[(x - \mu_x) + (y - \mu_y)]^2 = \text{var }(x) + \text{var }(y) + 2 E[(x - \bar{x})(y - \bar{y})]$$

When x and y are independent, $E[(x - \bar{x})(y - \bar{y})] = E(x - \bar{x})E(y - \bar{y}) = 0$ and the variance of the sum is the sum of the variances. The statistical approximation to this relation is

$$S_{x+y}^2 = S_x^2 + S_y^2 + 2 \text{ cov }(x, y)$$

‡ The extreme values for correlation coefficients are ± 1. When the relation is perfectly linear with a positive slope, the value of the correlation coefficient is $+1$. When the relation is perfectly linear and the slope is negative, the correlation coefficient is -1. The actual value of the slope is immaterial. The plausibility of this fact is easy to demonstrate. First the correlation coefficient measures degree of linearity. If $y = a + bx$, then cov $(x, y) = bS_x^2$ and $S_y^2 = b^2 S_x^2$. Therefore, $r = \pm 1$, depending on the sign of b, but not on its magnitude, since b cancels out in the ratio.

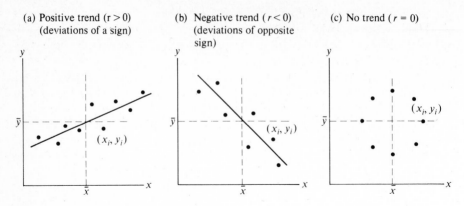

Figure 7.2 Correlation versus no correlation.

The data in Figure 7.2c facilitate an important observation. All data points are distributed in a perfect circle, suggesting that the variables x and y are related by the equation of a circle, $k^2 = x^2 + y^2$, where k is the radius. This is a perfect functional relation, so the variables cannot be independent; yet the correlation coefficient is zero, indicating no *linear* relation.* Clearly, a zero correlation coefficient does not imply that two variables are not functionally related; it only implies that they are not linearly related. On the other hand, if two variables are independent, their covariance must be zero.† Thus, independence is a sufficient condition for zero correlation, but zero correlation is merely necessary for independence.

The procedure for calculating the correlation coefficient from a set of data is demonstrated in Example 7.1.

Example 7.1: Correlating heights of offspring and parents Anthropologists suspect that a child's height at maturity is related to the average height of its parents. They have collected height data for a small island population and have asked their research assistant to determine the degree of correlation between average parental height and the height of adult offspring.

The research assistant first prepared a scatter diagram, as shown. She noticed an apparent trend in the data and calculated the correlation

---

* Notice, too, that since $E[(x - \bar{x})(y - \bar{y})] = 0$,

$$\text{var } (x + y) = \text{var } (x) + \text{var } (y)$$

even though the two variables are not independent. Consequently, the variance of a sum of variables is the sum of the variance whenever the covariance is zero; the variables need not be independent, although independence implies the same result.

† This follows from the fact that cov (x, y) is a statistical approximation for $E[(x - \mu_x)(y - \mu_y)]$ and for independent variables,

$$E[(x - \mu_x)(y - \mu_y)] = E(x - \mu_x)E(y - \mu_y) = 0$$

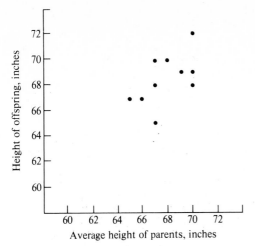

Scatter diagrams.

coefficient as a measure of the degree of linearity in the trend. Her calculations are tabulated below.

In her report to the anthropologists, the research assistant remarked that this preliminary analysis indicated a linear relation between the two variables and that the correlation coefficient was relatively large although there was some chance that it could be the result of statistical variability. She indicated that she would investigate this matter further and would also look into the possibility that height of male or female offspring correlated

Calculations for correlation coefficient

Observation number, i	Height data, inches		$x - \bar{x}$	$(x - \bar{x})^2$	$y - \bar{y}$	$(y - \bar{y})^2$	$(x - \bar{x})(y - \bar{y})$
	Parents, x	Offspring, y					
1	65	67	−2.8	7.84	−1.5	2.25	4.2
2	70	72	2.2	4.84	3.5	12.25	7.7
3	70	69	2.2	4.84	0.5	0.25	1.1
4	68	70	0.2	0.04	1.5	2.25	0.3
5	66	67	−1.8	3.24	−1.5	2.25	2.7
6	67	65	−0.8	0.64	−3.5	12.25	2.8
7	69	69	1.2	1.44	0.5	0.25	0.6
8	70	68	2.2	4.84	−0.5	0.25	−1.1
9	66	68	−1.8	3.24	−0.5	0.25	−0.9
10	67	70	−0.8	0.64	1.5	2.25	−1.2
Total	678	685	NA	30.0	NA	34.5	16.2
Average	67.8	68.5	NA	3.0	NA	3.5	1.62

$S_x^2 = 3.0$ cov $(x, y) = 1.62$

$S_y^2 = 3.5$ $r = 0.50$

better with the height of the parent of the same sex than with the average height of both parents.

The research assistant in Example 7.1 knew her business. Realizing that apparent correlations can be the consequence of pure chance, she planned to assess the validity of the observed correlation coefficient. To do this she would calculate a confidence interval around the estimate r and test the hypothesis that the true correlation coefficient ρ was really different from zero.

The procedure for assessing the accuracy and the significance of observed correlations is best explained in the context of regression. All computerized regression programs include provisions for these assessments. The F distribution is normally used in these programs although the Student's-t distribution can be used instead, and often is in mathematical texts. These are the topics of Section 7.2.

7.2 TWO-VARIABLE EQUATIONS

When two variables are linearly related, the value of either may be calculated from the value of the other. Linear equations relating two variables are always of the form

$$y = a_0 + a_1 x$$

Thus, one needs to know only the values of the constant a_0, which is the y intercept (the value of y when $x = 0$), and the constant a_1, which is the slope (the rate of increase, or decrease, in y as x is increased) to predict the value of y for a particular value of x. When y is written as a function of x, as it is above, y is called the *dependent variable*, and x is called the *independent variable*, because the value of y depends on the value of x inserted into the equation.* Notice also that

---

* Dependent and independent variables are also referred to as *endogenous* and *exogenous* variables, respectively.

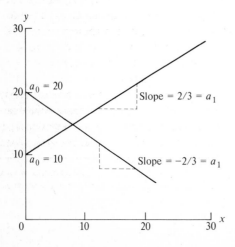

Figure 7.3 Two linear equations, $y = a_0 + a_1 x$.

whenever y can be written as a function of x, it is also true that x can be written as a function of y. In that case x becomes the dependent variable and y is chosen or determined independently. In Figure 7.3 linear equations for y as a function of x are plotted for different values of a_0 and a_1.

As suggested in Section 7.1 there are many situations in which two variables appear to be linearly related. Although statistical data may indicate variability in the relationship, one can often attribute observed variability to the accumulation of a multitude of external (exogenous) impacts which perturb the data. This situation suggests a model relating x and y of the type

$$y_i = a_0 + a_1 x_i + \varepsilon$$

where a_0 and a_1 are constants, x is not a random variable, and ε is a random variable which induces randomness in y. Because ε is assumed to represent the accumulation of a multitude of small perturbations in the value of y_i which occur independently of the value of x_i, the variable ε should be approximately normally distributed (according to the central limit theorem). It should have an expected value of zero $[E(\varepsilon) = 0]$ and a constant, but unknown, variance σ^2. With these assumptions it is theoretically possible to estimate the coefficients in the predictive equation and, using the statistical methods of Chapter 5, to assess their credibility.

To facilitate calculating the coefficients, the predictive equation is usually rewritten as

$$y_i = b_0 + b_1(x_i - \bar{x}) + \varepsilon$$

This form is obtained by adding and subtracting $a_1 \bar{x}$ in the original form of the equation and collecting terms so that

$$b_0 = a_0 + a_1 \bar{x} \qquad \text{and} \qquad b_1 = a_1$$

The change is superficial. The origin has been shifted for calculational convenience, while the slope $b_1 = a_1$ remains unaffected. If one wished to calculate the value of a_0 or a_1 after determining b_0 and b_1 one would simply observe that:

$$a_1 = b_1$$
$$a_0 = b_0 - a_1 \bar{x}$$

Henceforth in this chapter the modified form of the linear prediction equation is the one used.

The basic model is graphically displayed in Figure 7.4. The trend line represents expected (or predicted) values of y_i for each value x_i. That is,

$$E(y_i/x_i) \doteq E[b_0 + b_1(x_i - \bar{x}) + \varepsilon]$$
$$= b_0 + b_1(x_i - \bar{x}) + E(\varepsilon)$$
$$= b_0 + b_1(x_i - \bar{x})$$

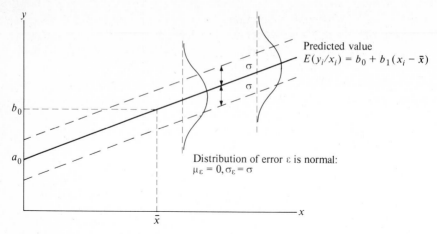

Figure 7.4 Simple linear model with error term.

The distributions of the actual y values about the trend line are represented as normal curves of constant variance. They are centered on the trend line. These distributions indicate how the ε term influences the value of each y_i and thereby represent the distribution of deviations one would expect to observe between the true values y_i and the corresponding values $E(y_i/x_i)$ predicted by the linear model.

The principal features of the linear model are summarized in Table 7.1 on page 148.

The objective of the least-squares method is to estimate the coefficients b_0 and b_1 from sample data. The approximate equation is commonly written as

$$\hat{y}_i = \hat{b}_0 + \hat{b}_1(x_i - \bar{x})$$

The value $\hat{y}_i$ is an estimate for $E(y_i/x_i)$ determined from actual data. If one had a sample of n paired data points (x_1, y_1) through (x_n, y_n), one could plot them in a scatter diagram. On the same diagram one could sketch in a linear equation for y_i in terms of x_i and indicate the deviations of the true values y_i from the predicted values by d_i, as shown in Figure 7.5. The d_i values would then represent the actual values assumed by the error variable ε for each sample value x_i. The sum of the squares of the deviations would be as small as possible if the coefficients of the predictive equation were calculated as follows.

The deviations d_i represent the differences

$$d_i = y_i - [\hat{b}_0 + \hat{b}_1(x_i - \bar{x})]$$

Therefore, the sum of the squared deviations D is

$$D = \sum_{i=1}^{n} [y_i - \hat{b}_0 - \hat{b}_1(x_i - \bar{x})]^2$$

$$= \sum_{i=1}^{n} [(y_i - \hat{b}_0)^2 - 2\hat{b}_1(x_i - \bar{x})(y_i - \hat{b}_0) + \hat{b}_1^2(x_i - \bar{x})^2]$$

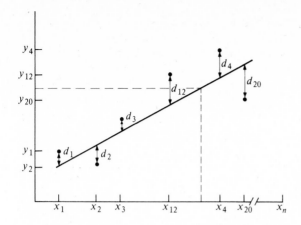

Figure 7.5 Predictive equation and deviations from data.

The sum of the terms $(y_i - \hat{b}_0)^2$ is minimized when $\hat{b}_0 = \bar{y}$ because then each term represents the deviation of y from its expected value.* The second term in the summation may now be written as the sum

$$-2\hat{b}_1(x_i - \bar{x})y_i + 2\hat{b}_1(x_i - \bar{x})\bar{y}$$

The second of these sums to zero, because the sum of deviations $(x_i - \bar{x})$ is zero. Therefore, to finish the process of minimizing D, the sum

$$\sum_{i=1}^{n} [-2\hat{b}_1(x_i - \bar{x})y_i + \hat{b}_1^2(x_i - \bar{x})^2]$$

must be minimized. This is accomplished by setting†

$$\hat{b}_1 = \frac{\sum_{i=1}^{n}(x_i y_i - \bar{x}y_i)}{\sum_{i=1}^{n}(x_i - \bar{x})^2} = \frac{\frac{1}{n}\sum_{i=1}^{n} x_i y_i - \bar{x}\bar{y}}{S_x^2} = \frac{\text{cov}(x, y)}{S_x^2}$$

which can be written more conveniently by multiplying the numerator and denominator by S_y^2 to obtain

$$\hat{b}_1 = \frac{\text{cov}(x, y)S_y^2}{S_x^2 S_y^2} = r\frac{S_y}{S_x}$$

When pieced together, the predictive equation for y_i as a linear function of x_i, determined by minimizing the sum of squared deviations, is simply

$$\hat{y}_i = \bar{y} + r\frac{S_y}{S_x}(x_i - \bar{x}) \qquad \text{estimated predictive equation}$$

*, † One way to find the values of $\hat{b}_0$ and $\hat{b}_1$ which minimize D is to set the derivatives of D with respect to $\hat{b}_0$ and $\hat{b}_1$ equal to zero and solve these equations for the coefficients. An alternative algebraic approach is explained in Mosteller's book which is referenced in Section 7.6.

As always, one also needs an estimate for the variability in data. The variability in the y_i values from the expected values $E(y_i/x_i)$ is measured by σ^2. The variability in the data values y_i from predicted values $\hat{y}_i$ is measured by S^2, the sum of squared deviations divided by the sample size. Actually S^2 is an estimate for σ^2. Thus,

$$S^2 = \frac{D}{n} \quad \text{estimate for } \sigma^2$$

By substituting the previously determined expressions for $\hat{b}_0$ and $\hat{b}_1$ into D, one finds that

$$S^2 = S_y^2(1 - r^2)$$

which suggests that the variation in the y dimension is reduced by $(1 - r^2)$ when measured from the approximate linear predictive equation, as opposed to being measured around the average value $\bar{y}$. In technical jargon, r^2, the estimated squared correlation coefficient, measures (determines) the fraction of the variability explained by the regression; r^2 is often called the *coefficient of determination*. The number $S = \sqrt{D/n}$ is commonly referred to as the *standard error of estimate* for the regression.

Example 7.2 demonstrates the process of estimating a regression equation and the variance around it. The data are taken from Example 7.1, where r was calculated.

Example 7.2: Estimating a regression equation The research assistant helping the anthropologists in Example 7.1 decided to develop an approximate predictive equation from the data supplied. She had already determined that

$$\bar{x} = 67.8 \quad S_x^2 = 3.1 \quad r = 0.53$$

$$\bar{y} = 68.5 \quad S_y^2 = 3.5$$

Consequently, it was a simple matter to calculate the coefficients for the regression equation, namely,

$$\hat{b}_0 = \bar{y} = 68.5 \quad \hat{b}_1 = r\frac{S_y}{S_x} = 0.57$$

and write the equation

$$\hat{y}_i = 68.5 + 0.57(x_i - 67.8)$$

$$= 29.9 + 0.57x_i$$

From this equation she could predict the height of immature offspring from the population stock represented by her data, provided that the average height of the parents did not exceed the limits of her data, namely, average heights of 65 to 70 inches.*

* Too frequently the untrained are tempted to predict values of the dependent variable for values of x which exceed the range of x values used in developing the predictive equation. This is an invalid use of the technique. Regression is a technique for data description and may be used for prediction only within the limits of the data.

The last major step in a regression analysis is to test the validity and accuracy of the correlation coefficient. The methods employed are those of Chapter 5. The ratio

$$F_{1,\,n-2} = \frac{r^2 S_y^2/1}{S^2/(n-2)} = \frac{(n-2)r^2}{(1-r^2)}$$

is an F random variable with parameters of 1 and $n-2$. To assess the validity of the conclusion that two variables are correlated one would ordinarily test $H_0 : \rho = 0$ against $H_1 : \rho \neq 0$ and reject the null hypothesis only if the estimate r were outside the range specified by the preselected error limit. However, this is equivalent to testing $H_0 : \rho^2 = 0$ against $H_1 : \rho^2 > 0$ and rejecting the null hypothesis only if r^2 were outside some critical range. In this form, the decision rule is

$$\text{Reject } H_0 \text{ if } \quad F_{1,\,n-2} = \frac{(n-2)r^2}{(1-r^2)} > k$$

The number k is determined from the $F_{1,\,n-2}$ table, according to the condition

$$P(F_{1,\,n-2} > k) \leq \text{error limit*}$$

If the specific estimate $(n-2)r^2/(1-r^2)$ is larger than k, one concludes that the likelihood that the two variables are *not* linearly related is no greater than the specified error limit; the results are said to be significant at that error limit.†

The next example completes the analysis of the height prediction theory postulated by the anthropologists in Example 7.1.

Example 7.3: Significance of the correlation coefficient The anthropologists' research assistant previously estimated the correlation coefficient (coefficient of determination) for the height data as $r = 0.53$ ($r^2 = 0.28$). She knows that $S^2 = (1-r^2)S_y^2$, so that from her calculations ($S_y^2 = 2.5$ and $S_x^2 = 3.1$) she now forms the ratio

$$\frac{r^2 S_y^2/1}{S^2/(n-2)} = \frac{(n-2)r^2}{(1-r^2)} = \frac{8(0.53)^2}{(1-0.53^2)} = 3.1$$

---

* The numbers k and c^2 are related by the expression

$$c^2 = k \left| \frac{S^2/(n-2)}{S_y^2/1} \right| \qquad \text{reject } H_0 \text{ if } r^2 > c^2$$

† This conclusion is equivalent to the conclusion that $b_1 = \rho \sigma_y/\sigma_x$ is not zero, because the ratio

$$\frac{b_1^2 S_x^2/1}{S^2/(n-2)}$$

is also an F variable with parameters 1 and $n-2$, since $b_1 = 0$ only when $\rho = 0$. The significance of both $\hat{a}_1 = \hat{b}_1$ and $a_0 = \bar{y} - \hat{b}_1 \bar{x}$ are usually assessed in computer algorithms, and their standard errors of estimate are printed out. The format of computer printouts vary, but the **SPSS** package organizes the results as follows:

$$\hat{y}_i = \hat{a}_0 + \hat{a}_1 x \qquad r^2, S^2$$

$$(S_{\hat{a}_0})(S_{\hat{a}_1}) \qquad \text{significance level}$$

Table 7.1 Simple linear regression summary

Theoretical model

$$y_i = b_0 + b_1(x_i - \bar{x}) + \varepsilon: N(\varepsilon/\hat{y}_i; \sigma)$$

$$E(y_i/x_i) = b_0 + b_1(x_i - \bar{x}): \text{var}(y_i/x_i) = \text{var}(\varepsilon) = \sigma$$

Estimation equation

$$\hat{y}_i = \hat{b}_0 + \hat{b}_1(x_i - \bar{x})$$

$$\hat{b}_0 = \bar{y}; \hat{b}_1 = rS_y/S_x$$

Coefficient of determination (proportion of variance explained)

$$r^2 = \left[\frac{\frac{1}{n^2} \sum\limits_{i=1}^{n} x_i y_i - \bar{x}\bar{y}}{S_x S_y} \right]^2$$

Standard error of estimate (approximation for σ)

$$S = \sqrt{S_y^2(1 - r^2)} \qquad \text{or} \qquad S^2 = Sy^2(1 - r^2)$$

and compares it to the values $k = 11.26$ and $k = 5.32$, which she determines from the F tables with parameters 1 and 8 and error limits of 1 percent and 5 percent, respectively. Since the ratio $F_{1.8}$ is smaller than either k value, she concludes that the chances are greater than 5 percent that the apparent correlation is a statistical artifact; she cannot reject the null hypothesis at either the 1 percent or 5 percent level.

For convenience of reference the highlights of simple linear regression are summarized in Table 7.1.

7.3 MANY-VARIABLE EQUATIONS

Seldom is a simple linear equation a good predictor. Crop yields, for example, depend on rainfall as well as fertilizer. And academic performance in the first year of college depends on many factors in addition to performance in high school. Consequently, a linear predictive equation or linear regression model which includes many independent variables $x_1, x_2, \ldots, x_k$ is generally more useful than a two-variable model.

Multivariable models are generalizations of the simple linear model. They are written

$$y_i = a_0 + a_1 x_{1i} + a_2 x_{2i} + a_3 x_{3i} + \cdots + a_k x_{ki} + \varepsilon$$

The coefficient a_0 represents the value of y when each of the x values is zero, as it did in the simple linear model. The coefficients $a_1, a_2, \ldots, a_k$ represent rates of change in y as the x values are varied individually, just as a_1 represented the slope of the linear predictive equation in the two-variable model. The dependent variables have two subscripts for accounting purposes. The second subscript i references the data point or observation number. The first subscript basically provides a name for the dependent variables so they can be identified easily. The x_{ji} are not random variables, but the error variable ε is. As in the two-variable case, ε is assumed to be the accumulation of a multitude of relatively small, independently acting factors and to have a normal distribution with zero mean and constant variance σ^2. Consequently,

$$E(y_i/x_{1i}, x_{2i}, \ldots, x_{ki}) = a_0 + a_1 x_{1i} + \cdots + a_k x_{ki}$$

In the multivariate model some assumptions are required which were not apparent in the simple linear model. First, one must assume that none of the dependent variables are exact linear functions of any group of the others. If they are so related, they are said to be collinear. This results in double attribution of the predictive values of variables and inflates the significance of the correlation. Second, one must ensure that the number of observations exceeds the number of coefficients to be estimated. In other words, if the coefficients to be estimated are $a_0, \ldots, a_{20}$, one must have at least 21 data points $(y_1, x_{11}, \ldots, x_{20})$. Without sufficient data the coefficients cannot be calculated and their accuracy assessed.

The objective of the least-squares method is to estimate $E(y_i/x_{1i}, \ldots, x_{ki})$ by finding least-squares estimates of the coefficients a_0 through a_k. To facilitate the calculations the predictive equation is rewritten

$$y_i = b_0 + b_1(x_{1i} - \bar{x}_1) + \cdots + b_k(x_{ki} - \bar{x}_k) + \varepsilon$$

as in the simple linear model. The deviations between the actual values of y_i and the predicted values are calculated and their squares are summed to get

$$D = \sum_{i=1}^{n} (y_i - \hat{y}_i)^2$$

$$= \sum_{i=1}^{n} (y_i - \hat{b}_0 - \hat{b}_1 x_{1i} - \hat{b}_2 x_{2i} - \cdots - \hat{b}_k x_{ki})^2$$

The derivatives of this equation are taken with respect to each of the coefficients and set equal to zero. The resulting equations are then solved to obtain expressions for the coefficients in terms of the sample data. As in the simple linear model the coefficients have the form:

$$b_0 = \bar{y}$$

$$b_i = \left(\begin{array}{c} \text{a correlation} \\ \text{coefficient} \end{array} \right) \times \left(\begin{array}{c} \text{ratio of standard} \\ \text{deviations} \end{array} \right)$$

However, the correlation coefficients have a special meaning and are called partial coefficients.* *Partial correlation coefficients* measure the strength of the linear relation between the dependent variable y_i and one independent variable when the remaining variables are held constant.

The correlation coefficient between y_i and all the dependent variables together (that is, between y_i and $\hat{y}_i$ predicted on the basis of all the variables) is called the multiple correlation for the regression. The *multiple correlation coefficient* corresponds to the correlation coefficient determined in the two-variable model. Its square represents the proportion of the variation in y which is explained by the regression; *the multiple correlation coefficient squared is the coefficient of determination.* It is estimated from the data using the relation

$$S^2 = S_y^2(1 - r^2)$$

which implies that

$$r^2 = 1 - \frac{S^2}{S_y^2}\dagger$$

If the multiple correlation coefficient is zero, so must be all the partial correlation coefficients and, consequently, the coefficients b_j in the regression equation. To test the hypothesis that all the b_j are zero,

$$H_0 : r^2 = 0 \quad \text{or} \quad b_1 = b_2 = \cdots = b_k = 0$$

the F statistic is used;

$$F_{k,\,n-k+1} = \left[\frac{n-k+1}{k}\right]\frac{r^2}{1-r^2}$$

* If four dependent variables are included in the regression, the partial correlation coefficient for the x_2 term is estimated by

$$r_{y,\,2/1,\,3,\,4} = \frac{\text{cov}\,(\hat{y}_{1,\,3,\,4}\hat{x}_{2i:\,1,\,3,\,4})}{S_{1,\,3,\,4}S_{x_2:\,1,\,3,\,4}}$$

where $\hat{y}_{1,\,3,\,4}$ = predicted value of y_i with only x_{1i}, x_{3i}, and x_{4i} in the predictive equation

$$\hat{y}_{i:\,1,\,3,\,4} = \hat{C}_0 + \hat{C}_1(x_{1i} - \bar{x}_1) + \hat{C}_3(x_{3i} - \bar{x}_3) + \hat{C}_4(x_{4i} - \bar{x}_4)$$

$\hat{x}_{2i:\,1,\,3,\,4}$ = predicted value of x_2 with only x_{1i}, x_{3i}, and x_{4i} in the equation

$$\hat{x}_{2i:\,1,\,3,\,4} = \hat{A}_0 + \hat{A}_1(x_{1i} - \bar{x}_1) + \hat{A}_3(x_{3i} - \bar{x}_3) + \hat{A}_4(x_{4i} - \bar{x}_4)$$

$S_{1,\,3,\,4}$ = standard error of estimate for the regression of y on x_1, x_3, and x_4
$S_{x_2:\,1,\,3,\,4}$ = standard error of estimate for the regression of x_2 on x_1, x_3, and x_4

† When the number of data points is not much larger than the number of variables in the equation, r^2 must be adjusted. The corrected form is

$$r'^2 = r^2\left[\frac{n-1}{n-k-1}\right] - \left[\frac{k}{n-k+1}\right]$$

The ratio $(n-1)/(n-k-1)$ which appears in this relation also appears in the F statistic used to test the significance of r^2. This ratio would be infinitely large if $n = k+1$, which is impermissible; hence the assumption at the outset that the number of data points exceeds the number of coefficients to be estimated.

The decision rule is to reject H_0 if the calculated value exceeds the value of F determined from the F table with $v_1 = k$, $v_2 = n - k + 1$ and a specified error limit. That is,*

$$\text{Reject } H_0 \text{ if } \quad \left[\frac{n - k + 1}{k}\right] \left[\frac{r^2}{1 - r^2}\right] > F$$

Fortunately, multiple-regression algorithms are available on most computer systems. The SPSS and SAS packages previously mentioned are commonplace these days. Even some hand-held calculators have multiple-regression capabilities. If these were not the facts, multiple regression would hardly ever be undertaken, because the calculations are beyond reason when more than a few dependent variables are included in a model. As it stands, multiple regression is one of the most frequently used tools in data analysis.

Example 7.4: Multiple regression The anthropologists of the previous examples decided to expand their analysis of influences on the height of children; they decided to include nutrition in addition to the average height of parents. They measured nutrition in terms of the average daily protein intake per child. Their predictive equation now contained two variables, namely, x_1, the average height of the parents, and x_2, the average daily protein consumption of the child, and had the form

$$\hat{y} + \hat{a}_0 + \hat{a}_1 x_1 + \hat{a}_2 x_2$$

Another research assistant, being unfamiliar with the technique of multiple regression, approached the problem indirectly. He reasoned that when x_2 was not included in the analysis, the predictive equation had the form

$$\hat{y} = \hat{d}_0 + \hat{d}_1 x_1$$

Therefore, the difference $z = y - \hat{y}$ (called the *residual* by statisticians) should vary with x_2, and perhaps other variables, but not with x_1, since the x_1 component of y is removed by subtracting $\hat{y}$.

Furthermore, he reasoned, diet in one generation correlates to some extent with diet in the previous generation, so x_1 and x_2 should be related. Hence one might be able to predict x_2 from x_1 using a simple regression equation of the form

$$\hat{x}_2 = \hat{b}_0 + \hat{b}_1 x_1$$

The residual, $w = x_2 - \hat{x}_2$, would represent the part of x_2 which is independent of x. Consequently, one ought to be able to predict z from w by a simple linear regression equation of the form

$$z = \hat{c}_0 + \hat{c}_1 w \qquad \hat{c}_0 = 0 \qquad \text{since the expected residual or error is zero}$$

---

* Previously we used k to represent the number found in the F tables. We are using F here because k has a different meaning in this instance.

From these observations he concluded that he ought to be able to calculate the coefficients $\hat{a}_0$, $\hat{a}_1$, and $\hat{a}_2$ from the simple regression coefficients $\hat{d}_0$, $\hat{d}_1$, $\hat{b}_0$, $\hat{b}_1$, and $\hat{c}_1$, as follows:

$$z = \hat{c}_1 w = \hat{c}_1(x_2 - \hat{x}_2)$$
$$= \hat{c}_1 x_2 - \hat{c}_1 \hat{b}_0 - \hat{c}_1 \hat{b}_1 x_1$$

substituting $z = y - \hat{y}$ and solving for y,

$$y - \hat{y} = y - \hat{d}_0 - \hat{d}_1 x_1 = -\hat{c}_1 \hat{b}_0 - \hat{c}_1 \hat{b}_1 x_1 + \hat{c}_1 x_2$$
$$y = (\hat{d}_0 - \hat{c}_1 \hat{b}_0) + (\hat{d} - \hat{c}_1 \hat{b}_1)x_1 - \hat{c}_1 x_2$$

where
$$\hat{a}_0 = \hat{d}_0 - \hat{c}_1 \hat{b}_0$$
$$\hat{a}_1 = \hat{d}_1 - \hat{c}_1 \hat{b}_1$$
$$\hat{a}_3 = \hat{c}_1$$

Reflecting on his observations he noted that the coefficient $\hat{a}_1$ characterized the relation between y and x_1 when the influence of x_2 was removed, whereas $\hat{a}_2$ characterized the relation between y and x_2 when the influence of x_1 was removed (by regressing the residuals). Satisfied that his reasoning met the requirements of the multivariate analysis the anthropologists wanted, he calculated the coefficients of the simple regressions and from these found the multiple-regression coefficients:

$$\hat{d}_0 = 29.9, \hat{d}_1 = 0.57 \qquad \hat{b}_0 = 30, \hat{b}_1 = 0.43$$
$$\hat{c}_1 = 0.3, \hat{a}_0 = 20.9, \hat{a}_1 = 0.44, \hat{a}_2 = 0.3$$

Hence, he predicted the value of y from x_1 and x_2 as

$$\hat{y}_i = 20.9 + 0.44x_{1i} + 0.3x_{2i}$$

Going one step further he calculated the multiple r^2 for his analysis, using the expression

$$S^2 = S_y^2(1 - r^2) \qquad r^2 = 0.8$$

where
$$S_y^2 = 3.5 \qquad S^2 = \frac{1}{n} \sum_{i=1}^{10} (y_i - \hat{y}_i)^2 = 0.7$$

The research assistant had reasoned correctly. His two-variable predictive equation was identical to that which would have been obtained using standard multiple-regression computer programs—although the computer programs are calculationally more efficient in that they calculate $\hat{a}_0$, $\hat{a}_1$, and $\hat{a}_2$ directly. Had he carried his reasoning a bit further, he would have observed, also, that the partial correlation coefficients for y and x_1 when x_2 is held constant and for y and x_2 when x_1 is held constant must be related to $\hat{a}_1$ and $\hat{a}_2$ as follows:

$$\hat{a}_1 = \hat{d}_1 - \hat{c}_1 \hat{b}_1 = r_{y,\,1/2} \frac{S_y}{S_{x_1}} \qquad x_2 \text{ constant}$$

$$\hat{a}_2 = \hat{c}_1 = r_{y,\,2/1} \frac{S_{x_2}}{S_{x_1}} \qquad x_1 \text{ constant}$$

7.4 STEPWISE INCLUSION OF VARIABLES

Computerized algorithms have made it possible to selectively include variables in a multiple-regression equation. These algorithms proceed by calculating individually the simple correlation coefficients for the independent variable and each dependent variable. The dependent variable having the largest r^2 is included as the first variable in a predictive equation,

$$y = a_0 + a_1 x_k$$

The algorithm selects the next variable for inclusion by calculating the coefficient of determination for two-variable models including the variable already selected and each of the remaining variables being considered,

$$y_i = a_0' + a_1' x_k + a_j' x_i \qquad j \neq k$$

In each case when a' is significant, the algorithm calculates the coefficient of determination r^2 for the two-variable models. The new variable selected for inclusion in the multiple regression is the one which increases r^2 by the largest amount when it is included. This process continues until a predetermined number of variables are included in the model or until no remaining variable is found to have a significant coefficient when included in the next step.

Stepwise inclusion of variables is a very powerful procedure and should be practically mastered by those who wish to use multiple regression for data analysis. Perhaps the best sources of information are the handbooks for specific algorithms, which are published by the computer software companies (see Section 7.7 for references).

Example 7.5: Stepwise inclusion of variables The results of a stepwise multiple regression for the following data are shown on page 154 in the form of the SAS printout. The independent variable is I. Notice that particulates (PART) was the first variable entered—it offered the largest r^2 (R^2 in SAS notation) of the two individual variables. However, the r^2 increased dramatically from 0.14167975 to 0.67047426 when both variables PART and NOX were included.

S T A T I S T I C A L   A N A L Y S I S   S Y S T E M   1

STEPWISE REGRESSION PROCEDURE FOR DEPENDENT VARIABLE INDX

STEP 1   VARIABLE PART ENTERED   R SQUARE = 0.14167975

	DF	SUM OF SQUARES	MEAN SQUARE	F	PROB>F
REGRESSION	1	6.44642857	6.44642857	0.99	0.3581
ERROR	6	39.05357143	6.50892857		
TOTAL	7	45.50000000			

	B VALUE	STD ERROR	TYPE II SS	F	PROB>F
INTERCEPT	8.05357143				
PART	0.33928571	0.34092648	6.44642857	0.99	0.3581

STEP 2   VARIABLE NOX ENTERED   R SQUARE = 0.67047426

	DF	SUM OF SQUARES	MEAN SQUARE	F	PROB>F
REGRESSION	2	30.50657895	15.25328947	5.09	0.0623
ERROR	5	14.99342105	2.99868421		
TOTAL	7	45.50000000			

	B VALUE	STD ERROR	TYPE II SS	F	PROB>F
INTERCEPT	16.62500000				
PART	0.94078947	0.31407153	26.90657895	8.97	0.0303
NOX	-2.10526316	0.74322890	24.06015038	8.02	0.0366

Data

Pollution index I	12	8	9	14	6	11	10	8
NO_x	4	4	5	5	6	6	7	7
Particulates	3	0	5	7	3	8	6	8

The column labeled DF provides the degrees of freedom for the F statistic. Since $n = 8$, these are $k = 1$ for the numerator and $n - k - 1 = 6$ for the denominator of F in Step 1. In Step 2 they are $k = 2$ and $n - k - 1 = 5$, respectively. The F statistic is obtained by dividing the mean square for the "regression" (row 1) by the mean square for the "error" (row 2). At step 1, $F = 0.99$. At step 2, $F = 5.09$. The last column specifies the probability that the F variable will exceed the calculated value. At step 1, PROB $> F$ is 35.81 percent, indicating that the relation between I and PART is not statistically significant (unless one is willing to accept a 35.81 percent chance of being wrong in concluding that r is significantly different from zero).

The B values for each step are the coefficients of the regression equations.

At step 1: $\hat{y} = \hat{b}_0 + \hat{b}_1(x_1 - \bar{x}_1)$:

$$\hat{b}_0 = 8.05357143 \qquad \hat{b}_1 = 0.33928571$$

At step 2: $\hat{y} = \hat{b}_0 + \hat{b}_1(x_1 - \bar{x}_1) + \hat{b}_2(x_2 - \bar{x}_2)$:

$$\hat{b}_0 = 16.62500000 \quad \hat{b}_1 = 0.94078947$$

$$\hat{b}_2 = -2.10526316$$

(Since each coefficient is, in fact, a random variable, standard errors and F statistics are calculated for these and presented, as well. See an SAS manual or an advanced text on regression for an explanation of those items.)

7.5 REGRESSIONS ON TIME

Although a substantive discussion of time series data and their analysis is beyond the level of this book, it is worth noting that regression techniques may be applied to such data. One may treat time as the independent variable and regress on it to estimate the coefficients of the equation

$$y_t = \alpha_0 + \alpha_1 t$$

or one may regress on previous values of the variable of interest. This is called *autoregression* and the general expression for it is

$$y_t = \alpha_0 + \alpha_1 y_{t-1} + \alpha_2 y_{t-2} + \cdots + \alpha_n y_{t-n}$$

where y_t is the predicted or estimated value for time t and y_{t-k} is the value of the variable k periods, earlier.

One can calculate the degree of correlation of values of y at any time t with values at a previous time $t - 1$ by calculating the autocorrelation coefficient,

$$\text{Autocorrelation} = \frac{\text{cov } (y_t, y_{t-1})}{\sigma_{y_t} \sigma_{y_{t-1}}}$$

when sufficient data are available. If autocorrelation is weak, one can proceed with regression analyses exactly as previously discussed. If autocorrelation is significant, the assumptions of independent observations and normally distributed errors with zero mean and constant variance do not hold. This means that hypothesis testing on r^2 and the regression coefficients as discussed previously is not possible and special techniques must be employed.

7.6 PROBLEMS

7.1 From a random sample of pigs the following data were gathered:

Length, inches	35	37.5	39	33.5	35.5	35	34	38	34	34.5	35
Weight, pounds	175	198	156	180	178	182	160	204	167	169	162

(a) Construct a scatter diagram.
(b) Find the regression equation of weight on length.
(c) Is r significant? At what level?
(d) For a pig of length 74 inches, what should be its weight? Comment.

7.2 For the following data do a stepwise multiple regression and interpret the results. Present the results concisely and on a separate sheet of paper (no computer printouts).

y	13	7	9	15	7	12	9	8
x_1	5	5	6	6	7	7	8	8
x_2	4	1	6	8	4	9	7	9

7.3 Egberts Apple Orchards, Inc., is about to set apple prices for customers who pick their own. Experience has shown that the quantity picked is related to price (in constant dollars), as given below. Egberts intends to use these data as part of the input to price determination. If they regressed quantity on price, they would in effect have a demand equation. This would be a first step in the process of price determination. Calculate a demand equation for Egberts, using the data supplied.

Pounds picked	50	65	85	95	90	100	80	105	125	115	120
Price, cents	99	80	80	70	75	65	75	60	55	55	50

7.4 Over the past 10 decades the cost of property destroyed in floods along a particular stretch of the Mississippi River has increased according to the following schedule. At the same time the ratio of average river width to average river depth has been altered due to dredging and the construction of retaining walls, as indicated in the table below.

(a) Is the apparent correlation between the two variables significant?

(b) Write the linear predictive equation for annual flooding costs as a function of the ratio of river width to river depth. Interpret this equation.

(c) Write the linear predictive equation for the ratio of river width to river depth. Interpret this equation.

(d) Discuss the other factors which ought to be considered in any analysis of annual cost of flooding. Which of these might be the most important factors and how could they be defined for inclusion in a multivariate regression?

Decade	Average annual cost, $ (thousands)	Average width to depth ratio
1871–1880	210	5
1881–1890	219	5
1891–1900	225	4.5
1901–1910	Unknown	4.5
1911–1920	244	4.2
1921–1930	255	4.1
1931–1940	266	4.0
1941–1950	271	4.0
1951–1960	290	3.3
1961–1970	205	3.1

7.5 Identify and employ four methods learned in this book by which one could estimate the average annual cost for the decade 1901–1910 in Problem 7.4. Discuss the accuracy of each approach; explain how and why they differ and make a reasoned argument for adopting one over the other.

7.6 The cost of collecting trash in the city has increased steadily over the past decade. Identify at least four variables which you believe have influenced cost. Define these variables in a form which would make them usable in a multivariate regression analysis of costs.

(a) How could you use a computerized regression algorithm to determine which of the variables describe the most important factors?

(b) How could you use the results of a regression analysis to estimate the marginal reduction in cost which might result from adjusting one of the factors appropriately through managerial actions?

7.7 At Locks and Dam No. 27, the average delay time per tow (barges plus towboat) has increased dramatically over the past decade. Possible reasons are:

- More tows are sufficiently large to require two lockages to process them through.
- New barges are sufficiently wide that when secured side by side, they cannot be processed through together.
- There is increasing noncommercial traffic using the locking system.
- There are more tows on the river each year.

(a) Write a (general) linear expression for average delay time as a function of these four factors.

(b) Explain the physical meaning of the coefficients.

7.8 The Department of Conservation needs a simple technique for estimating populations of small animals without doing an elaborate census. Conservation specialists have hit upon the idea of using high fatalities for rabbits and other small animals as an indicator of population density for well-defined neighborhoods along roads. They estimate population density by a regression equation developed on the basic census data at particular periods in time for specific neighborhoods and road fatalities for these same neighborhoods during the same periods. Data for one neighborhood are given below.

(a) Calculate the coefficients for a linear predictive equation, and, of course, determine the level of significance for the equation.

(b) What action will be required to ensure that this equation remains valid over an extended period of time?

(c) What other factors might it be wise to include in the equation? What might this do to the cost of maintaining the estimation procedure?

Year	Live rabbits/square mile (census)	Fatalities/mile (summer)
1	300	28
2	290	27
3	230	28
4	220	26
5	205	23
6	150	20
7	150	22
8	250	25
9	280	24
10	195	24

7.9 For the past 5 years agricultural production and farm population have diverged as indicated.

Year	1972	1973	1974	1975	1976
Production	15	17	18	20	22
Population	6	5	3	2	1

(a) Write the regression equation for production as a function of population.

(b) Extrapolate the population trend back to 1968.

(c) For the value obtained in part b, regress production back to 1968. How does this compare to the actual value of 10.9 for 1968? Name a better method of extrapolation.

7.10 A new soil moisture measurement procedure for use in the field has been invented. Its advantage is that samples need not be sent to a laboratory for analysis. Now farmers and contractors can do the test themselves. The new procedure has been calibrated against the standard laboratory procedure, producing the summary statistics.

$$\bar{x} = 13.8 \qquad \bar{y} = 13.8 \qquad n = 67$$

$$\sum y_i^2 = 13{,}291$$

$$\sum x_i^2 = 13{,}260 \qquad \sum x_i y_i = 13{,}145$$

(a) Calculate regression coefficients from those numbers, using the expressions (equivalent to those in the text):

$$\hat{b}_0 = \bar{y} \text{ and } \hat{b}_1 = \frac{\sum x_i y_i - n\bar{x}\bar{y}}{\sum x_i^2 - n\bar{x}^2}$$

(b) Is $\hat{b}_1$ significant at the 5 percent level?

Hint:

$$nS_x^2 = \sum_{i=1}^{n} x_i^2 - n\bar{x}^2 \qquad nS_y^2 = \sum_{i=1}^{n} y_1^2 - n\bar{y}^2$$

and

$$r = \frac{\hat{b}_1 S_x^2}{S_y^2}$$

7.11 For the data below answer the following questions:

(a) Which pair is more strongly correlated: (y, x) or (y, z)?

(b) For which pair should the fraction of variance explained be largest?

(c) For which pair should the slope of the predictive equation be steepest?

(d) What is the value of b_0 for each pair?

(e) What trick might you employ to improve the correlation for the worst of the pairs?

(f) Will this trick modify the value of b_0?

(g) Data

y	1	2	3	4	5
x	6	7	8	9	10
z	1	4	9	16	25

7.12 Intelligence is presumably related to both environmental and genetic factors. Scientists have constructed indices for each and collected data to investigate the strength of these relations. Use the following data to develop a multiple-regression equation to predict IQ and discuss your results.

IQ	120	80	90	140	60	110	100	80
Environment	100	80	100	99	50	90	90	50
Genetic	90	70	75	99	70	90	80	99

7.7 SUPPLEMENTAL READING

Barr, Anthony J., et al.: *SAS User's Guide*, SAS Institute, Inc., P.O. Box 10066, Raleigh, N.C. 27605, latest edition.

Benjamin, Jack R., and C. Allin Cornell: *Probability, Statistics and Decision for Civil Engineers*, McGraw-Hill Book Company, New York, chap. 4.

Chisholm, Roger K., and Gilbert R. Whitaker, Jr.: *Forecasting Methods*, Richard D. Irwin, Inc., Homewood, Ill., 1971.

Daniel, Cuthbert, and Fred S. Wood: *Fitting Equations to Data*, Interscience Publishers, New York, 1971.

Hays, William L., and Robert L. Winkler: *Statistics, Probability, Inference, and Decision*, Holt, Rinehart and Winston, Inc., New York, 1971.

Hoel, Paul G.: *Elementary Statistics*, 3d ed., John Wiley & Sons, Inc., New York, 1971, chap. 9.

Kilpatrick, S. James, Jr.: *Statistical Principles in Health Care Information*, University Park Press, 1977, chap. 11.

Mosteller, Frederick, Robert E. K. Rourke, and George B. Thomas, Jr.: *Probability with Statistical Applications*, Addison-Wesley Publishing Company, Inc., Reading, Mass., 1961.

Nie, Norman H., et al.: *Statistical Package for the Social Sciences, SPSS*, McGraw-Hill Book Company, New York, latest edition.

Tukey, John W.: *Exploratory Data Analysis*, Addison-Wesley Publishing Company, Inc., Reading, Mass., 1977.

EIGHT

APPLICATIONS OF STATISTICAL DATA ANALYSIS AND A CASE FOR ANALYSIS

The two applications discussed in this chapter are estimates of insulin supplies, by the consumer safety statistics staff of the Food and Drug Administration (FDA), and my own critique of the process, used by the FDA for determining carcinogenicity of food additives, as exemplified by the 1977–78 saccharin decision. Each example is treated in a separate section. The decision scenarios are described at the outset. Then methods of analysis are critiqued and the role of analysis in the decision process is examined. The first example is straightforward, requiring little commentary. Not so for the saccharin case. It provides a unique opportunity to emphasize the importance of the scientific method, contrast the classical procedures of Chapter 5 with bayesian procedures of Chapter 6 in the public policy setting, and illustrate the important but limited role of purely statistical data analysis in public decision making.

The case for analysis focuses on one aspect of a major policy question that has been debated for decades, namely, should Locks and Dam No. 26 on the Mississippi River near Alton, Illinois, be replaced or repaired? The data included permit one to analyze traffic flow through it and two sister locks, one immediately upriver, the other immediately downriver, and draw policy-relevant conclusions.

8.1 ESTIMATES OF INSULIN SUPPLIES NEED IMPROVEMENT

The FDA keeps track of supply of and demand for important drugs. Isolated shortages of insulin have been reported since 1974. In 1977, the FDA decided to investigate the potential for widespread drug shortages in the future.* The investigation pointed out the possibility of widespread shortages as early as 1982 under certain (pessimistic) conditions, identified technical factors which strongly influence supply, and indicated weaknesses in the quality and availability of basic data required to more accurately estimate demand and supply. The methods of analysis employed in the study were elementary. More sophisticated methods were not necessary, nor were data sufficiently accurate to justify more than crude projections. This study had important impacts on public policy, which reemphasizes an important theme in this book, namely, that the key to good policy studies is appropriate analysis, not methodological sophistication.

Jack E. Baer of the consumer safety statistics staff did the study. To estimate demand he used published works describing the characteristics of diabetics.† United States diabetic population estimates ranged from a reported 4.8 million to about 10 million in 1975, depending on the assumed rate of underreporting. About 30 percent of reported cases were insulin users. Baer assumed that the lion's share of unreported cases were mild and could be managed with controlled diets, so he used the lower number. This biased his demand estimates downward and, although he did not test the sensitivity of his final projections on this assumption, he should have. Using these reported data on diabetics per capita for 1964 as well as 1973 and 1975 U.S. Bureau of the Census population projections, Baer projected demand to 1984 with the simple proportional relation

$$\text{Insulin users} = 0.3 \text{ users per diabetic} \times \tfrac{4.8}{20.5} \text{ diabetics}$$
$$\text{per capita} \times \text{projected population}$$

His projections are shown in Figure 8.1.

Insulin is produced from the pancreas of cattle and hogs. Using Department of Agriculture statistics on hogs and cattle slaughtered each year from 1961 through 1976 in slaughterhouses which supply pancreas for insulin production and estimates for the average weights of hog and cattle pancreas, Baer estimated gross weight of pancreatic material for each year through 1976 and projected availability through 1984.‡ Pancreas availability, shown in Figure 8.2, is related to slaughter rates by

$$\text{Availability} = \tfrac{1}{3} \text{ lb/hog} \times \text{hogs slaughtered}$$
$$+ \tfrac{1}{2} \text{ lb/cow} \times \text{cattle slaughtered}$$

* Jack E. Baer, *Estimates of Insulin Supplies*, OPE Study 39, Consumer Safety Statistics Staff, Office of Planning and Evaluation, Office of the Commissioner, Food and Drug Administration.

† His two principal sources were: National Center for Health Statistics, *Characteristics of Persons with Diabetes, United States*, July 1964–June 1965; Laurentius O. Underdahl, "Classification of Diabetes Mellitus According to Therapeutic Requirements," in G. J. Hamuri and T. S. Danoaki, (eds.), *Diabetes Mellitus: Diagnosis and Treatment* (no date given); and unpublished data from the National Center for Health Statistics for the years 1973 and 1975.

‡ The procedure used to project slaughter rates is not specified in the report. Baer plotted cyclic patterns although linear projections would have sufficed. He may have used a simple method of averaging increases between peaks and decreases between valleys in his data and modifying the trend line accordingly. He could have used either linear extrapolation, linear regression over time, or exponential smoothing.

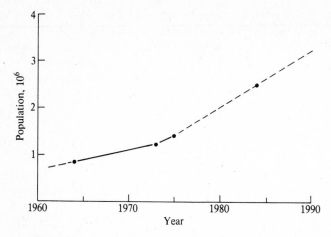

Figure 8.1 Projected insulin-using diabetic population. (*Source*: Baer.)

To estimate insulin yields, Baer sought advice from scientists at the Bureau of Drugs and at Eli Lilly and Company, the largest United States manufacturer of insulin. Eli Lilly scientists estimated yields at 700 to 3000 units per pound of pancreas, depending on conditions including age, weight, and height of the animal at slaughter; handling of the pancreas at slaughter; and processing conditions. The Bureau of Drugs expert estimated 1140 units per pound as a reasonable average based on extensive experience.* Baer did not attempt to modify his calculations to account for possible future imports, exports, or competing uses of pancreatic material, although he recognized these issues, for he had no data at all with which to work.

* There is no evidence that this number is any more than an educated guess. To compensate for uncertainty, Baer used the extreme values as well as the average estimate in his calculations. There was no need for him to apply the more sophisticated judgment quantifying procedures of Chapter 6 for the purposes of his study.

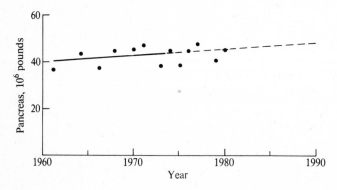

Figure 8.2 Projected pancreas availability (linear approximation on Baer's projection).

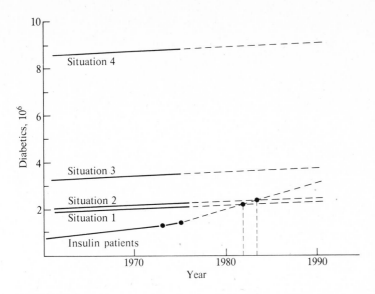

Figure 8.3 Projected demand and availability of insulin (linear approximations to Baer's curves).

Four plausible situations formed the basis for all comparisons of supply and demand in the future. These were:

1. One pound of insulin per 16,000 pounds of pancreas. Will supply 750 diabetics for 1 year (Eli Lilly).
2. Yield is 700 units per pound of pancreas. Average patient requires 40 units per day (FDA).*
3. Yield is 1140 units per pound of pancreas. Average patient requires 40 units per day.
4. Yield is 3000 units per pound of pancreas. Average patient requires 40 units per day.

Projections to 1990 are shown in Figure 8.3. Crossover points on supply and demand curves signal potential shortages.

In his report, Baer did not point out the limits of his projections, but anyone reading it could see them immediately because of his effective, concise writing style. He did point out the uncertainties in his basic data. He strongly emphasized his opinion that more accurate data for more refined projections were badly needed within a few years. He based his opinion on his finding that under particularly negative circumstances, serious shortages could occur by 1982. His recommendation for improved data collection and further analysis was simultan-

---

* Food and Drug Administration, Office of Planning and Evaluation, "Inflation Impact Assessment of Oral Hypoglycemic Drugs; Proposed Labeling Requirements and Public Hearing," January 1976.

eously a safe and reasonable position to take. It was effective, too. He would have been overstepping the limits of credibility for his work and might have been ignored had he claimed to have actually forecast future shortages.

8.2 TESTING FOR CARCINOGENICITY OF FOOD ADDITIVES: THE SACCHARIN EXAMPLE

In 1958 Congress amended the Federal Food, Drug, and Cosmetic Act in an attempt to guarantee that no known carcinogenic substances would be used as food additives. The amendment, sponsored by Congressman James Delaney of New York, has become known as the Delaney clause. In effect Congress rejected the notions of threshold levels below which carcinogenic substances are tolerated in the human body and took the position, in the absence of proof to the contrary, that substances carcinogenic for lower animals are carcinogenic for human beings as well. The amendment required the Secretary of Health, Education, and Welfare to establish appropriate testing and evaluation procedures for screening food additives suspected of being carcinogenic and to ban all additives found to be carcinogenic in human beings or animals. Operating responsibility was assigned to the FDA.

In keeping with the Congressional intent of seeking out and banning carcinogenic substances, without regard to dose rate or species of test animals, the FDA established laboratory procedures in which massive doses of suspected carcinogens are fed to rats or mice over extended periods. Experimental animals are divided into matched groups. The two groups are raised under identical conditions except that the test group consumes the suspected substance as a specified proportion of its diet.

The National Cancer Institute (NCI) has published guidelines for testing suspected carcinogens in laboratory animals. Salient points include:

Guidelines

Groups of 50 animals of one sex and one strain should be started on the experiment at 6 weeks or weaning. Control groups should also contain 50 animals. (In practice, 100 animals, 50 male, 50 female, should be used at each dose.)
The chemical should be administered by a route that mimics human exposure.
At least two doses, MTD (maximum tolerated dose) and MTD/2 or MTD/4, should be administered.
Treatment should be continued long enough (in practice, generally 24 months) to produce a maximum response.
Animals should be sacrificed (usually at 24 months) and necropsied according to detailed pathology procedures.
Tests should be conducted in two species, and the results of the more sensitive one should be given greater consideration.

Eventually all the animals are killed, or sacrificed, as the scientists like to say, and tumor rates for each group are determined pathologically. The substance is labeled carcinogenic if the tumor rate for the test group is significantly higher than for the control group in a sufficient number of independently conducted tests under equivalent conditions.

On the surface, the Delaney clause seems perfectly reasonable. In effect the Congress took the position that cancer is a dread disease and the public must be protected against the insidious threat posed by carcinogenic food additives. In fact, the amendment is both naive and presumptuous, which makes operationalization a bear of a problem. It naively suggests that one can determine when a substance is a carcinogen. The fact is, there is no way to prove carcinogenicity. The most one can do is make a reasoned judgment on the basis of available evidence. One can apply the hypothesis testing methods of Chapter 5 to data generated by scientifically designed laboratory tests to estimate the strength of that evidence. One can apply the bayesian methods of Chapter 6 to estimate directly the likelihood of inducing cancer under circumstances approximating those represented by available data. But proof is out of the question. Furthermore, the strength of carcinogenicity is a critical issue as are the relative benefits and costs of banning a substance. The Delaney clause notwithstanding, neither the public nor its elected or appointed officials are foolish enough to ban a product without exploring alternatives and weighing carefully the consequences of each before acting. It was certainly presumptuous on the part of the Congress to mandate a single course of action reflecting its own temporal views in 1958, without regard to changing attitudes or the risks and benefits specific to individual decision situations in the future.

The saccharin decision, a major public issue in 1977 to 1978, is an excellent case in point. Numerous tests for carcinogenicity had been conducted since 1948. But until 1977, evidence against saccharin was clearly not compelling. Then in 1977 a Canadian group announced results which convinced many that it should be banned. This triggered heated public debate. At one point the FDA proposed a ban but withdrew it under great political pressure until a more thorough review of the scientific evidence and evaluation of the consequences of a ban could be completed. As of January 1980 a final decision still had not been made. In fact sufficient evidence was not yet in when the National Academy of Sciences had been assigned the task of soliciting position papers from all camps and holding an open hearing in June 1979 at which the scientific evidence was critiqued and risks and benefits identified and evaluated to the extent possible. Clearly the question of carcinogenicity was still unresolved. But more importantly, the public and its officials were uncertain regarding the degree of carcinogenic risk associated with saccharin and the distribution of benefits and costs of a ban to users and producers. Until they had a clearer view of the nature and magnitude of these, they would not decide.

Consider the test data. The major saccharin studies, judged technically valid by the scientific community, are listed in Table 8.1.* The dose rates for these

---

* This judgment is based on a review study published in draft form by the Office of Technology Assessment: "Cancer Testing Technology and Saccharin (Draft)," U.S. Congress, June 1977.

Table 8.1 Selected tests in chronological order*

Year	Testing agency	Test animals
1948	FDA	Rats
1949	B. Lessel	Rats
1973	FDA	Rats
	Litton Bionetics, I	Rats
	Litton Bionetics, II	Rats
	Bio-Research Consultants	Rats
	Bio-Research Consultants	Mice
	Bio-Research Consultants	Rats
	Wisconsin Alumni Research Foundation	Rats
1974	Wisconsin Alumni Research Foundation	Rats
1975	I. C. Munro, et al.	Rats
1977	Canadian	Rats
1977	Canadian	Humans

* This list of studies was prepared by Dr. William P. Darby of Washington University. All tests listed were judged technically valid, Office of Technology Assessment, "Cancer Testing Technology and Saccharin (Draft)," U.S. Congress, June 1977.

tests were 5 percent saccharin, meaning that 5 percent of the diet for test animals consisted of saccharin. These are the basic data upon which the carcinogenicity issue turned. Scattered tests at 0.5, 1.5, and 7.5 percent dose rates shed no light on the issue, although a series of tests at 1 percent on male rats did yield indistinguishable malignancy rates among control and test populations, suggesting that carcinogenicity might not be a problem at low-dose rates. Test results are recorded in Table 8.2 for these studies.

Biostaticians favor the Fisher's exact method of (hypothesis) testing for differences in proportions, because sample sizes in carcinogenicity tests tend to be small. This method is based on the true, or exact, probability model for the test situation, thereby eliminating error introduced when tests are based on the normal approximation as they are for large samples. The hypotheses tested are of the form:

$$H_0: p_1 = p_2 \quad \text{not carcinogenic}$$

$$H_1: p_1 > p_2 \quad \text{carcinogenic}$$

The symbols p_1 and p_2 represent theoretical probabilities or propensities for tumors in test animals and control animals, respectively. Since test and control animals are selected randomly from the same population, tests for which the experimental proportion $\hat{p}_1$ is greater than $\hat{p}_2$ tend to incriminate saccharin. The strength of the evidence depends, of course, on the significance level of the test. Tests which are significant only at the 5 percent level are less convincing than tests which are significant at the 1 percent level, because greater differences in proportions are required for 1 percent significance.

Table 8.2 Test results for selected tests

5% dose rates

Year of test	Test group* Male n	Male k	Both sexes n	Both sexes k	Control group* Male m	Male l	Both sexes m	Both sexes l	Significant* at least at 5% level $\hat{p}_1 > \hat{p}_2$	$\hat{p}_1 < \hat{p}_2$
1948	—	—	17	9	—	—	20	0	B‡	
1949	—	—	20	2	—	—	20	0	NS	
1973†	21	1(1)	49	1(1)	25	1(1)	49	1(1)	NS	
1973	26	0	52	0	20	0	40	0	NS	NS
1973	26	0	52	1(0)	20	0	40	0		NS
1973	26	2(1)	—	—	30	2(2)	—	—		NS
1973	15	0	33	0	19	1(0)	36	1(0)		NS
1973	19	2(0)	37	2(0)	19	1(0)	36	1(0)	NS	
1973	—	—	20	4	—	—	17	0	NS	
1974†	15	8(7)	35	8(7)	12	0	28	0	M, B	
1975	54	0	108	0	57	1(1)	113	1(1)		B
1977†	45	12(8)	96	14(10)	42	0	89	0	M, B	

* Numbers in parentheses refer to malignant tumors, while other numbers refer to total benign and malignant tumors of the urinary bladder.

† Studies of offspring of first-generation test animals. Parents were fed saccharin from a few weeks of age. Offspring consumed it from the time of conception.

‡ B symbolizes combined male and female. M symbolizes males alone. NS means not significant.

Significance at the 5 percent level or better is recorded in Table 8.2.* The pattern of evidence is clearly inconclusive. The 1948 FDA study which started it all strongly suggested carcinogenicity, but none of the successive eight tests supported that hypothesis. In fact, two of them, taken out of context, weakly

* To carry out the test, first organize the data in tabular form such that the upper left cell represents the situation where the greatest portion of tumors is expected for an experiment. Thus,

	Tumors	No tumors	Totals	
Test group	k	$n - k$	n	$\hat{p}_1 = k/n$
Control group	l	$m - l$	m	$\hat{p}_2 = l/m$
Totals	K	$N - K$	N	

If k/n is greater than or equal to l/m, ($\hat{p}_1 > \hat{p}_2$), select a significance level and refer to tables for the hypergeometric distribution to find the largest value of l for which significance is realized. If the experimental value of l is no larger than this critical value, conclude that the test is significant at the chosen level. Otherwise the test is not significant at that level. If it should happen that l/m is greater than k/n, test the hypothesis that $\hat{p}_1 < \hat{p}_2$ by interchanging the rows and proceeding as before. Refer to Chapter 5 of this text or Chapter 8 of *Distribution-Free Statistical Tests*, by James V. Bradley, Prentice-Hall, Inc., Englewood Cliffs, N.J., 1968.

The hypergeometric tables are available in book form. See *Tables for Testing Significance in a 2 × 2 Contingency Table*, compiled by D. J. Finney, R. Latscha, B. M. Bennett, and P. Hsu, with an introduction by E. S. Pearson, Cambridge University Press, England 1963.

suggested that saccharin might be a cancer suppressant. The 1974 Wisconsin study on second-generation rats was the first corroborating evidence against saccharin and the 1977 Canadian study, again on second-generation rats, supplied additional strong evidence. However, between these two studies Munro reported results which cast serious doubt on the incriminating evidence and once again raised the question of whether saccharin might in fact be a cancer suppressant.

It was in the face of this patchwork of inconclusive and partially conflicting evidence that public officials wisely encouraged extensive public debate on the adequacy of the data and existing interpretations of it. The position argued by Dr. Umberto Saffiotti is particularly noteworthy, because it accentuates the need for careful reasoning in interpreting statistical results. Dr. Saffiotti argued that once positive results are obtained, carcinogenicity has been demonstrated and no number of negative tests before or after can negate that conclusion.* If he understood the implications of his argument, he would not likely have taken that position; for if decisions were made on that basis, practically every additive ever tested would eventually be banned. Here is why.

The phrase "positive results" refers to statistically significant results. Significance is the theoretical probability of obtaining tumor rates for test animals in excess of rates for control animals by at least the specified minimum, even though the test substance is not carcinogenic. In other words a 5 percent significance level indicates that there is a 5 percent chance of significant test results even though an additive is not carcinogenic. On this basis, 20 is the average number of tests required until the first positive test, according to the geometric model, because the expected number of trials until the first success is simply

$$E(N) = \frac{1}{p}$$

So long as the significance level selected is greater than zero, there is always a chance that positive results will be obtained early in a testing sequence, as indicated in Table 8.3. Note, too, that there is nearly a 10 percent chance of positive results in the first 10 tests, if 1 percent significance is the decision criterion, and much higher chances exist at the other significance levels. At the 5 and 10 percent levels, it is a practical certainty that within 50 tests any product, carcinogenic or not, can be banned with (apparent) scientific validity.

Scientists have long appreciated this statistical problem. Embedded in the scientific method is the principle that one significant result is insufficient evidence for a scientific conclusion. Strong corroborating evidence from independent researchers is always a condition for accepting results. In this regard, Dr. Saffiotti might counter that the 1974 test results corroborate the 1948 FDA

---

* This argument was made at a meeting entitled Saccharin: Scientific and Public Policy Issues, Society for Occupational and Environmental Health, Washington, D.C., September 1977. Related by Dr. William P. Darby.

Table 8.3 Cumulative probability of false positive tests within series of experiments

Test number	Significance level		
	0.01	0.05	0.10
1	0.010	0.050	0.100
2	0.020	0.098	0.190
3	0.030	0.143	0.271
4	0.039	0.185	0.344
5	0.049	0.226	0.410
6	0.059	0.265	0.469
7	0.068	0.302	0.522
8	0.077	0.337	0.470
9	0.086	0.370	0.613
10	0.096	0.401	0.651
15	0.140	0.537	0.794
20	0.182	0.642	0.878
30	0.260	0.785	0.958
40	0.331	0.871	0.985
50	0.395	0.923	0.995

Note: From the geometric model, with p equal to significance. Expected number of tests until the first positive result is the reciprocal of the significance level.

results; but his argument would still be weak, for Table 8.3 indicates almost a 34 percent chance of corroborating results within eight tests after the FDA study, even for noncarcinogenic substances.*

The strongest evidence in the whole series of tests is the 1977 Canadian study taken in conjunction with the 1974 Wisconsin study. These were equivalent studies on second-generation rats, and the significance levels for both were 1 percent or better. According to the binomial distribution with p taken as the significance level, the chance of two such positive results within three tests is about 3 in 10,000, which is very strong evidence that saccharin is a carcinogen in second-generation rats.†

Hypothesis testing is merely one way to judge the strength of evidence and not the most direct way. Bayesian methods have several advantages for regulatory decision making, especially regarding the balancing of benefits and costs.

The bayesian process corresponds to the inferential process employed by decision makers. The process is chronological so that historical context and

---

* The geometric model applies to the interval between positive tests as well as to the interval to the first positive test (see Chapter 4).

† The 1973 FDA study on second-generation rats is the first one in the series.

information are preserved. Preliminary evidence is systematically modified by new test results, and the total of accumulated evidence is used in the decision process. Classical hypothesis testing, on the other hand, explicitly ignores prior tests or experimental evidence except insofar as it is used to formulate a hypothesis for examination.

The artificiality of yes or no conclusions from test results, which characterizes classical methods, is avoided. So, too, is the conflict between false positive and false negative rates, which also characterizes classical methods.

The products of bayesian analysis are likelihood estimates. As test results accumulate for experimental and control groups, one sequentially estimates the likelihoods that similar populations will develop tumors (malignant or benign). The difference in likelihoods for the two groups represents the differential induced tumor rate due to the experimental agent (saccharin, in this case).

Probability estimates are precisely what decision makers need for explicit benefit-cost evaluations of decision alternatives. The expected number of additional tumors can be calculated for specific consumption rates for the population at risk. This cost may be compared directly to the expected consequences (costs) of restricting the use of the product. Table 8.4 illustrates the point symbolically.

The bayesian process is elegantly simple. One begins with parameters for the likelihood distributions which reflect previous information. For the beta distribution, which is appropriate here, these are t and r. New test results on n test animals, k of which developed tumors in the course of the study, modify these parameters as follows:

$$\text{New value of } t = \text{old value of } t + n$$

$$\text{New value of } r = \text{old value of } r + r$$

The expected value for the likelihood that similar groups will develop tumors

Table 8.4 Symbolic bayesian decision table

	Consequences		
Action	Expected number tumors avoided*	Induced medical problems†	Economic impacts
Ban saccharin	Maximum	Maximum	
Partial ban			Magnitude and Affected parties
No action	0	0	

* Expected number of tumors avoided is calculated by modifying differential propensities toward tumors for test and control groups according to human consumption rates and multiplying by the size of the at-risk population.

† Expected consequences to multiple at-risk groups, such as diabetics, are calculated individually and summed.

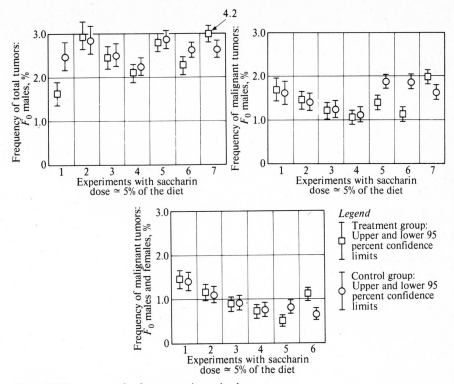

Figure 8.4 Tumor rates for first-generation animals.

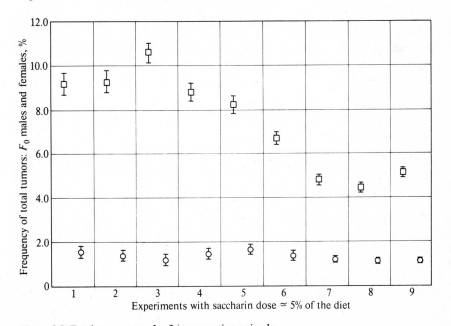

Figure 8.5 Total tumor rates for first-generation animals.

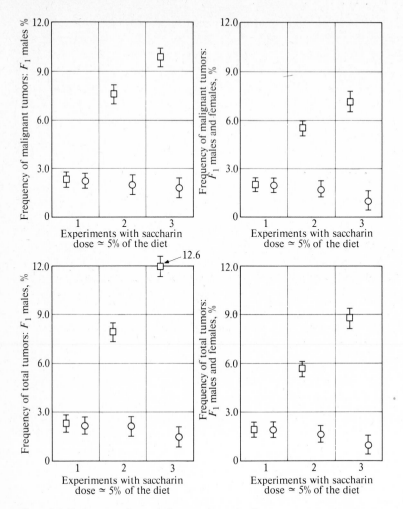

Figure 8.6 Tumor rates for second-generation animals.

under similar conditions, namely, $E(p) = t/r$, serves as one point estimate. The mode is another. The variance $V(p) = r(t - r)/t^2(t + 1)$ provides a measure of uncertainty. Confidence intervals for p are calculated directly from the beta distribution specified by the revised parametric values.

As test data accumulate, the parametric values t and r become increasingly stable and further testing has less and less impact on likelihood estimates. The usefulness of additional testing becomes obvious when $E(p)$ is plotted for sequential tests.

Figures 8.4 through 8.6 illustrate the bayesian approach to interpreting the body of data available for deciding the saccharin question. Data for both first- and second-generation studies on males only and males and females are combined. Results for total-tumor and malignant-tumor incidence rates only are presented.

Only a few of the technical issues relating to regulatory decisions regarding food additives in general and saccharin in particular were covered in this section. These are the key points from the perspective of this book. Anyone who digs into the subject will be astounded to learn how uncertain is almost every step in the regulatory process and how logically crude are some of the highly technical arguments which support decision making.

8.3 LOCKS AND DAM NO. 26 REPLACEMENT: A CASE FOR ANALYSIS

The inland waterways system of the United States is a major component of our commercial transportation system. The major artery of the waterways system is the Mississippi River, in the heartland of the nation. Commodity traffic on the upper Mississippi and a tributary, the Illinois River, alone has increased from about 11 million tons annually in 1950 to more than 54 million tons in 1970.

Since the 1930s the upper Mississippi has been altered to improve navigation and control flooding. A series of dams from St. Louis to Minneapolis operated by the U.S. Army Corps of Engineers has converted the upper 669 miles into essentially a stairway of water, as depicted in Figure 8.7. The dams

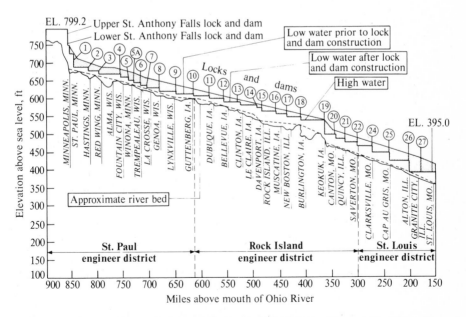

Figure 8.7 "Stairway of water" makes navigation possible between Minneapolis, Minnesota, and St. Louis, Missouri. Redrawn from U.S. Army Corps of Engineers, *The Upper Mississippi River Nine-Foot Channel*, U.S. Government Printing Office, Washington, D.C., 1974.

create slack-water pools which are maintained at about 9 to 10 feet in depth for navigation, by dredging. Locks at each dam permit movement of vessels from one level to another.

Increasing traffic on the river coupled with economic and technologic pressures for longer tows are straining the locking system at some points, causing long delays. Delays at Locks and Dam No. 26 at Alton, Illinois, are exceptionally long. It is not uncommon for tows to "backwater" in midstream for more than a day while waiting to process through. Delays are expensive and the waterways transportation industry, including the Corps of Engineers, has proposed replacing Locks and Dam No. 26 by a new facility with increased capacity to reduce congestion.

The Corps formally recommended replacing Locks and Dam No. 26 in a document entitled, "Report on Replacement, Locks and Dam No. 26, Mississippi River, Alton, Illinois," which was approved by the Secretary of the Army on July 14, 1969. As authority for its action, the Corps cited Section 6 of the Rivers and Harbors Act of 1909, which reads:

> Provided, that whenever, in the judgment of the Secretary of War, the condition of any of the afore-said works is such that its entire reconstruction is absolutely essential to its efficient and economical maintenance and operation as herein provided for, the reconstruction thereof may include such modifications in plan and location as may be necessary to make the reconstruction work conform to similar works previously authorized by Congress and forming a part of the same improvement, and that such modifications shall be considered and approved by the Board of Engineers for Rivers and Harbors and be recommended by the Chief of Engineers before the work of reconstruction is commenced.

This action followed upon a sequence of studies, technical reviews, and public hearings on the topic beginning as far back as September 1956. Numerous engineering design studies, an environmental impact statement, and a benefit-cost study have been done since.

As of January 1980 the old locks and dam has still not been replaced. Special interest groups including the rail industry, which competes with the waterways transportation industry, and environmental groups dispute the cost estimates attributed to replacement by the Corps. A major issue, and the one you are to address, is the impact on delay times expected for alternative replacement designs.

Assignment Consider yourself a policy analyst on the staff of the Secretary of the Army or a private consulting analyst. Estimate delay times at Locks and Dam No. 26 as it now operates, and estimate reductions expected for alternative replacement designs. Think in terms of Poisson arrivals and Poisson service at the locks. You will have to project traffic statistics, estimate locking capacities (service rates), and make numerous reasonable assumptions regarding locking procedures and tow configurations. The essential data are included below.

Prepare a technical report covering the details of your analysis and documenting your assumptions and conclusions. Include a carefully written execu-

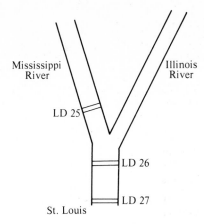

Figure 8.8 Geography.

Table 8.5 Lockages per month

| | | LD 25 | | LD 27 | | |
| | | Chamber, 600 ft | | Chamber 1, 1200 ft | Chamber 4, 600 ft | |
Year	Month	Singles	Doubles	Singles	Singles	Doubles
1975	N	110	140	425	—	—
	D	130	50	0	0	0
1976	J	40	10	250	95	105
	F	30	20	0	125	355
	M	100	150	0	290	340
	A	115	180	0	220	390
	M	100	205	130	180	400
	J	105	180	120	170	390
	J	110	160	0	250	390
	A	100	140	0	300	350
	S	85	135	310	250	140
	O	100	150	510	200	0
	N	100	145	505	200	0
	D	—	—	—	—	—
1977	J	10	10	200	50	0
	F	10	10	275	75	0
	M	70	80	600	160	0
	A	80	190	625	225	0
	M	80	205	675	225	0
	J	70	160	575	210	0

Notes: Chamber 1 at LD 27 processes no double tows. Chamber 1 was closed from February through April, and in July and August 1976.

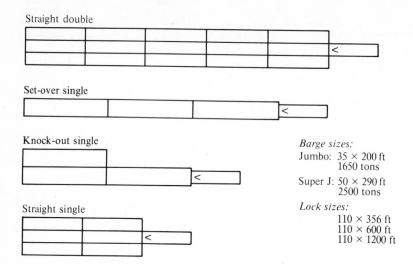

Straight double

Set-over single

Knock-out single

Barge sizes:
Jumbo: 35 × 200 ft
        1650 tons
Super J: 50 × 290 ft
        2500 tons

Straight single

Lock sizes:
110 × 356 ft
110 × 600 ft
110 × 1200 ft

Figure 8.9 Tows versus locks.

tive summary in place of a preface. Busy executives demand thorough, concise, clear, nontechnical summaries. These are their decision documents and must be of very high quality. They seldom read technical material. Put your best foot forward to maximize your influence.

System description The geography of the locks and dams system in the St. Louis region is illustrated in Figure 8.8. Ignore the rest of the system. Locks and Dam No. 27 has two locking chambers. The main chamber, No. 1, is 1200 feet long. The auxiliary chamber is 600 feet long. Locks and Dam No. 26 upriver, which processes nearly the same volume of traffic, also has two chambers; but the largest, No. 1, is only 600 feet long and the auxiliary is just 360 feet long. Locks and Dam No. 25, above the confluence of the Illinois River, has only one chamber, which is 600 feet long. All locking chambers are 110 feet wide.

Tows may consist of 1 to 15 barges. Costs incurred by towing companies are relatively independent of tow size; fuel consumption increases very little with tow size and crew complements are not variable. Consequently towing companies prefer to operate larger tows which are more profitable. Navigation is the primary constraint on tow size. The so-called straight double tow, depicted in Figure 8.9 with other common configurations, is practically the upper limit on tow size and is used primarily to move jumbo barges loaded with solid commodities. The set-over single configuration is commonly used to move loaded super jumbo barges which are designed to carry liquid commodities. The other, smaller configurations are most often used simultaneously to move empty and loaded barges. Each tow configuration represents a logistical decision reflecting technological, economic, and customer service constraints.

Data Traffic data for Locks and Dam Nos. 25, 26, and 27 for the period November 1975 through December 1977 are tabulated in Tables 8.5 through 8.7. These data are highly seasonal. They are also strongly affected by circumstances surrounding the operations of the locks, including weather and repairs, as indicated. Exercise care in drawing conclusions from them.

The data on lockage types especially require special interpretation. Any tow which locks through without reconfiguration is referred to as a "straight single" by analogy with the six-barge complement of the same name, depicted in Figure 8.9, which just fits into a 600-foot chamber. A double lockage represents a tow that is locked through in two distinct pieces. Thus, the 15-barge complement referred to as a "straight double" in Figure 8.9 is counted as a single lockage at a 1200-foot chamber, a double lockage at a 600-foot chamber, and a multiple-cut lockage at a 360-foot chamber.

The data are clearly incomplete. This is typical in policy studies. In particular, the data for Locks and Dam No. 26 are largely missing. You will have to make do with the data provided and the following information:

Locks and Dam Nos. 26 and 27 operated "normally" during June 1977. About nine straight double tows per day processed through chamber 1 at No. 27 while eight singles moved through the smaller chamber. During the same period,

Table 8.6 Millions of tons and thousands of barges processed

Year	Month	LD 25		LD 26		LD 27	
		Tons	Barges	Tons	Barges	Tons	Barges
1975	N	2.3	2.1	5.3	5.2	5.7	5.7
	D	1.2	1.2	3.6	4.2	3.9	4.5
1976	J	0.1	—	2.9	3.2	3.5	3.8
	F	0.2	0.5	4.2	5.0	5.0	5.5
	M	2.0	2.2	4.6	5.1	5.2	5.6
	A	2.2	2.5	5.0	5.5	5.9	6.0
	M	2.9	3.0	6.0	6.6	6.7	7.2
	J	2.9	3.0	5.8	6.1	6.1	6.5
	J	2.6	2.9	4.9	5.5	5.8	6.2
	A	2.2	2.3	4.5	5.0	5.2	5.8
	S	2.2	2.1	4.6	4.8	5.5	5.7
	O	2.8	2.7	5.1	5.3	5.9	5.8
	N	—	—	—	—	—	—
	D	1.1	1.1	3.9	4.1	4.2	4.7
1977	J	—	0.1	—	0.9	—	1.0
	F	0.1	0.1	0.7	2.2	0.8	6.1
	M	0.1	1.6	1.3	—	2.0	6.3
	A	2.8	2.5	6.0	6.2	6.7	7.0
	M	3.1	3.1	6.2	6.5	6.9	7.4
	J	2.8	2.4	5.7	5.8	6.3	6.2

Table 8.7 Delay and processing times

		Processing times, minutes						Average delay, hours		
		LD 25		LD 27						
		C 1		C 1		C 4				
Year	Month	S	D	S	D	S	D	LD 25	LD 26	LD 27
1975	N							1.8	11.0	12.0*
	D	30	100	46		—	—	0.5	3.4	1.7*
1976	J	34	100	50	153			0.1	13.5	6.6†
	F	40	99	—	—			0.3	13.2	20.6‡
	M	30	104	—	—			0.9	6.7	11.1‡
	A	28	100	—	—			0.9	27.0	14.0‡
	M	27	99	42	—			1.2	10.0	17.4†
	J	28	108	36	—			1.2	5.4	14.4†
	J	27	105	—	—			1.3	3.9	16.5‡
	A	26	107	—	—			1.4	3.9	6.8‡
	S	28	108	40	—			5.3	5.0	4.2†
	O	29	111	50	—			1.1	4.1	1.7
	N	30	109	40	—			0.9	3.5	0.3
	D	—	—	60	—			0.9	4.0	0.6
1977	J	50	153	41	97			1.6	1.8	0.8
	F	—	—	34	—			0.3	8.1	0.6
	M	27	104	39	—			0.7	14.8	0.5
	A	27	104	38	—			1.7	6.8	0.5
	M	29	106	42	—			2.0	10.4	0.5
	J	29	109	40	—	35	80	1.4	5.5	0.8

* LD 27, chamber 4 closed and chamber 1 constricted to 70 feet in width.
† LD 27, chamber 1 constricted to 70 feet.
‡ LD 27, chamber 1 closed.

64 percent as many double tows and 94 percent as many singles moved through No. 26; thus about 33 single-equivalent tows moved through No. 26.*

Processing times at No. 27 are 40 minutes at either chamber. At No. 26, processing times per single-equivalent tow are about 50 and 150 minutes for the 600- and 390-foot chambers, respectively.

* L. Icerman, J. K. Gohagan, and D. Culler, "Energy Impact of Modification of a River Transport System: The Upper Mississippi," *Energy*, vol. 4, June 1979, pp. 401–413.

THREE

BENEFIT-COST ANALYSIS

In this book, the term *benefit-cost analysis* refers to a class of methodologies in which the positive and negative impacts of competing public investment opportunities are systematically assessed and, to the extent possible, quantified and explicitly included in comparative evaluations of the alternatives. The two major subclasses are

- *Economic benefit-cost studies*, in which both benefits and costs are measured in monetary units, and
- *Cost-effectiveness studies*, in which benefits are measured in nonmonetary units, *or* are assumed to be comparable in magnitude and type for competing alternatives and need not be considered explicitly in the analysis.

Unsystematic and political evaluations are not included.

The purposes of benefit-cost studies are to help answer two fundamental questions:

- Does the net gain in public welfare anticipated from a new public project or program (an investment) justify the expenditures to be incurred?
- Is the anticipated distribution of benefits desirable?

Make no mistake about it, the tools of analysis are inadequate to provide definitive answers, except, perhaps, in absolutely trivial cases. However, thoroughly conducted benefit-cost studies provide essential information and insight into the relative merits of competing public investments.

When all benefits and costs can be adequately characterized in monetary units, the tools of analysis are quite effective in answering the first question. One simply evaluates the time streams of benefits and costs for competing alternatives using the methods of Chapter 9 and compares them using the decision rules and measures of economic efficiency of Chapter 10. If uncertainty is an important element, and it almost always is, it may be dealt with using the methods of Chapter 11.

The same tools of analysis apply when benefits are measured in nonmonetary terms. Unfortunately the decision rules developed for economic studies are no longer applicable and must be replaced by ad hoc rules suitable to the situation. Because of this difficulty economists have labored hard to develop techniques for translating nonmonetary benefits into roughly equivalent monetary sums in the belief that it is better to apply universal decision criteria to the latter than to apply variable decision rules to accurately portrayed nonmonetary impacts. These are the topics of Chapter 12.

The issue of equity, or the distribution of benefits and costs, is even more difficult to deal with. There are no analytical methods, no decision rules, for determining whether a distribution or redistribution of public welfare is good or bad. The tools of benefit-cost analysis can only help determine who will get what under what circumstances. However, this information itself can be enormously important in the decision-making process. How one might deal with the issue also is discussed briefly in Chapter 12.

Over the years welfare economists (and before them, philosophers, to be sure), have struggled with the question of what constitutes an acceptable public investment. In the early nineteen-hundreds, Vilfredo Pareto, an engineer turned economist, suggested that public projects were justifiable only if no one would suffer a net loss in personal welfare, as perceived by the individual, while at least someone would gain. One problem with the Pareto criterion is that if there were an individual in society who would lose only slightly while the rest of society would gain enormously from a public project, the project would not be justifiable; this seems to suggest that society at large would be the loser. Another problem is that there is no practical means of determining the views of all persons who might be impacted by a public project.

In the 1940s N. Kaldor and J. R. Hicks proposed a modified version of the Pareto criterion. They argued that if some persons could be made so much better off by a public project that they could compensate the losers and still have something left over, then the project was justifiable. They did not require that compensation actually be proffered. Basically, this is the concept employed in economic benefit-cost studies when net aggregated benefits are estimated to be positive, and on that basis a project is recommended.

Finally, in 1957 I. M. D. Little proposed that an economic (monetary) change was desirable if it caused a good redistribution of wealth (equity) and if potential losers could not profitably bribe potential gainers to oppose the change. His view was the broader one in that he included the issue of equity as one of the decision criteria.

Clearly, none of these three criteria provides sufficient guidance for public decision making. None of them deals adequately with the issues of nonmonetary impacts and equity. Although many economists would argue that these consider-ations have no place in benefit-cost studies, the fact remains that, in practice, decisions often turn on these considerations and analysts have an obligation to deal with them explicitly.

Public managers and political figures, not to mention the general public, are well aware of the limitations of economic analysis. To help compensate, the concept of benefit-cost analysis has been extended dramatically to include specifically natural and urban environmental considerations as well as technolo-gical issues. The National Environmental Policy Act of 1970 requires written statements for all major government projects. These statements tend to be volu-minous documents which qualitatively and in some cases quantitatively describe possible consequences to the natural environment. There is now talk of extend-ing this concept to require " Urban impact statements," where appropriate.

The concept of technology assessment also represents an extension of benefit-cost analysis, although followers of the idea might object to this perspec-tive. In this case the impact of new technology is the issue. The idea is to forecast the kinds of impacts one might expect from new technological developments in order to curtail or set controls for the undesirable ones. Although the idea has great appeal, it is not operational except superficially. One cannot forecast the future. That is a fact. Consequently, technology assessments have been little more than vapid guessing games. Good science fiction writers could do much better.

NINE

TIME VALUE OF MONEY

This chapter develops the fundamental technical tools for evaluating cash flows, whether they represent public projects or private activities. The most important concept to grasp is the relation between the discount rate (or interest rate) and the time value of money, especially as it relates to public projects. This is the subject of Sections 9.1 and 9.6. The remainder of the chapter develops the necessary tools of analysis and explores the impact of compounding frequency and inflation on cash flow evaluation.

The material in this chapter forms the building blocks for the economic evaluation of public projects.

Numbers in this chapter were calculated using a hand-held calculator and the interest tables supplied in Appendix 3.

9.1 DISCOUNT RATES

The value of money is not constant in time. We tend to value present consumption more highly than consumption postponed to the future. This may be due to our uncertainty about what the future holds, our expectation that prices will be higher in the future, our view that consumption foregone for a period is a cost to us in terms of pleasure, needs, or whatever, and many other factors. Opportunities lost because capital is not available at appropriate times also impose a time value on money. Higher utility for current consumption and competition for limited capital at particular points in time result in our willingness to pay a premium for money in the present. This premium is a measure of the relative worth of money over time. It is called the *discount rate* or *interest rate* and is not to be confused with inflation, which is discussed in Section 9.5.

The discount rate in the private sector varies according to the demand for and supply of money and with the quality of investment opportunities. High-quality corporate bonds may be desirable at 7 percent while riskier corporate bonds may have to offer 10 percent to attract buyers. Savings and loan associations may offer 5 percent on passbook accounts, $5\frac{3}{4}$ percent on 90-day accounts, and $6\frac{1}{2}$ percent on longer-term certificates of deposit in order to entice investors to commit their money for longer periods. Similarly, banks charge higher rates on shorter-term loans.

Understanding the terminology of interest or discount rates is essential. When an investment earns i times 100 percent per period and the earnings are immediately reinvested at the same rate, there is a compounding effect on total earnings. The interest rate in this case is referred to as the *compound interest rate*. It is the most commonly employed interest rate concept for economic studies. A more elementary concept is that of *simple interest*. An investment of 100 dollars which earns 20 dollars over some period, say 3 years, has associated with it a simple interest rate of 20 percent for 3 years. On the other hand, the same earnings will accrue from an annual compound interest rate of about 6 percent.

When the government spends money, it acquires that money by taxation or by increasing public debt (bonds, Treasury notes, etc.). In either case it is withdrawing money from the private economy. This tends to reduce immediate consumption and curtail investment for future productivity and economic expansion. Consequently, there is also a time value of money associated with public expenditures. It represents a societal cost in opportunities lost over time due to diverting private dollars into public expenditures or public consumption. This opportunity cost is reflected and, it is hoped, captured in what is called the *social discount rate*.

Although the discount rate employed by various public agencies when evaluating proposed public projects varies with the agency, in principle there ought to be a uniformly applicable social discount rate. A theoretical discussion of this issue is deferred to Section 9.6. Practically speaking, if public monies are derived 40 percent from issuing bonds (debt) and 60 percent from taxation, and the time value (opportunity cost) of the taxation portion is 8 percent and that of the borrowed portion is 6 percent, one might argue that the aggregated time value ought to be the weighted average of the two. In other words, the social discount rate or cost of capital ought to be about

$$i = 0.4 \times 0.06 + 0.6 \times 0.08 = 0.72 \; or \; 7.2\%$$

for all agencies. Basically this approach has been taken by the U.S. Bureau of the Budget (now called the Office of Management and Budget). The Director, Robert S. Mayo, issued a letter to all federal agencies on July 1, 1969, setting $4\frac{7}{8}$ percent as the minimum rate. Then in 1970 the Bureau set 10 percent as the appropriate social discount rate for all but water projects agencies, which were permitted to use lower rates. A 10 percent social discount rate is still considered generally appropriate, although some agencies use lower rates such as $6\frac{1}{2}$ percent.

9.2 PRESENT, ANNUAL, AND FUTURE WORTH FACTORS

When you deposit 100 dollars in your savings account at 5 percent per year, your action indicates that you prefer to postpone consumption of 100 dollars worth of goods. From your perspective, 100 dollars worth of consumption now is less desirable than 105 dollars worth of consumption 1 year from now. Another way of saying this is that your present 100 dollars has a future value, for this investment, of 105 dollars 1 year hence and you view the 5 dollars earned as ample compensation for deferring consumption for 1 year.

Comparisons of alternative investment schemes are facilitated by the availability of mathematical relationships between the present and future worth of money invested at specific interest rates. Suppose one wanted to know the future worth 10 years hence of 100 dollars invested now at 5 percent per year, assuming all earnings were reinvested immediately at 5 percent (compounded annually). The mathematical relationships between *present worth* $P = \$100$ and *future worth F* at the end of years 1 through 10 are shown in Table 9.1. All earnings or payments have been treated as if they occurred at the end of the year. This is a conventional practice. Abstracting from Table 9.1, the future worth F of P dollars invested now at i times 100 percent for n years, compounded annually, is simply

$$\left(\frac{f}{p}\right)_n^i P = (1 + i)^n P$$

The factor $\left(f/p\right)_n^i$ is tabulated in tables such as those reproduced in Appendix 3, for convenience.*

---

* In many books discounting factors such as $\left(f/p\right)_n^i$ are printed as shilling fractions whether they appear in text or in a displayed equation. In this book they are printed as shilling fractions in text and as built-up fractions in displayed equations. The intent is to remind the reader that discounting factors are indeed fractions, the magnitudes of which depend on the discount rate i and the project lifetime n.

Table 9.1 Future worth of 100 dollars

End of year	Relationship	Future value, $
(Now) 0	$F = P = 100$	100.00
1	$F = 100 + 0.05(100)$	105.00
2	$F = 100(1 + 0.05)^2$	110.25
3	$\vdots$	115.76
4	etc.	121.55
5	$\vdots$	127.62
6	$\vdots$	134.00
7	$\vdots$	140.71
8	$\vdots$	147.74
9	$\vdots$	155.13
10	$F = 100(1 + 0.05)^{10}$	162.89

By exactly the reverse argument, the present worth of a future sum is given by

$$P = \frac{1}{(1 + i)^n} F$$

The factor in this case is abbreviated

$$\left(\frac{p}{f}\right)_n^i = \frac{1}{(1 + i)^n}$$

Note that

$$\left(\frac{p}{f}\right)_n^i = \frac{1}{(f/p)_n^i}$$

Example 9.1: Present worth If a person promises to pay you 100 dollars in 10 years for the use of P of your dollars for the period and the interest rate available to you through alternative investments is 5 percent compounded annually, you can estimate your required investment level P to yield 100 dollars in 10 years. It is

$$P = \left(\frac{p}{f}\right)_{10}^5 100$$

$$= \$61.39$$

Clearly you should be willing to loan 61.39 dollars for the return of at least 100 dollars in 10 years. If this person wants more than 61.39 dollars for his or her future promise of 100 dollars, you would decline on a purely economic basis, although you might agree to do so for some other reasons.

Annuities, or uniform annual series, are important to economic analyses, also. It is often more convenient to compare capital expenditure projects on an annual basis than on either a present or future worth basis, particularly when project lifetimes differ for projects being compared. The basic idea of an annuity is simple. Suppose you had 100 dollars to invest now at 5 percent compounded annually and you wanted to withdraw equal amounts A at the end of each of the next 3 years, thereupon exhausting your account. What annuity A must you withdraw? The cash flow diagram for the problem is given in Figure 9.1. Proceed as follows:

1. At end of year 0 you invest 100 dollars.
2. At end of year 1 you have $100(1 + 0.05) - A$.
3. At end of year 2 you have $[100(1 + 0.05) - A](1 + 0.05) - A$.
4. At end of year 3 you have $[100(1 + 0.05)^2 - A(1 + 0.05) - A](1 + 0.05) - A$, which must equal zero.
5. Solve for A using the equation in step 4.

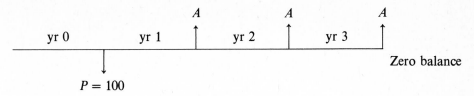

Figure 9.1 Annuity cash flow diagram.

The solution

$$A = \frac{100(1 + 0.05)^3}{(1 + 0.05)^2 + (1 + 0.05) + 1} = 36.72$$

also shows that annual receipts of 36.72 dollars for three years are equivalent to an initial lump sum receipt of 100 dollars which can be invested at 5 percent.

That is,

$$100 = 36.72\left(\frac{1}{(1 + 0.05)} + \frac{1}{(1 + 0.05)^2} + \frac{1}{(1 + 0.05)^3}\right)$$

The determination of the general relation between uniform annual series and present sums is straightforward.* They are

$$P = A\frac{(1 + i)^n - 1}{i(1 + i)^n} = A\left(\frac{p}{a}\right)_n^i$$

and

$$A = P\frac{i(1 + i)^n}{(1 + i)^n - 1} = P\left(\frac{a}{p}\right)_n^i$$

The factor $(p/a)_n^i$ is called the *present sum of a uniform series* factor, and $(a/p)_n^i$ is called the *uniform series of a present sum* factor or *capital recovery* factor. Notice that $(a/p)_n^i = 1/(p/a)_n^i$.

---

* The general equation would be

$$P(1 + i)^n - A(1 + i)^{n-1} - A(1 + i)^{n-2} - \cdots - A(1 + i) - A = 0$$

Solving for P yields

$$P = \frac{A}{1 + i}\left(1 + \frac{1}{1 + i} + \cdots + \frac{1}{(1 + i)^{n-1}}\right)$$

The bracketed term is the $(n - 1)$st partial sum for the series $1 + x + \cdots + x^{n-1} + \cdots$, where $x = 1/(1 + i)$ is less than 1. The compact form of this partial sum is

$$S_{n-1} = \frac{1 - x^n}{1 - x} = \frac{(1 + i)^n - 1}{i(1 + i)^{n-1}}$$

Thus,

$$P = A\frac{(1 + i)^n - 1}{i(1 + i)^n}$$

Table 9.2 Discount factors and forms

Ratio	Factor	Form
P/F	$(p/f)_n^i$	$1/(1 + i)^n$
F/P	$(f/p)_n^i$	$(1 + i)^n$
P/A	$(p/a)_n^i$	$[(1 + i)^n - 1]/i(1 + i)^n$
A/P	$(a/p)_n^i$	$i(1 + i)^n/[(1 + i)^n - 1]$
F/A	$(f/a)_n^i$	$[(1 + i)^n - 1]/i$
A/F	$(a/f)_n^i$	$i/[(1 + i)^n - 1]$

It is a simple matter to relate annuities and future worth by multiplying appropriate factors. Since

$$F = P(1 + i)^n \quad \text{and} \quad P = A\frac{(1 + i)^n - 1}{i(1 + i)^n}$$

clearly,

$$F = A\frac{(1 + i)^n - 1}{i} \quad \text{and} \quad A = F\frac{i}{(1 + i)^n - 1}$$

$$= A\left(\frac{f}{a}\right)_n^i \qquad\qquad = F\left(\frac{a}{f}\right)_n^i$$

Finally, notice for future reference that

$$\left(\frac{a}{f}\right)_n^i + i = \frac{i + i[(1 + i)^n - 1]}{(1 + i)^n - 1} = \frac{i(1 + i)^n}{(1 + i)^n - 1} = \left(\frac{a}{p}\right)_n^i$$

These discount factors and their mathematical forms are summarized in Table 9.2. The value of any factor may be found either in tables of the type reproduced in Appendix 3 or computed using a calculator or electronic computer.

Example 9.2: Evaluating a bond You own a 1000 dollar, 4 percent bond which will mature in 10 years. You have been offered 950 dollars for your bond. You must decide whether to accept the offer and invest the 950 dollars at an annual rate of 5 percent.

Assuming all bond dividends are paid at the end of the year, the cash flow diagrams for your two options are:

1. Keep the bond until maturity

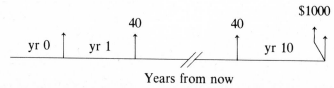

Years from now

2. Sell the bond

The future worth of keeping the bond, assuming you invest your dividends at 5 percent, is

$$F = 1000 + \left(\frac{f}{a}\right)^5_{10} 40 = \$1503.12$$

Selling the bond and reinvesting the 950 dollars at 5 percent yields

$$F = 950 \left(\frac{f}{p}\right)^5_{10} = \$1547.45$$

Clearly, unless the 5 percent investment is too risky, it would pay to sell the bond and reinvest your money.

Example 9.3: Evaluating cash flows The city is considering repaving a section of highway. The money for the job will come 40 percent from bonds issued at 5 percent and 60 percent from tax revenue. The opportunity cost of capital for the two sources is assumed to be 5 percent and 10 percent, respectively. Two surface materials are under consideration. Asphalt type 1 costs twice as much to lay, 10,000 dollars per mile, but lasts 10 years, or twice as long as type 2, with less maintenance. Yearly maintenance costs are 100 dollars for type 1 and 150 dollars for type 2. Which surface would be the most economical over the 10-year period? The appropriate discount rate is

$$i = 0.4 \times 0.05 + 0.6 \times 0.1 = 0.02 + 0.06 = 0.08$$

or 8 percent.

Cash flow diagrams for the options are

Type 1

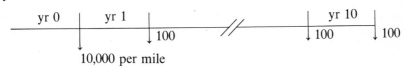

Type 2

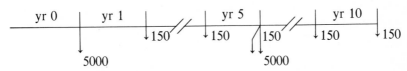

Comparison in present equivalent terms requires equalizing the periods, as in the diagrams. The present equivalent costs are:

Type 1
$$P = 10{,}000 + \left(\frac{p}{a}\right)_{10}^{8} 100$$

$$= \$10{,}671.00$$

Type 2
$$P = 5000 + \left(\frac{p}{a}\right)_{10}^{8} 150 + \left(\frac{p}{f}\right)_{5}^{8} 5000$$

$$= \$9409.43$$

Comparison on an annual basis is more direct. The annual equivalents can be calculated without equalizing project periods.

Type 1
$$A = 100 + \left(\frac{a}{p}\right)_{10}^{8} 10{,}000$$

$$= \$1590.29$$

Type 2
$$A = 150 + \left(\frac{a}{p}\right)_{5}^{8} 5000$$

$$= \$1402.28$$

The same numbers result from taking the products:

Type 1
$$\left(\frac{a}{p}\right)_{10}^{8} 10{,}671.00 = 1590.29$$

Type 2
$$\left(\frac{a}{p}\right)_{10}^{8} 9409.43 = 1402.28$$

Clearly type 2 is preferable, assuming other considerations are equivalent. Either annual or present accounting methods may be used, depending on which is most convenient for the problem at hand.

Example 9.4: Another cash flow Slightly altering the cash flows of Example 9.3 makes calculations more involved. Assume no maintenance is performed in the year in which resurfacing is to be done. The cash flows are

Type 1

Type 2

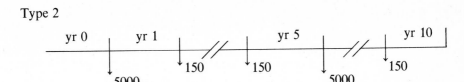

In this case it is best to concentrate on present equivalent costs.

Type 1
$$P = 10,000 + \left(\frac{p}{a}\right)_9^8 100$$

$$= \$10,624.69$$

Type 2
$$P = 5000 + \left(\frac{p}{a}\right)_4^8 150 + \left(\frac{p}{f}\right)_5^8 \left[5000 + \left(\frac{p}{a}\right)_4^8 150\right]$$

$$= \$9237.86$$

9.3 SPECIAL DISCOUNT FACTORS

In some cases cash flows are likely to increase regularly over time. Then special discount factors are required.

Arithmetic increases It sometimes happens that reasonable estimates of future operating costs (e.g., labor, materials, etc.) for a project are best estimated by adding on an increment k in succeeding years, as in Figure 9.2.

The value of the incremental growth alone is calculated directly as a future value most easily. Each increment k is an annuity running for a different number of years. Therefore,

$$F = k\left[\left(\frac{f}{a}\right)_{n-1}^i + \left(\frac{f}{a}\right)_{n-2}^i + \cdots + \left(\frac{f}{a}\right)_1^i\right]$$

$$= \left(\frac{k}{i}\right)[(1+i)^{n-1} + (1+i)^{n-2} + \cdots + (1+i) - n + 1]$$

$$= \left(\frac{k}{i}\right)\left[\left(\frac{f}{a}\right)_n^i - n\right]$$

Figure 9.2 Cash flow diagram for arithmetic increases.

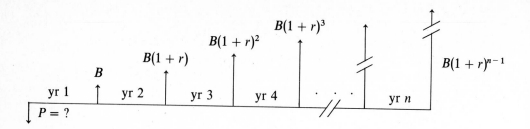

Figure 9.3 Cash flow diagram for geometric growth.

The complete value of the series is, of course,

$$\left(\frac{k}{i}\right)\left[\left(\frac{f}{a}\right)_n^i - n\right] + \left(\frac{f}{a}\right)_n^i A$$

Geometric increases When future flows are growing (or decreasing) at a constant rate r per year, the cash flow diagram is as shown in Figure 9.3.
The present worth of this series is given by

$$P = B\left[\frac{1}{1 + i} + \frac{1 + r}{(1 + i)^2} + \cdots + \frac{(1 + r)^{n-1}}{(1 + i)^n}\right]$$

The solution depends on the relative values of the growth rate r and the discount rate i. There are three cases:

Case 1. If $r > i$, then $(1 + r)/(1 + i) = 1 + \varepsilon$, $\varepsilon > 0$, so

$$P = \frac{B(1 + \varepsilon)}{1 + r}[1 + (1 + \varepsilon) + (1 + \varepsilon)^2 + \cdots + (1 + \varepsilon)^{n-1}]$$

or $\qquad P = \dfrac{B}{1 + i}\left(\dfrac{f}{a}\right)_n^\varepsilon$

Case 2. If $r < i$, then $(1 + r)/(1 + i) = 1/(1 + \varepsilon)$, $\varepsilon > 0$, so

$$P = \frac{B}{1 + r}\left[\frac{1}{1 + \varepsilon} + \frac{1}{(1 + \varepsilon)^2} + \cdots + \frac{1}{(1 + \varepsilon)^n}\right]$$

or $\qquad = \dfrac{B}{1 + r}\left(\dfrac{p}{a}\right)_n^\varepsilon$

Case 3. If $r = i$, then $(1 + r)/(1 + i) = 1$, so

$$P = \frac{nB}{1 + r} = \frac{nB}{1 + i}$$

9.4 COMPOUNDING PERIOD AND DISCOUNT RATES

Although compounding periods of 1 year are most common, monthly, daily, and even continuous compounding of earnings are not uncommon, and one must be careful to distinguish among them. Fortunately, one compounding frequency is as easy to employ as another and if one prefers to work with *equivalent annual interest rates*, the conversion is quite simple.

When compounding is more frequent than annual, a special name is given to the product of the interest rate per period with the number of periods in a year. It is called the *nominal annual interest rate* and is to be distinguished from the equivalent annual interest rate. For example, if the compound monthly interest rate is 1 percent, the nominal interest rate is 12 percent; the equivalent annual rate is 12.68 percent.

An investment of 100 dollars earning i times 100 percent per month has a year-end worth of

$$F = (1 + i)^{12}100$$

For some annual equivalent rate i_a the same value would be calculated from

$$F = (1 + i_a)100$$

Equating these two expressions and solving for i_a, one finds

$$i_a = (1 + i)^{12} - 1$$

The general result for m compounding periods per year is

$$i_a = (1 + i)^m - 1$$

Clearly, the period rate can be calculated from the annual rate by

$$i = (i_a + 1)^{1/m} - 1$$

If the compounding period is daily or less, the annual equivalent rate is approximately the same as for continuous compounding. For continuous compounding, i_a is found by observing that

$$i_a = (1 + i)^{im/i} - 1 \qquad \text{compounding frequency } m$$

has a limiting value of

$$i_a = e^{im} - 1 \qquad \text{continuous compounding}$$

as m approaches infinity while i approaches zero (such that im, the nominal rate, remains constant, which it must be for a specified annual equivalent rate).* The nominal interest rate may be calculated by taking the natural logarithm of this relation

$$im = ln e^{im} = ln(i_a + 1)$$

* The number e has an approximate value of 2.71828. Its actual value cannot be calculated; it is an indefinite number. Since the limiting value of $(1 + i)^{1/i}$ is e, as i approaches zero the limiting value for $(1 + i)^{im/i}$ is e^{im}.

Table 9.3 Nominal and period rates for $i_a = 10\%$

Periods per year, m	Period rate, i	Nominal rate, im
2	$(i_a + 1)^{1/2} - 1 \quad = 0.049$	0.098
4	$(i_a + 1)^{1/4} - 1 \quad = 0.024$	0.096
6	$(i_a + 1)^{1/6} - 1 \quad = 0.016$	0.096
12	$(i_a + 1)^{1/12} - 1 \quad = 0.008$	0.096
365	$\begin{cases} (i_a + 1)^{1/365} - 1 = 0.0002612 \\ \ln(i+1)/m \qquad = 0.0003 \end{cases}$	0.0953 0.0953

Table 9.4 Period and annual equivalent rates for $im = 12\%$ (0.12)

Periods per year, m	Period rate, i	Equivalent annual rate, i_a
2	0.06	0.1236
4	0.03	0.1255
6	0.02	0.1261
12	0.01	0.1268
365	$\begin{cases} 0.00328 \\ \text{Continuous approximately } e^{0.12} - 1 \end{cases}$	0.1274 0.1275

Discounting cash flows where the compounding period is more frequent than annual may be done directly, using the period interest rate i and the total number of periods m with the standard tables of discount factors, provided the number of periods does not exceed the extent of the tables. Otherwise, one simply converts period rates to annual equivalent rates and then refers to the tables. Hand-held calculators with discount factors built in and computerized discounting programs eliminate this problem entirely.

The nominal and period interest rates for different compounding periods are tabulated in Table 9.3 for $i_a = 10\%$.

For comparison, Table 9.4 shows how i and i_a vary with compounding period when im is held fixed.

Example 9.5: Sell or keep a bond A 1000 dollar bond pays 4 percent annual interest (nominal) and dividends are paid semiannually. You could sell the bond for 950 dollars today and reinvest your money at 5 percent compounded annually. Should you sell or keep the bond, which comes due in 10 years?

1. Keep the bond until maturity

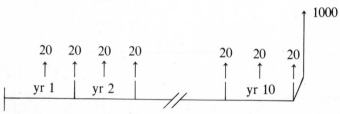

2. Sell the bond

The equivalent annual annuity for option 1 is calculated as $i_a \times 1000$. Since

$$i = \frac{0.04}{2} = 0.02$$

then
$$i_a = (1 + i)^2 - 1 = 0.0404$$

The annual annuity is therefore

$$A = \$40.40$$

The future worths of options 1 and 2 are

1. $F = 1000 + \left(\frac{f}{a}\right)^5_{10} 40.40 = \$1508.16$

2. $F = \$1547.45$

Selling appears to be the best option.

9.5 INFLATION

Inflation requires special attention. An inflation rate of 4 percent per year means that 1 dollar today will be worth 96 cents a year from now, and so forth. Higher inflation rates discourage investment at fixed rates of return. That is, a 5 percent savings account might look good when the inflation rate is 1 percent but not so good when the inflation rate is 4 percent. Clearly, inflation alters the time value of money and must, therefore, be accounted for in economic evaluations of alternative capital investments.

One dollar invested at 5 percent yields 1.05 dollars in one year. If inflation is 4 percent, that 1.05 dollars will have a real value less than 1.01 dollars at that time. That is, the effective interest rate is less than 1 percent (actually 0.8 percent):

$$F = (1 + 0.05) \times 1 \qquad \text{ignoring inflation}$$

$$F = (1 + 0.05)(1 - 0.04) \times 1 \qquad \text{including inflation}$$

Should one insist on a "real" gain of i_r per year when the inflation rate is r, one would have to obtain an investment rate i large enough to ensure that

$$1 + i_r = (1 + i)(1 - r)$$

In other words, the investment interest rate required would be

$$i = \frac{i_r + r}{1 - r}$$

When r is small,

$$i \simeq i_r + r$$

is a commonly quoted estimate.

Example 9.6: Cost escalation The city accepted a bid of 100,000 dollars for a job to begin in 1 year. After working on the job for 1 year, the contractor claims the total cost will be about 150,000 dollars by the time it is completed at the end of the next year. Assuming an inflation rate of 12 percent annually, what is the real annual rate of increase in the cost of the job?

Refer to the cash flow diagram. The annual

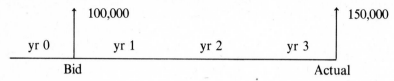

rate of cost escalation is found by interpolation using interest tables, as follows.

$$\left(\frac{f}{p}\right)_3^i = 1.5 \quad \text{actual}$$

$$\left(\frac{f}{p}\right)_3^{14} = 1.405 \qquad \left(\frac{f}{p}\right)_3^{15} = 1.521$$

So, $i = 14.896\%$

$$i_r = i - r - ir = 0.14896 - 0.12000 - 0.01787$$

$$= 0.01108 \ or \ 1.108\%$$

9.6 THEORETICAL GENERALIZATIONS

Two topics are treated in this section. One is a generalized view of the time value of money. The second is a theoretical view of when it makes economic sense to withdraw money from the private sector for investment in the public sector. The discussion is based on the work of Stephen A. Marglin (see Section 9.7).

Decision makers with mystic powers would have a clear view of the future. If they knew that 10 years hence would be a bumper year in terms of economic returns on existing investments yet 3 years hence would be relatively lean, they might search for investment opportunities paying more in year 3 than in year 10. That is, they might value an additional unit of return in year 3 more than an additional unit in year 10. Being mystics, they would be capable of placing relative weights on unit changes in net cash (or benefit) flows for all years into the future. Their weighting schemes, of course, would reflect their views of the time value of money year by year as compared to the present. Consequently, they could calculate the net present equivalent (NPE) of any new investment opportunity with costs C_j and returns R_j in year j ($j = 1, \ldots, n$) as

$$\text{NPE} = (R_0 - C_0) + w_1(R_1 - C_1) + \cdots + w_n(R_n - C_n)$$

Assuming that the weights w_j decrease as j increases, meaning that present consumption is valued more highly than future consumption, there are annual rates of change r_j in the weights such that

$$w_j = (1 + r_j)w_{j+1}$$

Because the mystics could foresee the future, the r_j values would differ from year to year. The weights, therefore, may be written in terms of the r_j values as follows:

$$w_1 = \frac{w_0}{1 + r_0}$$

$$\vdots$$

$$w_2 = \frac{w_1}{1 + r_1} = \frac{w_0}{(1 + r_0)(1 + r_1)}$$

$$w_n = \frac{w_{n-1}}{1 + r_{n-1}} = \frac{w_0}{(1 + r_0)(1 + r_1) \cdots (1 + r_{n-1})}$$

The NPE for the project now has the form

$$\text{NPE} = (R_0 - C_0) + \frac{R_1 - C_1}{1 + r_0} + \cdots + \frac{R_n - C_n}{(1 + r_0)(1 + r_1) \cdots (1 + r_{n-1})}$$

Notice that when one assumes the rates to be the same from year to year, this relation reduces to the form discussed in previous sections, namely,

$$\text{NPE} = (R_0 - C_0) + \frac{R_1 - C_1}{1 + r} + \cdots + \frac{R_n - C_n}{(1 + r)^n}$$

$$= (R_0 - C_0) + \left(\frac{p}{f}\right)^r_1 (R_1 - C_1) + \cdots + \left(\frac{p}{f}\right)^r_n (R_n - C_n)$$

when $r_j = r$ for all j. Usually one makes this assumption for practical reasons; no one is a mystic.

Now consider the problem of deciding whether to invest a dollar in the private versus the public sector. Suppose the social discount rate were known to be i. Suppose further that a dollar invested would earn a dollars per year forever in the private sector because of perpetual reinvestment, and b dollars per year for the project lifetime of n years in the public sector (see Figure 9.4).

1. Private

2. Public

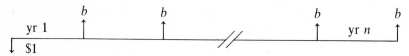

Figure 9.4

The present equivalents for each would be

1. Private

$$P_a = \left(\frac{p}{a}\right)^i_\infty a = \frac{a}{i}$$

2. Public

$$P_b = \left(\frac{p}{a}\right)^i_n b = \left(\frac{(1 + i)^n - 1}{i(1 + i)^n}\right) b$$

Since both calculations reflect the social value of the investment (i = social discount rate), the only reason one would divert a dollar from private to public investment would be if P_b were greater than P_a. That is, public investment would be economically rational if

$$b > \frac{a(1 + i)^n}{(1 + i)^n - 1}$$

In principle, b represents an annuity factor for a dollar invested in the public sector. Note, too, that there exists a synthetic discount rate s such that

$$\frac{a(1 + i)^n}{(1 + i)^n - 1} = \frac{s(1 + s)^n}{(1 + s)^n - 1} = \left(\frac{a}{p}\right)^s_n$$

namely,

$$s = \frac{(1 + i)a}{(1 + i) - 1}$$

Consequently, the factor $(a/p)^s_n$ may be compared directly with b to determine whether public investment is desirable.

The catch in all of this is that the appropriate rates a and b, and therefore s, can only be roughly approximated. Furthermore, this analysis ignores the political facts of life. Therefore, for practical purposes one must estimate social discount rates as discussed in earlier sections and test the sensitivity of analyses to the estimates. Public decision makers must still exercise their political judgment regarding the relative worths of public versus private investment on a project-by-project basis, weighing reasonable estimates of the economic value of the projects against other important factors.

9.7 PROBLEMS

9.1 Calculate present, future, and annual equivalents for the cash flow diagramed. Use $i = 10\%$.

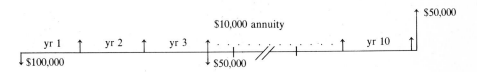

9.2 Labor costs are increasing at a rate of 3 percent per year in a certain industry. A 10-year project is being "costed out" for the purpose of bidding on the job. First-year labor costs are estimated at 50,000 dollars. What is the present equivalent cost of labor for the project if the discount rate is 15 percent?

9.3 The U.S. Army Corps of Engineers analysts are estimating the recreational benefits which might accrue if a proposed reservoir is built in a rural area of a Midwestern state. By comparing the proposed project with fairly similar projects in nearby states they have estimated that the average number of visitors per summer is a function of many variables including the recreational site, the ratio of motel and lodge-type accommodations to campsites, and the number of boat docks per acre of water. Data for the other projects are available:

    (*a*) What method(s) of analysis might the analysts use to convert the available data into a projection of demand or visitor rates for the proposed project?

    (*b*) If the analysts had found that the only really important variable in estimating demand was the average distance to the three largest nearby centers of population, how could they estimate a demand curve and calculate increased public welfare from the new project? Sketch such a demand curve with the approximate shape you would expect it to have for this situation.

9.4 The regional air pollution control board has hired a consulting company to estimate the economic merits of forcing a local smelting plant to install expensive pollution control equipment to reduce emissions. The company estimated the health benefits by applying public health data gathered in a nationwide study on the costs (income lost owing to morbidity and early death plus health care costs) from respiratory diseases as a function of air pollution concentrations. Criticize this estimation procedure and propose a better approach.

9.5 Operating costs for a public facility are expected to grow (arithmetically) at a rate of 5000 dollars per year. Base-year costs (year 1) are expected to be 100,000 dollars. User revenues, on the other hand, are expected to grow (geometrically) at a rate of 133 percent per year. Base-year revenues are 500 dollars. In what year would this public facility break even if these increases persist? Assume the discount rate is 10 percent.

9.6 Two investment opportunities are being considered. Opportunity A offers 6 percent compounded annually. Opportunity B offers $\frac{1}{2}$ percent compounded monthly. How much will 1000 dollars be worth in 15 months if invested in A, in B?

9.7 Calculate the effective discount rate i_r if the rate of inflation is 12 percent and the rate of return on an investment is 13 percent. What rate of return would be required to guarantee a "real" rate of return of 13 percent? Evaluate the real present worth of the cash flow diagramed, assuming these rates.

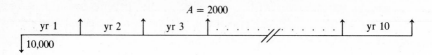

9.8 Develop a reasoned argument for allowing one agency, specifically a water resource agency, to use a social discount rate of, say, 4 percent while requiring other agencies to use, say, 10 percent.

9.9 You own a $7\frac{1}{2}$ percent, 10-year corporate bond with 1000 dollar maturity value. You paid 995 dollars for the bond just over one year ago. Assume your dividends are paid annually and you always invest them at your best alternative rate of return for similar quality investments. The current market value of your bond is 1050 dollars.

(*a*) If your best alternative investment rate of return for similar quality investments is 8 percent, should you sell your bond now and reinvest your money?

(*b*) At what alternative rate of return would the two options become equivalent?

9.10 Calculate the equivalent annual rate for a 1000 dollar certificate of deposit (CD) offered by a bank. A nominal rate of 4 percent is paid quarterly.

9.8 SUPPLEMENTAL READING

Haveman, Robert H., and Julius Margolis (eds.): *Public Expenditures and Policy Analysis*, Markham, Rand McNally, Chicago, 1970, chap. 10.

Marglin, Stephen A.: *Public Investment Criteria: Benefit Cost Analysis for Planned Growth*, MIT Press, Cambridge, Mass., chap. 2.

Merewitz, Leonard, and Stephen H. Sosnick: *The Budget's New Clothes*, Markham, Rand McNally, Chicago, 1971, chap. 7.

Riggs, James L.: *Economic Decision Models*, McGraw-Hill Book Company, New York, 1968, chap. 5.

Smith, Gerald W.: *Engineering Economy: Analysis of Capital Expenditures*, 2d ed., Iowa State University Press, Ames, 1973, chap. 4.

"Through the Looking Glass," *Wall Street Journal*, Apr. 13, 1977.

U.S. Bureau of the Budget, *A Conceptual Basis and Measurement Technique for Computation of a Social Discount Rate*, U.S. Government Printing Office, Washington, D.C., 1970.

TEN

ECONOMIC EVALUATION METHODS

Economic efficiency is one important consideration in public investments. Welfare economists define an efficient investment as one which increases the net economic welfare of society; that is, economic benefits outweigh economic costs. In this chapter measures of economic efficiency are defined and compared. Then the critical welfare economic concepts of efficiency benefits and costs are defined. Finally, proper methods of evaluating competing public investments are demonstrated for each measure of economic efficiency.

This chapter is quite technical and narrowly focused. Its central purpose is to thoroughly familiarize the reader with the methods of analysis employed in economic studies and to demonstrate their comparability and some pitfalls.

10.1 MEASURES OF ECONOMIC EFFICIENCY

Three basic measures of economic efficiency and two variations are discussed in this chapter. These are the *net present equivalent*, the standard and marginal *benefit-cost ratios*, and the comprehensive and marginal *rates of return* (also called *return on investment* and *internal rate of return*). The net present equivalent and rate of return concepts are applicable when both benefits and costs are expressed in dollar equivalents. The benefit-cost ratio can be quite useful even when benefits are not expressed in dollar equivalents (as demonstrated in Chapter 17), so long as they are all expressed in comparable units, such as lives saved, hospital days avoided, or other such measures. The major difficulty in the latter case is in determining at what rate to discount time streams of nonmonetary benefits. Table 10.1 summarizes these concepts in analytical form.

Table 10.1

---

1. Net present equivalent

   NPE = discounted benefits − discounted costs

2. Benefit-cost ratio
   (*a*) Standard

   $$B/C = \frac{\text{discounted benefits}}{\text{discounted costs}}$$

   (*b*) Marginal

   $$\Delta B/\Delta C = \frac{\text{increment in discounted benefits}}{\text{increment in discounted costs}}$$

3. Rate of return
   (*a*) Comprehensive

   ROR − discount rate for which NPE = 0

   (*b*) Marginal

   MROR − discount rate for which MNPE = 0

   where MNPE = increment in discounted benefits − increment in discounted costs

---

10.2 EFFICIENCY BENEFITS AND COSTS

Benefits expressed in monetary terms are usually referred to as *efficiency benefits*. The word "efficiency" refers to a state of the economy. An efficient economy is one in which it is impossible to reallocate resources among producers or goods and services among consumers so as to improve the welfare (or utility) of some producer or some consumer without reducing the welfare of another. If resources invested in a new public project produce an improvement in the aggregated level of public welfare, the economy is assumed to be more efficient and the benefits of the new project are defined to be the net increase in public welfare.*

Efficiency benefits should be measured from demand curves.† The procedure for calculating benefits from estimated demand curves is demonstrated in Figure 10.1. The demand curve shown is of a particularly simple form, namely, linear. For this discussion assume that the demand curve applies to the whole population to which the benefits might accrue; it is an aggregated demand curve. Assume also that there is no loss of public welfare elsewhere in the economy due to resource allocation. Then, for the limited number of individuals in the population who would consume Q_1 units at a price per unit to them of P_1 dollars, the

---

* If all losers were compensated, the economy would, in fact, be more efficient. If they were not compensated, one might argue that it would be less efficient than before.

† Unfortunately, demand curves of credible accuracy often cannot be constructed. In principle, they can be estimated using regression methods, but often the necessary data are not available.

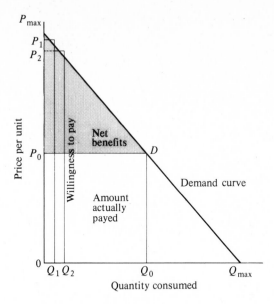

Figure labels within the image:
- P_{max}
- P_1
- P_2
- P_0
- Price per unit
- Willingness to pay
- Net benefits
- D
- Demand curve
- Amount actually payed
- 0
- $Q_1 Q_2$
- Q_0
- Q_{max}
- Quantity consumed

Figure 10.1 Demand curve and efficiency benefit estimation.

willingness to pay is approximately the rectangular area P_1 times Q_1. If the price were reduced to P_2, additional consumers would purchase the good and the net increase in willingness to pay would be the rectangular area P_2 times $(Q_2 - Q_1)$. The total willingness to pay for the consuming population would be approximately the sum of the two shaded rectangles. Proceeding in this manner, the aggregated willingness to pay for the quantity Q_0 at a price of P_0 would be the area with corners 0, P_{max}, D, and Q_0. On the other hand, the total cost to consumers of Q_0 units at price P_0 per unit would be the rectangle with corners 0, P_0, D, and Q_0. The net benefits then would be the triangular area with corners P_0, P_{max}, and D; economists call this *consumer surplus*.

Although analysts frequently assume that resources invested in new public projects or activities have no impact on public welfare derived from other public activities, there may be cases where this assumption is inappropriate. After all, the economy is a very complex system and adjustments (resource allocation) in one component of a system lead to reverberations in other components. For example, if resources were directed toward improvements in the Mississippi River locks and dam system, some of the water transportation benefits realized would be at the expense of the competing rail transportation system. In this case the change in public welfare would have to be estimated from demand curves for the river and rail systems separately and from these the net increase (or decrease) calculated. This approach is demonstrated in Figure 10.2. For the sake of exposition an unrealistic assumption is incorporated, namely, that although the locks and dam system is made more efficient, the demand curves are not affected.

Prior to the improvements, users of the river system paid an amount in dollars equal to the area $A_2 + A_4$ for a service they valued at the sum

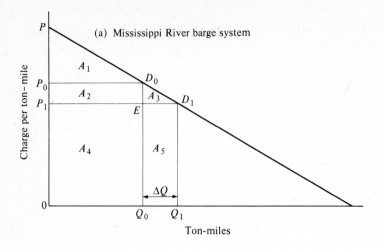

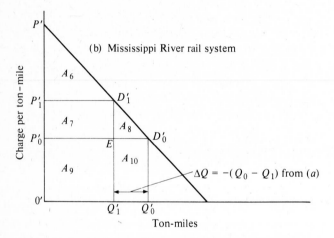

Figure 10.2 Demand curves for competing transportation systems.

$A_1 + A_2 + A_4$. Thus, their consumer surplus was A_1. In the equilibrium of competition, users of the rail system paid $A_9 + A_{10}$ for a service they valued at $A_6 + A_7 + A_8 + A_9 + A_{10}$. Their surplus was $A_6 + A_7 + A_8$. If the improvements in the river system reduce the price to users from P_0 to P_1, demand may be expected to increase by ΔQ to Q_1. In this case river users will pay $A_4 + A_5$ dollars for the service, and the consumer surplus will increase by $A_2 + A_3$, which is the net benefit gain expected. On the other hand, demand on the rail system should be reduced by ΔQ as a result of the waterway improvements, resulting in a net welfare (benefit) loss of $A_7 + A_8$. The net gain in public welfare expected to accrue only to users of the two systems, ignoring other impacted groups, would be the difference,

$$\text{Net change in benefits} = (A_2 + A_3) - (A_7 + A_8)$$

Table 10.2 Types of costs

Bid evaluation and contractual expenses
Capital invested
Contract management
Price incentives to contractor
Materials and labor furnished by the government
Operating costs including insurance, labor, maintenance
Depreciation

Table 10.3 Benefit-cost definitions

Efficiency (economic) benefits = willingness to pay − actual consumer fees
Costs = capital + start-up + operating costs to the agency + depreciation

Whereas benefit estimation is both conceptually difficult to understand and practically difficult to do, cost estimation presents only practical problems. Once the appropriate costs have been identified, only the accounting is difficult. The kinds of costs which should be included in estimates are listed in Table 10.2. These are costs to the public agency supplying the service or public good; costs to users are not to be included. The concepts of benefits and costs are summarized in Table 10.3.

10.3 NET PRESENT EQUIVALENT

When benefits and costs are measured in monetary units, the net present equivalent (NPE) is the best measure of economic efficiency to employ. It measures the magnitude of the excess of present equivalent benefits over present equivalent costs.*

The basic decision criteria applicable to NPE analyses are elementary. When an appropriate social discount rate is used, projects with positive NPEs are theoretically acceptable because discounted benefits to the public exceed discounted costs to the government. Projects with the largest NPEs are, of course, most desirable, because they offer the greatest net gain. Those with negative NPEs are economically unacceptable.† However, there is an important qualification. Benefits or costs of a nonmonetary nature cannot be included in NPE calculations because the units of measurement are different. Thus, when

---

* The NPE is consistent with the widely accepted concept of marginal economic analysis (see Section 10.5). That is, one should continue to invest in public projects until incremental benefits no longer exceed incremental costs, or until capital is exhausted.

† The trick is to select from among acceptable projects a group or portfolio so as to maximize the overall gain in social welfare when available investment capital is limited. This issue is discussed in Sections 10.6 and 10.7 and again in Chapters 11 and 14.

nonmonetary considerations are important, these decision criteria no longer suffice. They may be used only to identify programs which have acceptable economic implications, as a starting point in the decision process.*

Example 10.1: Comparing mutually exclusive projects Consider three projects requiring about the same initial capital investment and each having a lifetime of 10 years. Suppose an agency has sufficient initial capital to invest in only one of three projects. Project 1 has the characteristic of water management projects; large initial capital investments are required and benefits, though large, accrue only after many years. It also requires capital expenditures in the second and third years although operating costs are relatively small. Project 2 has nearly the same geometric growth of benefits although it requires less capital over time and has larger operating costs. Project 3 requires multiple investments of capital because project lifetime is half those of projects 1 and 2.

Project data
Thousands of dollars

	Project 1			Project 2			Project 3		
Year	Operating costs	Capital	Benefits	Operating costs	Capital	Benefits	Operating costs	Capital	Benefits
Now	0	100	0	0	100	0	0	100	0
1	0	100	0	28	0	0	18	0	55
2	10	50	0	28	0	0	18	0	55
3	15	0	0	28	0	0	18	0	55
4	15	0	5	28	0	5	18	0	55
5	15	0	10	28	0	10	18	100	55
6	15	0	20	28	0	20	18	0	55
7	15	0	40	28	0	40	18	0	55
8	15	0	80	28	0	80	18	0	55
9	15	0	160	28	0	160	18	0	55
10	15	0	320	28	0	320	18	0	55

Calculations of the NPEs are straightforward. Note that $\varepsilon = (1 + r)/(1 + i) - 1$, where r, the geometric growth rate, is 1 for projects 1 and 2 benefits.

Project 1

$$\text{NPE} = -100 - 100\left(\frac{p}{f}\right)_1^i - 60\left(\frac{p}{f}\right)_2^i - 15\left(\frac{p}{a}\right)_8^i\left(\frac{p}{f}\right)_2^i$$
$$+ \frac{5}{1 + i}\left(\frac{f}{a}\right)_7^\varepsilon\left(\frac{p}{f}\right)_3^i$$

* Nonmonetary benefits is the subject of Chapter 12.

Project 2

$$\text{NPE} = -100 - 28\left(\frac{p}{a}\right)^i_{10} + \frac{5}{1+i}\left(\frac{f}{a}\right)^\varepsilon_7\left(\frac{p}{f}\right)^i_3$$

Project 3

$$\text{NPE} = -100 - 100\left(\frac{p}{f}\right)^i_5 + 37\left(\frac{p}{a}\right)^i_{10}$$

The results for various discount rates are shown in the table and graph below.

NPE

Project	0%	4%	10%
1	255	100.08	−23.78
2	255	114.38	−0.74
3	170	117.91	65.26

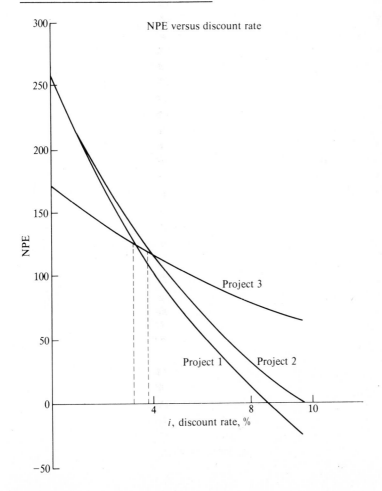

Notice how important is the choice of discount rates in the analysis. At rates below about 8.5 percent each project is acceptable, yet above 10 percent only project 3 is acceptable. In the range from 0 to 3.2 percent project 2 represents the best investment of the limited capital available, although project 1 is a close runner-up. Beyond about 3.7 percent project 3 is best. Nowhere is project 1 clearly preferable. One must always explore the implication of various discount rates in an analysis because, as was discussed in Chapter 9, the social discount rate can be only roughly estimated and such explorations provide important insight.

10.4 BENEFIT-COST RATIO

The standard benefit-cost ratio is the ratio of the total discounted benefits to the total discounted costs for a project or investment. It is a commonly used measure of economic efficiency. When benefits and costs are measured in monetary units and the proper social discount rate is employed, all projects having ratios greater than unity are acceptable, because discounted benefits exceed discounted costs.*

One must be careful in using the benefit-cost ratio as a basis for selecting projects. Those having the largest benefit-cost ratio(s) may not be the most desirable. Under conditions of limited budgets, where not all acceptable projects can be funded, it is not always true that funding those with the highest ratios maximizes social welfare. The reason is quite simple. A small investment of 5 dollars which yields a present equivalent of 10 dollars provides a benefit-cost ratio of 2. A larger investment of 20 dollars which yields 30 dollars provides a benefit-cost ratio of only $1\frac{1}{2}$. Yet the second investment leaves one economically better off than the first. Thus, if the 20 dollars were available for investment, and only one of the two opportunities could be selected, the (second) option with the lower benefit-cost ratio would be preferable.

However, the marginal benefit-cost ratio, if properly applied, does lead to the best economic decision. The idea is to order all acceptable investment opportunities according to capital requirements. One then compares the next most capital-expensive one to the least and so forth. The incremental capital outlay is justified by the incremental net benefit stream, if the ratio of the incremental benefits to incremental costs is greater than unity. When the marginal ratio is larger than unity and there is sufficient capital to invest in only one or the other project, the more capital-expensive alternative is preferable, because the net return is larger; that is, the NPE is larger.

One complication in applying the marginal ratio as just described comes to the fore when capital investment continues beyond year 0. One can either assume that extra capital will be available as needed in future years and rank

---

* Annual equivalents, present equivalents, or future equivalents for benefits and costs may be used in the ratios. The ratios will be identical.

projects according to initial capital required or one can work with present equivalent capital required. The first option is not particularly desirable. It justifies to projects having continuing capital demands, without a proper accounting of the opportunity costs of tying up future capital. The second option portrays project costs more accurately and accords consistent treatment to capital and operating costs in the future.

The different modes of analysis and their implications are best clarified by example.

Example 10.2: Improper treatment of capital costs Information for three projects follows. Only one project may be selected. The social discount rate for the analysis is 10 percent and each project has a lifetime of 10 years.

	Project A	Project B	Project C
Capital			
Year 0	$10,000	$17,000	$5,000
Year 1	—	—	7,000
Annual benefits	3,000	4,000	3,200
Annual operating costs	800	500	800

The standard benefit-cost ratios are calculated as follows:

Standard ratio:

Project A $\quad \dfrac{B}{C} = \dfrac{3000}{800 + 10,000(a/p)_{10}^{10}} = 1.24$

Project B $\quad \dfrac{B}{C} = \dfrac{4000}{500 + 17,000(a/p)_{10}^{10}} = 1.22$

Project C $\quad \dfrac{B}{C} = \dfrac{3200}{800 + [5000 + 7000(p/f)_1^{10}](a/p)_{10}^{10}} = 1.21$

All three projects are acceptable at 10 percent since all three ratios are greater than unity.

Project C requires the least initial capital outlay, followed by project A. The marginal benefit-cost ratios for the incremental investments to move from project C to project A or to project B are:

Marginal ratio:

From C to A $\quad \dfrac{\Delta B}{\Delta C} = \dfrac{-200}{[5000 - 7000(p/f)_1^{10}](a/p)_{10}^{10}} = 0.9$

From C to B $\quad \dfrac{\Delta B}{\Delta C} = \dfrac{800}{[12,000 - 7000(p/f)_1^{10}](a/p)_{10}^{10} - 300} = 1.3$

The incremental initial investment required for project A in lieu of project C is not justified by the marginal benefits anticipated, as indicated by the marginal ratio of 0.9, even though project A has the largest B/C ratio. The incremental investment to move to project B from project C is justified, however, since the marginal ratio is greater than unity.

The net annual equivalent (NAE) values for projects A, B, and C are 572.55, 733.33, and 550.62 dollars, respectively.* Notice that the NAE for project A is larger by 21.93 dollars than for project C. That is, moving from project C to project A would net 21.93 dollars annually. Obviously project A is preferable to project C. Why then did the analysis via marginal benefit-cost ratios not signal a move from project C to project A? (The answer is that when the 7000 dollars required in year 1 is accounted for, project C requires greater present equivalent capital investment than project A, 11,363.64 dollars compared to 10,000 dollars, and the marginal ratio analysis algorithm does not permit one to move backward to a less capital-demanding alternative.) The algorithm correctly signaled a profitable move from project C to project B, which has the largest NAE, because these two projects are properly ordered in terms of capital requirements.

The next example provides a proper marginal analysis of project alternatives A, B, and C.

Example 10.3: Proper treatment of capital costs Using the data from the last example, but recognizing that the 7000 dollars capital expenditure at the end of the first year of project C should be discounted to the present and added to the 5000 dollars before ranking alternatives by capital requirements, the correct marginal benefit-cost analysis follows.

Project	Present equivalent capital required, $	NAE, $	Standard B/C	Marginal $\Delta B/\Delta C$ over A	Marginal $\Delta B/\Delta C$ over C
A	10,000.00	572.55	1.24	—	—
C	11,363.64	550.62	1.21	0.9	—
B	17,000.00	733.33	1.22	1.19	1.30

Marginal ratios:

C over A

$$\frac{\Delta B}{\Delta C} = \frac{200}{(11{,}363.64 - 10{,}000)(a/p)_{10}^{10}} = 0.9$$

B over A

$$\frac{\Delta B}{\Delta C} = \frac{1000}{(17{,}000 - 10{,}000)(a/p)_{10}^{10} - 300} = 1.19$$

* Since NAE = $(a/p)_n^i$ NPE, both measures lead to identical decisions.

$$\text{B over C} \qquad \frac{\Delta B}{\Delta C} = \frac{800}{(17{,}000 - 11{,}363.64)(a/p)^{10}_{10} - 300} = 1.30$$

All three projects are acceptable. If only one is to be selected, and at least 17,000 dollars in capital is available, project B is best because it offers the largest annual equivalent increase in public welfare; 733.33 dollars. The marginal benefit-cost analysis leads step by step to the same conclusion. That is, starting with project A, which demands the least capital, marginal analysis indicates the undesirability of moving to project C, since $\Delta B/\Delta C = 0.9$. On the other hand, the analysis shows that it would be wise to select project B in lieu of project A, since $\Delta B/\Delta C = 1.19$ in this case.

10.5 RATE OF RETURN

The three terms, rate of return, return on investment, and internal rate of return are all phrases referring to the discount rate at which the NPE of a project has a value of zero. Private sector investors frequently contrast the rate of return for an investment option under consideration with a minimum rate which they consider acceptable. If the rate of return for the option is greater than the minimum acceptable rate, the option itself is judged acceptable. Public sector managers, many of whom hail from the private sector and are quite comfortable with the concept, often compare projects on their rates of return. Projects with higher rates of return are considered preferable.

The concept of rate of return can be quite useful. When one does not know the proper discount rate to use for project evaluation, one can establish a reasonable lower limit, by political or other means, and select among projects on that basis. By ranking projects according to capital required, as in Section 10.4, one can calculate rates of return on incremental capital investment. Thus, starting from the least capital-consuming alternative, one can move increment by increment to the project which offers the greatest return of discounted benefits for the capital invested, by selecting a more costly investment over a less costly one only if the rate of return on the margin is greater than the acceptable minimum. When marginal rate of return analyses are properly executed, they select the same alternative as do the NPE and marginal benefit-cost ratio methods.

Some caveats are in order, however. Overall rates of return on projects are not sufficient indicators of relative desirability. Projects with larger overall rates of return may, in fact, be less desirable economically than projects with lower rates of return when capital budgets are limited, for the same reason that projects with larger standard benefit-cost ratios are economic ventures which may be inferior to projects with smaller ratios. Overall rates of return only indicate whether projects are acceptable in the sense that their rates surpass an established minimum required; that is, they are merely indicators of acceptability. Marginal rates must be examined in order to choose among alternatives.

A more serious consideration is the nonuniqueness of the rate of return. The NPE is a polynomial in i and may be zero for many different values of i.* This may happen whenever accumulated end-of-year net cash flow changes sign more than once. If the sign changes only once, the rate of return for the project is unique.

Example 10.4: Multiple rates of return The NPE for the cash flow diagramed may be written as

$$\text{NPE} = \frac{1000[1.5 - (1 + i)][1.2 - (1 + i)][1.0 - (1 + i)]}{(1 + i)^3}$$

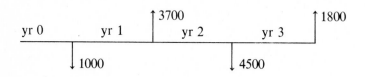

Clearly, NPE $= 0$ when $i = 0$, 0.2, and 0.5, as the sketch of NPE versus i indicates; only the numerically referenced points are accurate, the curve is roughed in.

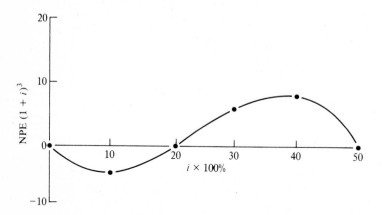

Example 10.5: Rates of return comparisons for mutually exclusive alternatives The data of Example 10.2 are used in this example. Projects A, B, and C are first ranked from least to most present equivalent capital required. The

---

* A polynomial in x is a function like $p(x) = a_0 + a_1 x + a_2 x^2 + a_3 x^3 + \cdots + a_n x^n$, where the a's are constants. A polynomial of degree 10, meaning the highest power is $n = 10$, may have as many as 10 real roots. That is, there may be as many as 10 values of x for which $p(x) = 0$. For example a polynomial which can be factored as follows has five real roots, $(c_1, c_2, c_3, c_4, c_5)$:

$$p(x) = (x - c_1)(x - c_2)(x - c_3)(x - c_4)(x - c_5)$$

NPEs for each are calculated for various discount rates, and overall rates of return are estimated by linear interpolation. Marginal rates are estimated similarly. The results are summarized below and NPEs are plotted against the discount rate.

NPE equations:

Project A $\qquad$ $\text{NPE} = -10{,}000 + 2200\left(\dfrac{p}{a}\right)^i_{10}$

Project B $\qquad$ $\text{NPE} = -17{,}000 + 3500\left(\dfrac{p}{a}\right)^i_{10}$

Project C $\qquad$ $\text{NPE} = -5000 - 7000\left(\dfrac{p}{f}\right)^i_1 + 2400\left(\dfrac{p}{a}\right)^i_{10}$

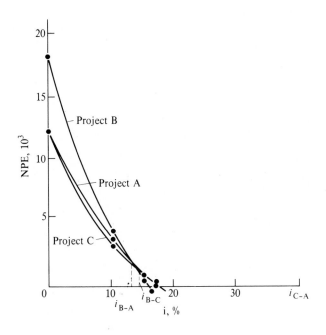

Net present equivalent:

Project	Discount rate, %					
	0	10	15	16	17	18
A	17,000	3518	—	—	248	−113
B	18,000	3384	566	−84	−695	—
C	12,000	4506	959	—	198	−146

Rate of return (comprehensive):

Project A $\qquad i \simeq 17 + (18 - 17)\left(\dfrac{248}{248 + 113}\right) = 17.7\%$

Project B $\qquad i \simeq 15 + \dfrac{566}{650} = 15.9\%$

Project C $\qquad i \simeq 17 + \dfrac{198}{344} = 17.6\%$

To calculate the marginal rates of return first subtract the benefit-cost cash flow for project A from that for project C to obtain the incremental flow. The discount rate for which the NPE of this marginal flow equals zero is the marginal rate of return for the incremental investment in project C over project A. In the same manner calculate the marginal rates for project B over project A and project B over project C.

Capital	C over A
Year 0	$5000 - 10{,}000 = -5000$
Year 1	$7000 - 0 \quad = 7000$
Annual benefits	$3200 - 3000 \quad = 200$
Annual costs	$800 - 800 \quad = 0$

$$\text{NPE} = 200(p/a)^i_{10} - 7000(p/f)^i_1 + 5000$$

The marginal rates of return may be calculated directly from these equations or, since NPE (incremental) = NPE (project C) − NPE (project A), one may estimate them from the graphs of overall NPE versus discount rate. The discount rate at which NPE lines for two projects cross is the marginal rate of return for the incremental flow of the more capital-demanding project over the other.

Project	Equivalent capital	Rate of return, %	Marginal rate of return (%) over project A	C
A	10,000.00	17.7	—	
C	11,363.64	17.6	40	
B	17,000.00	15.9	13	13.5

According to the rate of return criterion, if the minimum acceptable rate is less than 15.9 percent, all three projects are acceptable and B is the best, just as the marginal benefit-cost and NPE analyses indicated. If it is between 15.9 and 17.6 percent, only projects A and C are acceptable. If the minimum rate required is above 17.7 percent, none are acceptable.

Starting with project A and assuming the minimum rate required is less than 15.9 percent, the move to project C is acceptable since the incremental investment yields about 40 percent in return. The move from project C to project B is acceptable only if the required rate is less than 13.5 percent since that is the rate for the additional capital investment required.

The rate of return measure is not a substitute for the NPE or the marginal benefit-cost ratio measures. It can be useful when only a minimum discount rate is specified. Then by calculating the rates of return for marginal investments one can systematically select the best available alternative. On the other hand, using this minimum discount rate directly in NPE calculations would lead to the same choice.

10.6 PROJECTS PORTFOLIO

How should a public manager with an anticipated capital budget of, say, 100,000 dollars select among a multitude of project proposals totaling more than 100,000 dollars? Assume his primary objective is to maximize the increase in public welfare within his budgetary constraints.* What should he do if many of the proposed projects are economically superior to ongoing projects? Should he review the entire portfolio of ongoing projects annually and develop a new portfolio of old and new projects which maximizes the public good within the total budget of his agency? That is, should public agencies practice zero-based budgeting? These are critically important practical and political as well as economic issues. Only a rudimentary discussion of them is provided here.

Theoretically, it makes eminently good sense to require public managers to regularly review their entire projects portfolio and revise it so as to maximize public welfare. Practically, however, it is impossible, as advocates of the Planning Programming and Budgeting System (PPBS) initiated in the federal government in the late 1960s learned by experience. Even if one could reasonably estimate the true public welfare for public projects and knew how to equitably balance the distribution of benefits and costs to society, the labor, time, and dollar costs of comprehensive reviews make them impractical. Even within small corporations or individual households, the costs of comprehensive portfolio reviews are so high that they are seldom undertaken. Consequently, zero-based budgeting for public agencies is viewed by most managers as practically impossible. The practical alternative is to fund new projects with newly available capital and design the new projects portfolio to maximize the increase in social welfare to the extent possible.

---

* There are many other objectives embraced by public managers, from pleasing the boss, to responding to political pressures from special interest groups and constituencies, to capturing the limelight. These are extremely important factors in decision making but they are not within the scope of this section and so are not treated here.

A straightforward approach to maximizing the increase in public welfare is to develop an array of all possible portfolios of proposed projects which meet budgetary constraints. These then are mutually exclusive alternatives which should be evaluated and compared using the NPE or marginal benefit-cost methods or perhaps the rate of return method demonstrated in previous sections of this chapter. This is a practical approach if the number of possible portfolios is small or if electronic computers are employed for portfolio formation and comparison.

Example 10.6: Simple portfolio evaluation Data for three proposed projects are given. Assume that the public manager anticipates a capital budget of 100,000 dollars. Capital not invested in one of these projects will not be used elsewhere so it will generate zero return. The social discount rate is 10 percent. Project lifetimes are 5 years.

	Project A	Project B	Project C
Present equivalent capital, $	60,000	40,000	50,000
Annual benefits, $	20,000	15,200	15,500
Annual operating costs, $	1,000	1,200	500
$n = 5$			

All three projects have positive NPEs and benefit-cost ratios greater than unity. Obviously, their rates of return exceed 10 percent. Thus, all three projects are acceptable. Two additional portfolios, project A plus project B and project B plus project C, must be evaluated.

	Portfolio				
	A	B	C	A + B	B + C
Present equivalent capital, $	60,000	40,000	50,000	100,000	90,000
Annual benefits, $	20,000	15,200	15,500	35,200	30,700
Annual operating costs, $	1,000	1,200	500	2,200	1,700

Comparisons:

			Marginal $\Delta B/\Delta C$ over project			
Portfolio	NPE	B/C	B	C	A	B + C
B	13,071	1.20				
C	6,862	1.33	0.15			
A	12,025	1.13	0.95	1.59		
B + C	19,933	1.23	1.12	1.29	1.24	
A + B	25,096	1.21	1.18	1.32	1.29	1.43

Clearly the optimal portfolio is project A plus project B. It offers the greatest excess of discounted benefits over discounted costs, NPE = 25,096. It does not offer the largest standard benefit-cost ratio; but marginal benefit-cost analysis, of course, does indicate that the incremental capital required to move from portfolio B to portfolio (B + C) is justified, $\Delta B/\Delta C = 1.12$, and the move from portfolio (B + C) to portfolio (A + B) is desirable, $\Delta B/\Delta C = 1.34$, ultimately leading to the same selection as does the NPE method.

One might be tempted to construct a portfolio by accepting a single project, say, the one with the largest net present value or largest standard benefit-cost ratio, and step by step add the next most desirable project until capital is exhausted. Unfortunately, these approaches do not generally result in optimal portfolios. The points are easily demonstrated as follows.

Three projects are under consideration. The data are given in Table 10.4. The capital budget is fixed at 800 dollars. Project A has the largest NPE and B/C; but if chosen first for inclusion in the portfolio, it would exclude a better opportunity, namely, the combination of projects B and C.

Sophisticated methods of portfolio analysis have been developed by management scientists, operations researchers, and financial analysts. For example, integer programming methods allow one to maximize overall the NPE by selectively including projects in the portfolio subject to budgetary and other constraints. Details on these methods, which are quite valuable when many options exist and an array of constraints are in force, are published in the management-science and financial literature.

10.7 THEORETICAL CONSIDERATIONS

The net present value is the difference between discounted benefits and discounted costs. Clearly an infinite number of projects, each with different benefit-cost streams can have the same NPE. The rays having unit slope in Figure 10.3 symbolize all possible projects with positive benefits and positive NPEs. Notice that projects with benefits and costs (B_1, C_1) and (B_2, C_2) have the same ratio

Table 10.4 Project data
In present equivalents

	Project A	Project B	Project C	Projects B + C
Capital required, $	800	500	300	800
Benefits, $	1080	1060	1030	2090
Costs, $ (total)	1000	1000	1000	2000
NPE	80	60	30	90
B/C (standard)	1.08	1.06	1.03	1.05

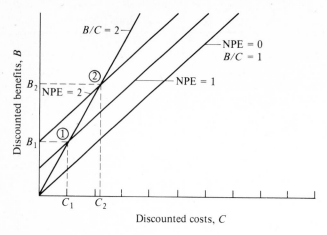

Figure 10.3 Rays of NPE and B/C.

rating, $B/C = 2$, yet the second project offers greater net return. This is a now-familiar phenomenon. Notice, too, that the slope of all NPE lines is unity. This is an important observation, for it helps explain why the NPE and marginal benefit-cost ratio methods of analysis lead to the same project selections.

The maximum NPE criterion could have been stated as follows. Continue to invest available capital so long as the increment in NPE is positive. The mathematical equivalent of this statement is: continue to invest available capital, so long as for the next increment

$$\Delta \text{NPE} = \Delta B - \Delta C > 0$$

or, equivalently,

$$\frac{\Delta \text{NPE}}{\Delta C} = \frac{\Delta B}{\Delta C} - 1 = \varepsilon$$

where ε is a positive number and B and C are discounted benefits and costs. Optimality (the best economic situation) is achieved when ε has been reduced to the extent possible. When an unlimited array of portfolio options is available and sufficient capital for any one exists, optimality is reached when $\varepsilon = 0$ (see Table 10.5).

Table 10.5 Conditions of optimality

- Unlimited array of portfolios and sufficient capital

  $\Delta \text{NPE} = 0$ or $\Delta B/\Delta C = 1$

- Limited capital or portfolio options

  $\Delta \text{NPE} = \varepsilon$ or $\Delta B/\Delta C = 1 + \varepsilon$ where ε is as small as possible

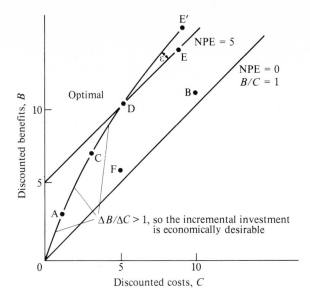

Figure 10.4 Portfolio selection.

Figure 10.4 portrays an array of possible portfolios. All projects indicated by points to the right of and below the NPE = 0 line have negative net equivalents and benefit-cost ratios less than unity, and thus have been excluded from the diagram as infeasible. Assume that portfolio A is the least capital-intensive, followed by the other portfolios in alphabetical order, and that there are sufficient funds to invest in A, B, C, or D but not E (or E').

Portfolio D is the optimal solution. Even if capital were available for portfolio E, the incremental investment would yield zero net benefits, ΔNPE = 0 and $\Delta B/\Delta C = 1$, so portfolio E would be inferior to portfolio D. Were portfolio E' available in lieu of portfolio E, portfolio D would still be optimal, because E' exceeds the capital budget, although

$$\Delta\text{NPE} = \varepsilon \qquad \text{and} \qquad \frac{\Delta B}{\Delta C} = 1 + \varepsilon$$

The positive number ε is called the *shadow price*. It represents the incremental net benefit expected from an additional dollar invested. For different operating units in the same government agency, the value of ε may be quite different. Capital should be diverted to the agency with the largest shadow price, for there is where the largest incremental gain in public welfare will accrue per dollar of capital expended. For example, if the shadow price for the Departments of Defense (DOD), Health, Education, and Welfare (HEW), and Agriculture (USDA) were $\varepsilon_1 = 0.2$, $\varepsilon_2 = 0.5$, and $\varepsilon_3 = 0.1$, respectively, preferential treatment in capital budgeting should be given to HEW. If capital budgeting were done optimally, the outcome would be that the shadow price for all these agencies would be equalized, say, to $\varepsilon = 0.3$.

10.8 PROBLEMS

10.1 Using the data in Example 10.1 calculate the standard benefit-cost ratio ($i = 10\%$) and the comprehensive rate of return for each project. Also calculate the marginal benefit-cost ratio and incremental rates of return. Use the marginal benefit-cost ratio method to select the best alternative. Using the concept of minimum required rate of return, explain the conditions under which different alternatives are optimal according to the rate of return criterion.

10.2 The director of the state department of environmental protection has a limited capital budget of 100,000 dollars. Four projects are under consideration. The estimated benefits and costs for these in thousands of dollars are

	\multicolumn{8}{c}{Project}							
	No. 1		No. 2		No. 3		No. 4	
Year	C	B	C	B	C	B	C	B
0	100	0	40	0	34	0	40	0
1	0	20	0	5	0	17	0	0
2	0	20	0	10	0	17	0	0
3	0	20	0	15	34	17	0	0
4	0	20	0	20	0	17	0	0
5	0	20	0	25	0	17	0	100

(a) Select the best portfolio using the NPE and benefit-cost criteria for $i = 0\%$, $i = 5\%$, and $i = 10\%$, assuming leftover capital is not invested elsewhere.

(b) Same as part a except assume that leftover capital is invested at 10 percent; at 20 percent.

10.3 Using the data from Problem 10.2 calculate the rate of return for each portfolio. If the minimum acceptable rate of return is 15 percent, which alternative is best? Which is best if the minimum required rate is 18 percent?

10.4 Suppose the minimum required rate of return for the environmental protection agency of Problem 10.2 is 21 percent. Then, none of the proposed projects meet the minimum. Suppose, however, that for political reasons, at least one project must be launched. Analyze the portfolios available and select the optimal one. Assume leftover capital yields no return.

10.5 Construct a numerical example countering the argument that public decision makers are exercising the best economic decision rule when they fund first the projects with the highest standard benefit-cost ratios to the limit of their capital budgets.

10.6 Allowing one agency to use a discount rate of, say, 4 percent while requiring others to use, say, 10 percent, has an impact on the concept of shadow prices as signals for distributing capital among competing agencies. Explore and explain this impact.

10.7 Consider the following investment opportunities:

	A		B		C		D	
Year	B	C	B	C	B	C	B	C
0	0	100	0	50	0	15	0	75
1	50	0	15	0	20	15	0	0
2	0	0	15	0	20	15	0	0
3	50	0	15	0	20	15	0	0
4	0	0	15	0	20	15	0	0
5	50	0	15	0	20	15	150	0

Social discount rate = 5%
  Available capital = $100

(*a*) Calculate PEB, PEC, NPE, *B/C* ratio, and rate of return for each investment opportunity. Which is the best investment and why?

(*b*) Suppose there were a fifth opportunity to invest your capital with a rate of return 1 percent higher than the highest rate calculated in part *a*, but you had to invest in one of the four original projects. Which would you choose and why?

(*c*) Calculate the marginal ROR and *B/C* ratio for the alternatives not picked in part (*b*). Would you invest in a different project than that picked in part (*b*)? Why or why not?

10.8 The regional health planning agency is reviewing a proposed capital expenditure by a local hospital that wishes to expand its nuclear medicine department. The economic information they have submitted is summarized here. As a staff analyst of the health agency you must evaluate the proposal from an economic perspective. Analyze these data to the extent possible, draft a memo laying out the additional information you need to complete your analysis, and/or recommend to your board whether or not to issue a permit to the hospital on a purely economic basis.

Financial and cost considerations

A. *Estimated total project development costs*
   Equipment and installation costs $122,750.00
B. *Source of capital investment funds* . . . .
   Hospital short-term investments and cash reserves* $122,750.00
C. *Unit "fixed capital expenses" (depreciation, interest, and amortization) and patient revenues*
   Estimated number of nuclear
   medicine department
   diagnostic procedures† 1977:   3578
                                                    1978:   3935
                                                    1979:   4328

   Estimated annual "fixed expenses" (depreciation, interest, and amortization)‡
first 3 operating years $13,819.00
   Estimated "fixed expense"
   per nuclear medicine
   diagnostic procedure 1977: $3.86
                                               1978: $3.51
                                               1979: $3.19

   Estimated gross revenue
   per nuclear medicine
   diagnostic procedure (patient
   charge) first 3 operating years $71.42
   Estimated unit "fixed expense"
   as percentage of patient charge 1977:   5.4%
                                                     1978:   4.9%
                                                     1979:   4.5%

* The application documents an excess of $122,750.00 in short-term liquid investments.

† The forecasted annual growth rate in departmental patient loads is 10.0 percent. The annual rate of increase for the department was 9.0 percent for the period 1974 through 1976.

‡ Depreciation-amortization estimates apply to two Pho/Gamma camera systems with an estimated economic useful life of 10 years.

10.9 Calculate the rate of return for the following cash flow. Is it unique?

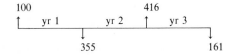

10.10 Agency A has proposed a new project. The marginal benefit-cost ratio is $\Delta B/\Delta C = 1.25$. Agency B has proposed a new project. The marginal benefit-cost ratio for this project is $\Delta B/\Delta C = 1.55$. You are the budget officer for both agencies. There is enough money in the budget to fund only one project.

(a) Which project is economically preferable?
(b) Which agency has the largest shadow price?
(c) What are the shadow prices for these agencies?
(d) What assumption have you made about discount rates used by the agencies?

10.11 You have two investment options. You can buy 100 shares of stock at 10 dollars per share or a 1000 dollar, 6 percent bond. The bond would mature in 5 years. The stock could be sold in 5 years. Both investments pay dividends at the end of the year only; the stock will pay 50 dollars per year. Assuming you will hold one instrument for 5 years, what selling price for the stock will make it an equivalent investment to the bond? The market interest rate for reinvestment is 10 percent.

10.12 Calculate the rate of return for the following cash flow.

yr 0	$-15,000$
yr 1	$4,000$
yr 2	$3,000 + 100$
yr 3	$2,000 + 200$
yr 4	$4,000 + 300$
yr 5	$7,000 + 400$

10.13 Washington University raised tuition for next year by 350, to 4300 dollars. Parents of incoming freshmen may pay 17,200 dollars in a lump sum in September to cover all 4 years or they may pay annually. If they pay annually, they may expect tuition to increase 350 dollars in each of the sophomore, junior, and senior years (gradient). Suppose you were a parent with alternative investment opportunities for your 17,200 dollars. At what rate of return would the two payment plans be economically equivalent?

10.9 SUPPLEMENTAL READING

Haveman, Robert H., and Julius Margolis (eds.): *Public Expenditures and Policy Analysis*, Markham, Rand McNally, Chicago, 1970, chap. 14.

Marglin, Stephen A.: Public Investment Criteria: Benefit Cost Analysis for Planned Growth, MIT Press, Cambridge, Mass., 1967, chap. 2.

McKean, Roland N.: *Efficiency in Government through Systems Analysis*, John Wiley & Sons, Inc., New York, 1958.

Merewitz, Leonard, and Stephen H. Sosnick: *The Budget's New Clothes*, Markham, Rand McNally, Chicago 1971, chaps. 6 and 8.

Smith, Gerald W.: *Engineering Economy: Analysis of Capital Expenditures*, 2d ed., Iowa State University Press, Ames, 1973, chap. 11.

U.S. Bureau of the Budget, *A Conceptual Basis and Measurement Technique for Computation of a Social Discount Rate*, U.S. Government Printing Office, Washington, D.C., 1970.

ELEVEN

INCORPORATING UNCERTAINTY

In Chapters 9 and 10 benefits and costs were treated as known entities. Unfortunately, it is a rare occasion when either is well known. At best, cost and benefit estimates, founded on accumulated data, may be projected into the future via extrapolation, regression, or some other method. At worst, they must be crudely estimated, almost guessed, with great uncertainty. Project lifetimes, too, were treated as known, when, in fact, they also can be estimated only roughly. These kinds of uncertainties are explicitly treated in this chapter and methods for including them in benefit-cost studies are explained. The main point is that a decision may change if the value of a parameter is varied, which implies that a bad estimate may lead to a bad decision.

11.1 SENSITIVITY ANALYSIS

One should always check the sensitivity of analytical results to variations in the parameters of analysis. As we observed in Chapter 9, the discount rate employed in a benefit-cost study may have a significant influence on project selection. For some values of the discount rate one project may appear economically preferable while for other values another project may look better. Clearly, the magnitudes of long-term expenditures and benefits, and rates of increase in costs and benefits over time, are also important parameters to vary in economic efficiency studies. Project lifetime is another possibly important parameter.

The ranges over which parameters should be varied depend very much on how confident one is that they fall within certain bounds. This, in turn, depends on how the parameters are estimated and the confidence one has in the data used. For example, if good statistical data on wage rates over a number of years were available and regression methods were applied to the data to estimate a trend line, one would want to test not only the predicted wage rates for the period, but the limits of the three standard deviation confidence bands as well. In other cases simply varying parametric values individually to find the crossover values at which project ranks change or projects become unacceptable economically may represent the best form of sensitivity analysis.

Example 11.1: Sensitivity The U.S. Army Corps of Engineers has proposed a new dam and reservoir on the Lovely River. Independent consultants have estimated the initial capital costs of the project to be in the range of 10 to 15 million dollars. The Corps has estimated operating costs, including labor and materials, by scaling up costs for other similar operations in the region. Using trend analysis methods the Corps projected annual operating costs as shown. The ranges are based on expert judgment and are at the very limits of plausibility, as the experts see it. Benefit estimates, too, were made by the Corps. The base value was estimated by a market survey and is considered accurate to within ± 1000 dollars. Annual growth rates were estimated from data for earlier similar projects adjusted for population and economic growth trends. The most likely rate of increase of benefits to the public is expected to be between 14 and 28 percent, as shown in the diagram. The social discount rate was assumed to be 10 percent.

The parameters to which the decision on whether to fund this project is likely to be sensitive are:

- Initial capital costs
- First-year benefit estimates
- Rate of increase in annual benefits
- Annual operating costs
- Project lifetime

A thorough sensitivity analysis would consist of varying each of these over the ranges indicated while holding all others constant. The results would then be summarized. However, there are five parameters and at least three values for each (minimum, maximum, and most likely or expected values); so there are 243 (3^5) separate conditions for which the NPE or benefit-cost ratio would have to be calculated. Thus, for the sake of exposition only two variations on initial capital and two on the benefits growth rate are considered here.

The NPE for this project is

$$\text{NPE} = -C_0 - 5000\left(\frac{p}{a}\right)_{20}^{10} - \frac{500}{0.1}\left(\frac{p}{f}\right)_{20}^{10}\left[\left(\frac{f}{a}\right)_{20}^{10} - 20\right] + \frac{100,000}{1.1}\left(\frac{f}{a}\right)_{20}^{\varepsilon}$$

where $\varepsilon = (1 + r)/1.1 - 1$ and r is the rate of increase in benefits.

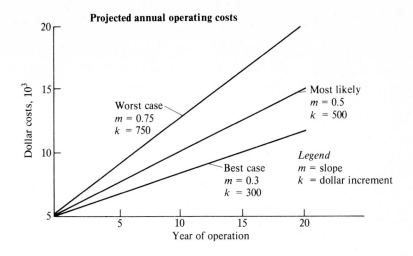

Projected annual operating costs

Worst case
m = 0.75
k = 750

Most likely
m = 0.5
k = 500

Best case
m = 0.3
k = 300

Legend
m = slope
k = dollar increment

Dollar costs, 10^3

Year of operation

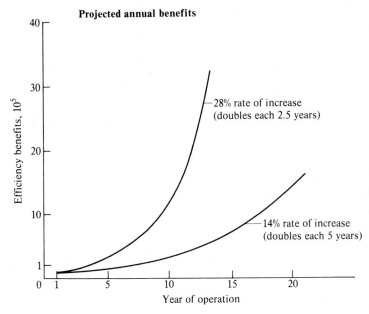

Projected annual benefits

28% rate of increase
(doubles each 2.5 years)

14% rate of increase
(doubles each 5 years)

Efficiency benefits, 10^5

Year of operation

NPE

		ε (%)	Initial capital, $ millions	
			10	15
Annual benefits	14	3.6	−7.5	−12.5
growth rate, %	28	16.4	+0.9	−4.1
$i = 10\%$ $n = 20\%$ $k = 500$				

Clearly, only if the growth rate for benefits is on the order of 28 percent annually and the initial capital costs are near the lower estimate of 10 million dollars does the project seem reasonable. Of course, the outcome could be improved if the project life were longer or the discount rate lower.

11.2 CONTINGENCY STUDIES

Quite often the time streams of benefits and costs for a project are contingent on the occurrence of events over which program managers have no control. For example, labor costs during the lifetime of a project may be significantly altered by future contract negotiations, materials costs may be subject to future regulatory or environmental legislation or to international trade agreements modifications, and capital availability for replacement equipment during the project lifetime may be subject to variations in the monetary policy of the Federal Reserve Board. When contingencies can be identified, they should be incorporated into the analysis explicitly. For each contingency projects should be evaluated separately. If time permits, one may wait to see which contingent event occurs and then select the project. Or, if immediate action is required, one might gamble on which contingency will actually be realized and choose the best action under those assumed circumstances. One might even calculate the expected (weighted average) NPE or benefit-cost ratio using the relative likelihoods of contingent events, and select the project which offers the best expected return when all contingencies are incorporated into the evaluation probabilistically.*

The last two approaches introduce additional parameters into evaluation studies, namely, relative likelihoods. For contingent events these must be estimated by statistical and judgmental methods combined, as discussed in previous chapters. One would expect project selection to be influenced by these likelihood estimates (otherwise why include the contingencies in the analysis at all); therefore sensitivity analysis should be applied to these parameters to determine the ranges within which certain alternatives are superior to others, as discussed in Section 11.1.

Example 11.2: Contingencies City Hospital is considering opening a free-standing ambulatory surgery facility. A private consulting firm, the Managerial Analytics Group, has carefully reviewed legislation, medical care practices, the needs of the target population, as well as the surgical demand statistics maintained by the hospital. The firm also developed capital and operating cost estimates and constructed appropriate fee schedules. Benefits

---

* The idea of incorporating contingencies into the analysis probabilistically and selecting a project on the basis of greatest expected return is central to the methodology called decision analysis, discussed in Chapters 18 through 21.

and costs estimates, including adjustments for shifting patient loads from the main hospital to the ambulatory unit, were prepared for two contingencies:

- Pending legislation expanding the range of procedures permitted within a freestanding ambulatory surgical unit passes this year
- The legislation does not pass this year

The summary report submitted to the board of directors included the following data, as well as many variations and background information.

Likelihood (unknown)	Law passes p	Law fails $1 - p$
Annual equivalent benefits and costs, $ (thousands)		
Capital required	30	20
Operating costs (net)*	400	300
Benefits (net)	450	300
Project life = 20 yrs		
Discount rate = 10%		

* Benefits reflect reductions in working days lost due to shorter hospital stays as well as benefits of improved scheduling from the perspective of patients.

On the basis of the data above it seems that the unit will provide positive net benefits to the community only if the new law passes. If the

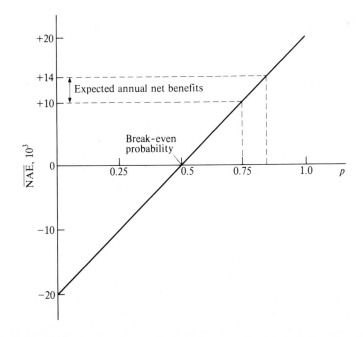

chances are good that the law will pass, opening the ambulatory unit would seem to be a good decision, otherwise it would seem to be a poor choice. The expected (average) NAE is

$$\overline{\text{NAE}} = 20{,}000p + (-20{,}000)(1 - p)$$
$$= -20{,}000 + 40{,}000p$$

The value $\overline{\text{NAE}}$ is larger than zero, indicating that the project is desirable, so long as p is larger than $\frac{1}{2}$.

The $\overline{\text{NAE}}$ was plotted as below for extreme ranges of benefit and cost estimates, permitting the board members to see how the value of p influenced the estimated desirability of the project. The board made a collective estimate that the value of p was in the range from 75 to 85 percent and opted to launch the venture.

11.3 BREAK-EVEN ANALYSIS

The concept of *break-even analysis* has numerous applications in economy studies, from facility capacity design and rate setting for profit maximization to go, no-go decision making.* One important application is in the treatment of uncertainty. Should it become obvious that project selection will be strongly influenced by certain parameters of analysis, one may wish to identify the parametric values which represent crossover points for decision making. In addition it might be important to decision makers that projects pay for themselves or at least become self-supporting. For example, one might be interested in how long it would take for annual benefits to match annual costs under certain conditions. Or, one might be concerned with how long a project would have to operate in order that equivalent annual capital expenditures be matched by equivalent annual net benefit flows (excluding capital).† The method called break-even analysis is designed to answer such questions.

Example 11.3 A municipality is considering two pool constructions for its new aquatic recreational facility. The longer-lasting standard design would cost about 67 percent more than the less-expensive experimental design. There is no reason to believe that public welfare would be enhanced more by one alternative than the other. Neither is there reason to believe that revenues would be affected by the choice. Hence, only costs need be considered.

Company S, which proposed the standard design at a cost of 100,000

---

* Any good text on engineering economy, such as those listed for supplemental reading, details many applications, especially to private sector decision making.

† In the private sector, where rapid recovery of capital invested is an important consideration, the concept of *payback period* is employed. The payback period is the number of years it takes for accumulated undiscounted net revenues to equal the capital invested.

dollars, is a well-known, reputable company. Company X, which proposed the experimental version for 60,000 dollars, is also well known and reputable. The companies agree that the lifetime for the standard design is about 25 years. How long must the experimental version last to make it a better economic alternative than the standard design?

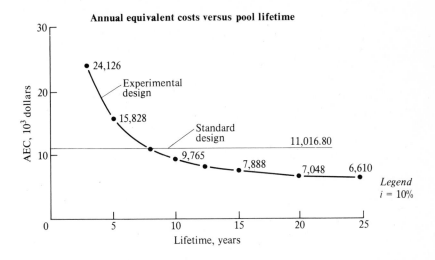

Annual equivalent costs versus pool lifetime

Lifetime, years

With a discount rate of 10 percent, the annual equivalent cost of the standard pool over a period of 25 years is 11,016.80 dollars. The annualized cost of the experimental construction plotted versus lifetime indicates that if the pool can be expected to last at least $8\frac{1}{2}$ years, it will be the economically superior choice. Now the problem is to determine the likelihood that the $8\frac{1}{2}$-year lifetime will be achieved.

On occasion the benefits of competing projects may be completely unknown, except that they are assumed to be nearly identical. When this is the case, benefit-cost studies become cost-effectiveness studies and projects are selected on the basis of least cost. The assumption of equal benefits should always be questioned. Break-even (or sensitivity) analyses on this assumption as well as on the costs and other parameters of analysis are important.

If for whatever reason one of the competing projects (or portfolios) must be selected, it is appropriate to ignore totals and compare the projects entirely on their differences. If the differences are uncertain, break-even analysis, sensitivity analysis, or contingency studies (described previously) may be applicable, as in the next example.

Example 11.4 Air pollution legislation is being written by Congress. Two alternate bills are being considered. The total costs and benefits of the bills are so uncertain that no one is seriously trying to estimate them. Instead, analysts are concentrating on estimating the magnitude of the differential benefits and costs of the two bills. At this point the best estimates of the differentials are as shown below. If the social discount rate is assumed to be 10 percent, which of the two bills appears preferable? The legislative impacts may be assumed to carry to perpetuity.

	Differential, $ (millions) Law 1 over law 2
Start-up capital	400 ± 10% error*
Annual equivalent	
Benefits	200 ± 20% error
Management costs	50 ± 10% error
Corporate operating costs	100 ± 20% error

* Error ranges represent the two standard deviation limits for the estimation distributions employed. Thus, the analysts are approximately 95 percent confident that the actual differences will be within these ranges.

The annual equivalents for these differences are plotted below. If the likelihoods are such that the benefit-cost differences can be expected to place the marginal NAE above the break-even line (in the box), law 1 would appear superior; otherwise, law 2 probably is best.

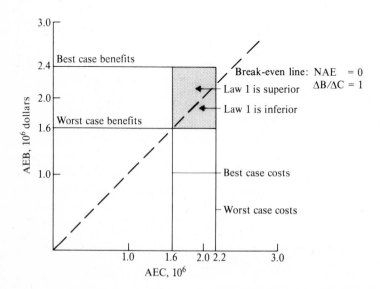

Of course, for such important decisions much reestimation of differential impacts and extensive sensitivity and contingency analyses are essential. One can never find *the* answer; one can only obtain better estimates and greater insight.

11.4 PROBLEMS

11.1 The Congress is considering legislation which would create a federally underwritten National Health Insurance Program. Staff analysts in the Library of Congress, Legislative Reference Service, and in the General Accounting Office have been assigned to estimate the costs of a comprehensive, full-coverage program and of a catastrophic- (economically) coverage-only program.

(*a*) Describe a reasonable plan of analysis which they might follow. In it identify objectives, tasks, data needs, and a plan of analysis.

(*b*) How might they estimate the differential benefits for the two options?

(*c*) Discuss the nature of the uncertainties the analysts should treat in their analyses. For example, what kinds of uncertainties might they find in the data bases they must use?

11.2 U.S. Army Corps of Engineers analysts are estimating the recreational benefits which might accrue if a proposed reservoir were built in a rural area of a Midwestern state. By comparing the proposed project with fairly similar projects in nearby states they have estimated that the average number of visitors per summer is a function of many variables including the cost and time to travel to the recreational site, the ratio of motel and lodge-type accommodations to campsites, and the number of boat docks per acre of water. Data for the other projects are available.

(*a*) What method(s) of analysis might be used to convert the available data into a projection of demand or visitor rates for the proposed project?

(*b*) Describe the uncertainties the analysts must face in estimating demand and how they might treat them in their analyses.

(*c*) If the analysts had found that the only really important variable in estimating demand was the average distance to the three largest nearby centers of population, how could they estimate a demand curve and calculate increased public welfare from the new project? Sketch such a demand curve with the approximate shape you would expect it to have for this situation.

11.3 The regional air pollution control board has hired a consulting company to estimate the economic merits of forcing a local smelting plant to install expensive pollution control equipment to reduce emissions. The company estimated the health benefits by applying public health data gathered in a nationwide study on the costs (income lost due to morbidity and early death plus health care costs) of respiratory diseases as a function of air pollution concentrations.

(*a*) Criticize this estimation procedure and propose a better approach.

(*b*) Discuss the uncertainties inherent in estimates of benefits for these kinds of public health programs. How would you propose to treat them if you were the analyst?

11.4 The city purchased equipment for 20,000 dollars. Half the purchase price was borrowed from a bank at 6 percent compounded annually. The loan was to be paid back with equal annual payments over a 5-year period. The equipment was expected to last 10 years, at which time it would have a salvage value of 4000 dollars. Over the 10-year period the operating and maintenance costs were anticipated to equal 4000 dollars per year. The social discount rate is 10 percent. What is the equivalent uniform annual cost for the investment? Check the sensitivity of annual equivalent cost (AEC) to lifetime and salvage value.

11.5 The city is planning to repave a major road. Two asphalt road surfaces, standard asphalt and experimental type Ex. 1, are under consideration. Each surface should provide equivalent benefits to residents, although no estimates of these magnitudes have been made. The experimental asphalt surface would be cheaper to lay down, but would probably not last as long as the more expensive, standard surface. Both surfaces would require about the same annual maintenance, although these costs have not been estimated for the experimental surface. The costs of laying the surfaces have been estimated by two contracting companies as shown.

	Resurfacing costs per mile, $	
	Company A	Company B
Experimental type 1 Lifetime unknown	20,000	18,000
Standard asphalt Lifetime average 8 years	40,000	42,000

The appropriate discount rate is between 5 and 10 percent.

(a) For each set of estimates determine the minimum lifetime required of the experimental surface to make it a better investment than the standard surface.

(b) Is your answer to part a sensitive to the discount rate used?

(c) Under what conditions should Company A be selected over Company B?

11.6 The Center for Disease Control has statistics which indicate that the likelihood of contracting disease D, a seasonal influenza, in a given year is distributed normally, with a mean of 30 percent and a variance of 15 percent for individuals not inoculated. For inoculated individuals, the likelihood is distributed normally, with a mean value of 14 percent and a variance of 9 percent. Last year 25 million people purchased the inoculation at a price of 4 dollars. A nationwide study indicates that 50 million people would probably have the inoculation this year if it cost about 2 dollars.

(a) What would be the anticipated consumer surplus, or increase in public welfare, if the inoculation were offered nationwide at 1 dollar per inoculation rather than 4?

(b) The government is considering subsidizing the distributor of the inoculation. Congress is considering allocating a yearly amount of 104 million dollars to the program. Economists estimate that the economic loss per annum to individuals who contract the disease is about 50 dollars on the average. At what price per inoculation would the benefits from economic savings and the demand-estimated increase in public welfare equal the cost of the government subsidy?

(c) Explore the uncertainty in this problem.

11.7 For a given operation there are certain fixed costs and certain variable costs. Fixed costs do not vary with the level of service. Variable costs vary with the level of service. Suppose that

C_F = fixed costs = 1.8 million dollars per year
C_V = variable costs = 3 dollars per unit of service
Q = units of service per year

(a) Find the total cost of service for

$$Q = 400,000 \text{ units of service}$$

(b) Write the total cost equation in terms of Q.

(c) What is the break-even level of service if benefits accrue at 4 dollars per unit of service?

11.8 The Municipal Trash Collection Agency disposes of waste at a cost of 2 dollars per ton. Half this waste (by weight) is recyclable glass. To recover all the glass waste, the agency can buy a sorting and glass recycling system for 25,000 dollars. Assume the equipment life is 10 years, salvage value is zero, operating costs are 5000 dollars per year plus 5 cents per ton of glass processed, and the required rate of return is 20 percent.

How much must the agency receive for the recycled glass in order to break even, if:

(*a*) Annual trash tonnage is 10,000 tons?

(*b*) Annual trash tonnage is 15,000 tons?

11.5 SUPPLEMENTAL READING

Fabrycky, W. J., and G. J. Thuesen: *Economic Decision Analysis*, Prentice-Hall, Inc., Englewood Cliffs, N.J., 1974, chap. 13.

Riggs, James L.: *Economic Decision Models*, McGraw-Hill Book Company, New York, 1968, chap. 2.

Smith, Gerald W.: *Engineering Economy: Analysis of Capital Expenditures*, 2d ed., Iowa State University Press, Ames, 1973, chap. 8.

TWELVE

NONMONETARY CONSIDERATIONS AND EQUITY

The decision rules employed in the previous three chapters break down when not all benefits and costs can be measured adequately in monetary units. How can one meaningfully apply the NPE concept when costs are measured in dollars and benefits are measured in disease-free days or reduced accident rates, or both? How could one calculate a rate of return in such circumstances? What is an appropriate decision rule to use when benefit-cost ratios for competing programs are given in different terms such as lives saved per dollar expended, for one, and books acquired per dollar expended, for the other? Is a ratio of 1 to 10,000 for the first better than 1 to 10 for the second? Even worse, how can one compare programs which produce psychological, moral, or other qualitative impacts to programs which produce measurable impacts?

Since important public investment decisions seldom turn on monetary considerations alone, the problem faced by the analyst is to find ways to incorporate nonmonetary impacts, both quantitative and qualitative, into the analysis so as to elucidate issues, alternatives, and potential impacts. Some insight into how one might treat benefits or costs which are measurable in units other than monetary is provided in Section 12.1. The basic idea is that multifaceted benefits can sometimes be adequately reduced to a single numerical measure or index. The idea of explicit accounting of both qualitative and quantitative impacts in tabular and scenario formats is discussed in Section 12.2. The issue of equity, which is really a decision criterion (although it may be thought of as an impact), is the topic of Section 12.3. The general theme of this chapter is that it is the responsibility of analysts to elucidate issues, alternatives, and potential consequences of all types so that the trade-offs implied by the selection of one alternative over another are explicit.

12.1 COMPARABLE MEASURABLE IMPACTS

Often, nonmonetary impacts can be measured in other units. Frequencies of events, such as deaths or morbid days avoided, can be counted. Pollution concentrations and animal population densities can be measured, too. In fact, many more impacts can be quantified than one might expect, and a good analyst can sometimes invent useful numerical measures for unique situations. So long as all costs and all benefits can be quantified in comparable terms, the benefit-cost ratio concept can be employed in both the standard and marginal forms, although the decision rule that projects are justifiable if the ratio is larger than unity is no longer applicable. When costs are measured in monetary units and benefits are measured in numbers of lives saved, for example, the benefit-cost ratio will indicate the number of lives saved per dollar expended. In other instances the ratio might represent the number of hospital days avoided or the number of accidents avoided per dollar expended. Although comparing similar projects using such ratios is easy and relatively objective, deciding that a project is worth initiating at all is purely a matter of judgment.

Deciding at what rate to discount nonmonetary benefits is a critical issue since it is not clear that the same rate is as applicable to these as to monetary benefits and costs. Perhaps the best thing to do is employ a variety of discount rates including zero as well as the rates at which economic costs are discounted in the study, and display the results so that decision makers can observe the implications of each selection.

Example 12.1: Nonmonetary benefits and monetary costs Data for two public safety projects are given. Benefits are measured in fatalities avoided, and costs are in terms of dollars. The social discount rate for dollar expenditures is 10 percent. Program periods are 10 years. The nature of fatalities is similar, so that quality aspects may be ignored.

	Program A	Program B
Initial capital	$100,000	$150,000
Annual flows		
Fatalities avoided	240 for 5 yrs; 300 thereafter	360
Operating costs	$10,000	$8,000

If future lives saved are valued at the same level as lives saved in the near term, fatalities avoided should not be discounted. If future lives are valued less than present lives, they should be discounted. Results for discount rates of zero and 10 percent in terms of benefit-cost ratios are given below:

	Standard B/C	Marginal $\Delta B/\Delta C$ over program A
Future lives not discounted		
Program A	0.017	
Program B	0.018	0.024
Future lives discounted at 10%		
Program A	0.010	
Program B	0.011	0.016

Whether or not lives are valued equally in the future, the marginal ratio indicates that if one or the other program is to be selected, program B is the best choice. The higher the rate at which future lives are discounted, the less difference there is between the two programs.

These results certainly would not represent reality if the quality of fatalities differed. For example, if program A were directed at infrequent but catastrophic and emotionally appalling accidents such as airline crashes or commuter train collisions and program B were directed at more frequent, but individualized motorcycle accidents, then program A might appear desirable for one or both discount rates used in the analysis.

In situations where benefits (or costs) are multifaceted and must be tallied in different units, acceptable rates of substitution of one type for another may sometimes be found. In this way multifaceted benefits may be represented in terms of a single unit of measurement, perhaps even in monetary units. If so, analytical methods previously discussed are applicable. However, many nuances and details tend to be obscured in the process of analysis. Unless the analyst takes pains to ensure that readers understand the implications of this reduction process, the analysis can be quite misleading. The treatment of multifaceted benefits is discussed more extensively in Chapter 20, where the concept of utility is introduced.

Example 12.2: Multifaceted benefits trade-off The Legislature is considering alternative bills; one would authorize a program to reduce highway accident rates, and the other would authorize a program of emergency care for persons who accidentally ingest toxic substances. Two measurable benefits expected from the highway program are reduced fatality rates and reduced rates of serious, nonfatal physical injury. From the emergency care program, reduced fatality rates are anticipated.

Using highway and vital statistics records, analysts have projected the impacts below.

	Highway safety program	Poison emergency unit
Initial capital	$10,000	$50,000
Annual flows		
Fatalities avoided	5,000	10,000
Serious injury avoided	5,000	
Operating costs	$5,000	$10,000

The Public Preference Survey Corporation conducted a survey designed to capture relative public attitudes toward fatalities versus serious injuries of a specified class from highway accidents and to estimate equivalence rates for auto fatalities and deaths from accidental poisoning. Their results are graphed here. They asked respondents to indicate their relative degree of aversion to auto fatalities compared to poisonings and auto injury.

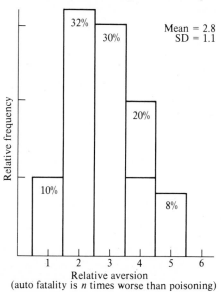

Histogram of relative aversion of an automobile fatality compared to a fatal poisoning

Mean = 2.8
SD = 1.1

Relative frequency

Relative aversion
(auto fatality is *n* times worse than poisoning)

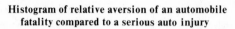

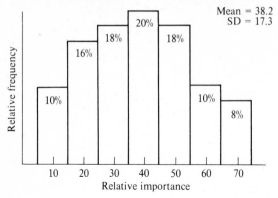

Histogram of relative aversion of an automobile fatality compared to a serious auto injury

On the basis of these data, analysts converted all benefits to equivalent highway fatalities avoided, as shown in the revised impacts table below. On this basis, if project lifetimes are assumed to be 10 years and the economic discount rate 10 percent, and if the value of human lives is constant over time, these data imply benefit-cost ratios of between 0.76 and 0.79 for the highway program and between 0.14 and 0.32 for the poison program, when the ratios are based on annual equivalents. The ranges are 1.24 to 1.28 and 0.23 to 0.52 for the respective programs when the ratios are based on the present equivalent. By either method, the highway safety program seems to be the better of the two. However, it is worth observing that for comparison with other kinds of programs the ratios of present equivalents place the highway program in the most favorable light; this anomaly is a result of discounting dollar expenditures but not lives.

	Highway safety program	Poison emergency unit
Initial capital	$10,000	$50,000
Annual flows:		
Equivalent highway fatalities avoided	5090–5250	2564–5882
Operating costs	$5000	$10,000

Not satisfied with the results obtained at that point, the analysts used automobile insurance data to estimate the value of a human life as implied by premiums paid. They reasoned that an aggregated demand curve, in terms of 1977 dollars, constructed from insurance premium data converted to standardized units of coverage, would provide a rough estimate of the average policyholder's view of the economic value of a life. However, generating the demand curve proved unreasonably complex; so they used the average premiums for bodily injury and major medical coverage paid in 1977 as a lower-bound estimate of the economic value ascribed by policyholders to injury and fatalities. Thus, they estimated annual equivalent values per serious physical injury and per traffic death as 35 and 40 dollars, respectively.

Using these rough estimates, the analysts then estimated the benefits of the alternative programs as:

	Highway safety program	Poison emergency unit
Annual benefits		
Fatalities avoided	$225,000	$115,380–$264,690
Serious injury avoided	175,000	
Total	$400,000	$115,380–$264,690

The respective benefit-cost ratios and NAEs were

	NAE	Standard B/C
Highway safety program	$393,372	60.0
Poison emergency unit	$97,242–$246,552	6.4–14.6

suggesting that both programs were desirable; but within the range of uncertainty considered, the highway safety program appeared to offer greater public welfare dividends.

The approach taken in Example 12.2 suggests a generalized approach to making the multifaceted benefits of one alternative comparable to those of its competitors. The idea was to devise relative weights for all measurable benefits of both alternatives and combine them into a single measurable index of the same numerical units for both. Explicitly, an index of equivalent auto deaths, call it I, was developed where

Highway safety project:

$$I = \text{deaths avoided} + \left(\frac{1}{38.2 \pm 17.3}\right) \text{injuries avoided}$$

Poison emergency unit:

$$I = \left(\frac{1}{2.8 \pm 1.1}\right) \text{fatal poisonings avoided}$$

In a more general case one might scale many benefits in this way to get an index

$$I = B_0 + W_1 B_1 + \cdots + W_n B_n$$

where B_0 is the basic unit of measure.

12.2 EXPLICIT ACCOUNTING

Multifaceted decision problems cannot always be reduced to single-faceted problems. Qualitative impacts are particularly difficult to treat. But even when all impacts can be quantified and substitution rates estimated, as in Example 12.2, important information can be obscured by numerical analysis. Consequently, an important aspect of benefit-cost studies is an explicit inventory of anticipated impacts for competing alternatives.

One format for displaying monetary and other measurable impacts is a simple matrix. In the matrix of Figure 12.1, competing alternatives are listed in the far left column and estimated aggregated impacts of specific kinds are associated with each. This is not a substitute for the kinds of analyses described previously; it is merely a supplement which, if accurately and tastefully presented, provides deeper insight regarding the nature and relative magnitude of costs and benefits.

Provided with this kind of information public managers can, in principle, at least, balance the merits of competing alternatives and make an enlightened selection. In effect this means that they roughly estimate, for example, the relative weights of lives saved versus books purchased, as described in Section 12.1. Although analysts may be incapable of developing explicit weighting schemes which adequately represent managerial preferences, good managers are skilled at factoring many variables into the decision process and the analyst has a responsibility to provide the necessary information in a useful form.

Qualitative impacts can sometimes be described in the same matrix with quantitative impacts by adding a third column. However, the matrix format offers only limited space for explanation, which encourages shallow treatment of

Quantitative annual impacts matrix

| Project or program | Monetary, $ | | | Other quantitative benefits | | | |
	Benefits	Capital costs	Operating costs	Lives saved	Accidents avoided	Books purchased	Acres preserved
A	10,000	100,000	1000	800	2000		
B	—	150,000	500	—	—	15,000	
C	20,000	250,000	900	500	—	—	2000

Figure 12.1

the qualitative impacts. It is often better to describe qualitative aspects in narratives which may also relate some of the quantitative impacts with a discussion of who gets (or pays) what in what proportions and in what time frame.

Keep in mind that an explicit accounting of impacts is intended to help clarify issues. Thus, accuracy and efficiency of presentation are essential. Analysts must studiously avoid drawn out, complex comparisons or discussions, since they will tend to confuse readers and may encourage them to ignore the analysis and make their selection without the benefit of the insight analysis can provide.

12.3 EQUITY

Public investments tend to redistribute public welfare. Thus, even if a project is desirable on the basis of expected net gain in public welfare, it could produce undesirable distributional effects and be judged undesirable on that basis. Highway and airport development projects are prime examples. The truth is, equity is a decision criterion, and benefit-cost studies which do not clarify distributional impacts are incomplete.

What is equitable is a matter of judgment. There can be no universal decision rule, such as NPE, which will provide a clear indication of which alternative is best. On the other hand, all the analytical techniques of benefit-cost analysis demonstrated in previous chapters can be brought to bear to estimate who will win and who will lose how much and what. Knowing the potential distributional impacts makes it possible to identify alternatives which meet economic or other efficiency criteria and also satisfy some subjective standard of equity.

Example 12.3: Equity in NPE distribution Two projects have been proposed. Both projects meet the economic acceptability criterion of positive NPEs when a social discount rate of 10 percent is assumed. All benefits and costs are in monetary units. Distributional data are tabulated as follows:

Project	NPE, $	Distribution of NPE to	
		Group 1, %	Group 2, %
A	100,000	70	30
B	80,000	30	70

Without knowledge of the distributional impacts, project A would appear best on the basis of the NPE criterion. However, there could be mitigating circumstances making the distributional aspects of project B more desirable; then one might favor project B.

When benefits (or costs) are multifaceted, distributional impacts are more difficult to assess. Even if the distribution of all quantitative impacts can be anticipated, public managers and analysts alike have no rationale for assigning relative preference weights to different kinds of benefits for impacted groups. Options open to managers include guessing group preferences, laying the issues open to public discussion, or ignoring the difficulty and acting on their own preferences.

Example 12.4: Multifaceted benefits Two routes have been proposed for a new highway. Route A is through a nice residential area along an unused railroad right-of-way. It would offer shorter driving times and greater safety to drivers than the alternative, route B. However, it would seriously affect the privacy of residents whose yards abut the old railroad right-of-way, because the road would be so wide that a buffer line of trees between the alley and the road would be impossible. Moreover, highway noise would be a serious problem.

Route B would be more circuitous but would cost less to build and maintain. Unfortunately, it would merge for a distance with an existing highway along a particularly busy segment, so the accident rate could be quite high.

Data below were tabulated for analysis.

	Distribution	
	To drivers	To residents
Annualized impacts	Route A over route B	Route A over route B
Marginal benefits		
Fuel savings ($)	50,000	—
Lives saved (adults)	15 ± 5	—
Marginal disbenefits (negative impacts to public)		
Noise (average dBA now is 40)*	—	25
NO$_2$ Exposure (average now is 80 micrograms per meter cubed)†	—	100
Pedestrian lives lost (children)	—	5
Other aesthetics		Qualitative
Annualized costs		Route A over route B
Capital required		50,000
Operating costs		200,000

* Average decibel levels above 50 dB on the A scale are considered harmful; 65 dBA is serious.

† Average annual nitrous oxide concentrations above 100 micrograms per cubic meter are considered harmful; 180 micrograms per cubic meter is serious.

The capital differential is offset by fuel savings. Here are the tough questions. Net lives saved by route A are estimated at 10 ± 5. However, is a child's life worth more or less than the life of a driver or adult passenger? If the net monetary value of a life saved were greater than 13,500 dollars, it would give the edge to project A; but how can one place a value on a child's life which has credibility with the parents and other residents?

Now what about the 50 noise pollution units and the 80 nitrous oxide pollution units? How can they be weighed into the analysis? Are they worth 1 life or 10 lives, and are those adult or juvenile lives? Also, what about the quality of environment the residents will lose when the wide strip of grass and shrubs is converted into concrete?

Clearly, the drivers and residents have different preferences for all these impacts. Perhaps one could find appropriate weights for all impacts for both groups and could create an index which characterized the value of each alternative for each group. Then one could select the best alternative on that basis. Perhaps one could estimate how much drivers would be willing to pay for the fuel savings and lives saved and the amount residents would accept in compensation for the disbenefits they would incur from project A. Then, if willingness to pay exceeded required compensation, project A would be optimal. But do either of these approaches adequately compensate for the equity problems imposed? That is a question well worth pondering.

The discussions and examples in this section have barely scratched the surface of the question of equity. This difficult issue cannot be resolved by analytical methodologies. However, well-executed analyses which carefully document distributional impacts implied by alternative courses of action provide the foundation for considerations of equity. In fact, informed judgment depends on the facts and insight produced by systematic analysis of quantitative data and qualitative issues.

12.4 PROBLEMS

12.1 Prepare a critical analysis of Example 12.2. Discuss the economics, the methods by which substitution rates were determined, the treatment of uncertainty, and the conclusions. Explain how you would have liked to have seen the analysis done and what the advantages would be.

12.2 Devise an analysis plan for converting all benefits (and disbenefits) into monetary units. Analyze the data in Example 12.4 after converting each to a monetary equivalent according to your plan. Which option looks best? Qualify your analysis and recommendation. Pretend that you have been commissioned by the county to complete the study.

12.3 The city has a single piece of ground suitable for two competing projects. A new general hospital is much needed and has been proposed for the site. The hospital would serve primarily underprivileged people. The competing proposal is a solar energy research facility which the federal government would like to build. Data for the two projects are given as below. The social discount rate is 10 percent.

Hospital

Costs (thousands of dollars)
  Capital: 50% city, 50% federal

Capitalized initial expenditures (contracts, etc.), $	500
Annual equivalent investment (40 yrs), $	1000
City operating costs	
(depreciation, maintainence, other), $	8000
Land value, $	5000

Benefits (annualized)

New jobs	70
Health care quality	Qualitative
Image of the city	Qualitative

Solar research facility

Costs (thousands of dollars)
  Capital: 80% federal

Capitalized initial expenditures, $	200
Annual equivalent investment (40 yrs), $	1,800
Federal operating costs, $	1,000
Land value, $	10,000

Benefits (annualized)

New jobs	200
Research results	Qualitative
Image of the city	Qualitative
Nonsalary expenditures in the city	(20% of operating costs)

Analyze these data as thoroughly as possible from an economic perspective. Find a way to quantify the qualitative aspects in monetary terms, if possible. Test the sensitivity of your economic analysis to the values placed on qualitative factors, to uncertainty in cost and numerical benefit estimates, and to the discount rate used. What other possible impacts should have been included in the analysis, and how?

12.4 Describe a complex public investment problem with which you are familiar. Identify the economic and other measurable impacts and assign reasonable numerical values. Describe the qualitative aspects of the problem, including equity issues and uncertainties. Do as thorough a job as you can, because someone else will have to analyze the problem and formulate a recommendation for action.

12.5 City council members are discussing the pros and cons of a policy which would require all city contracts and purchases to be awarded to the low bidder. One council member states that there must be some evaluation of the bidders themselves, with a preference given to those who have provided good work in the past and who have "stood behind" their work and products. Another member states that any departure from the objective low-bidder policy introduces politics into the selection and ultimately wastes public money. What policy do you recommend, and why?

12.6 An investor is deciding whether to buy bonds or stocks. He estimates that the probabilities of inflation and recession are, respectively, 0.7 and 0.3. For his anticipated investment of 10,000 dollars, he believes his choice of stocks would gain 10 percent per year during a period of inflation and would lose 5 percent of their value in a recession. His selection of bonds would show a gain of 4 percent regardless of inflation or recession.

  (a) Which alternative would the investor choose using the expected value criterion?

  (b) At what probability of recession would he be indifferent to the two alternatives?

12.5 SUPPLEMENTAL READING

Arrow, K. J., and A. C. Fisher: "Environmental Preservation, Uncertainty, and Irreversibility," *Quarterly Journal of Economics*, vol. 88, 1974, pp. 312–319.

Darcy, Douglas C., R. E. Kuenne, and M. C. Mcguire: "Approaches to the Treatment of Incommensurables in Cost-Benefit Analysis," Paper P-966, Institute for Defense Analysis, Program Analysis Division, Contract NSF-C-764, July 1973.

Lind, Robert C.: "The Analysis of Benefit-Risk Relationships: Unresolved Issues and Areas for Future Research," in *Perspectives on Benefit-Risks Decision Making*, National Academy of Engineering, Washington, D.C., 1972.

Merewitz, Leonard, and Stephen H. Sosnick: *The Budget's New Clothes*, Markham, Rand McNally, Chicago, 1971, chaps. 2 and 6.

Mishan, E. J.: "Evaluation of Life and Limb: A Theoretical Approach," *Journal of Political Economy*, July–August 1971, pp. 687–705.

Okun, Arthur M.: *Equality and Efficiency: The Big Tradeoff*, paper, The Brookings Institution, Washington, D.C., 1975.

Pauker, Stephen G., and Jerome P. Kassirer: "Therapeutic Decision Making: A Cost-Benefit Analysis," *New England Journal of Medicine*, vol. 293, no. 5, July 31, 1975.

THIRTEEN

APPLICATIONS OF BENEFIT-COST ANALYSIS AND A CASE FOR ANALYSIS

Two benefit-cost studies are described in this chapter. Both represent efforts to evaluate proposed large expenditures of public funds over a long period of time. Your case for analysis pertains to the polio vaccine development program.

A *cost-effectiveness study* of the incipient Uniformed Services University of the Health Sciences, versus a scholarship program, to obtain physicians for the military services is described in Section 13.1. This study was done by the General Accounting Office (GAO), which is part of the legislative branch of the United States government. It was initiated because two previous studies, including one by the Department of Defense (DOD), had drawn conflicting conclusions using different measures of effectiveness. The GAO recommended against the university program; but its conclusions are suspect, since it assumed the two programs to be mutually exclusive approaches to procuring physicians, which they were not. Also, it did not account for the different goals of the programs and for the unique benefits to be derived from each. Congress voted in May 1977 to continue funding for the university, perhaps accepting the argument advanced by the Department that the two programs were not competitors.

A *benefit-cost study* of a proposed satellite system which would survey and monitor natural resources of the earth is the subject of Section 13.2. This study, one of three major studies within a 5-year period, was commissioned by the U.S. Department of the Interior, at the request of the Office of Management and Budget (OMB), Executive Office of the President. It was conducted by the Earth Satellite Corporation (Earth Sat.) and Booze, Allen and Hamilton over a 3-year period. The other two were commissioned by the National Aeronautics and Space Administration (NASA) and were conducted by ECON Corporation.

Whatever its faults, this is a fine example of a major effort to evaluate the benefits and costs of a major public program in the face of political infighting.

The three studies produced conflicting findings. The OMB used the relatively negative findings of the Earth Sat. study as evidence to support its argument that the program was unjustifiable. The second ECON study was used as evidence by NASA and congressional supporters that the program was a wise investment of public funds. To help resolve conflicting evidence, the National Science Foundation (NSF), at the request of the OMB, commissioned a comparative analysis and critique of the 1975 ECON study, which contained the encouraging evidence, and the more negative Earth Sat. study. The principal investigators were Daryll E. Ray and Kendell Keith of Oklahoma State University. They concluded that the benefits projected for the agricultural sector were seriously overestimated in both studies. This was a serious charge, since the justification for the program depended strongly on agriculture. Although conflicting implications of the various studies remain unresolved, legislation to operationalize the system was reintroduced in Congress in 1977.

13.1 UNIFORMED SERVICES UNIVERSITY OF THE HEALTH SCIENCES*

In 1972 the Uniformed Services Health Professions Revitalization Act (PL92-426) authorized two programs by which the Department of Defense could procure physicians. The Department could provide up to 5000 scholarships to medical, dental, and other health trainees at accredited schools; as of 1975, 4730 scholarships had been granted. Recipients would be trained in civilian schools and would ordinarily be required to serve a minimum of one year in the military for each year of training. In some cases the service requirement could be met in civilian institutions. This program was intended to meet the short-term needs of the uniformed services for medical personnel.

The Department was also authorized to create and operate the degree-granting Uniformed Services University of the Health Sciences, consisting of medical, dental, veterinary, and allied health professions schools, and to construct the facilities to house the university. The university was to be within 25 miles of the District of Columbia. It was required to graduate at least 100 physicians annually starting no later than 1982. The faculty were to be military and civilian professors. Students were to be commissioned as 2d lieutenants and would be required to serve the military or other federal institutions for a minimum of 7 years after graduation. The university was intended to meet the long-term workforce needs of the uniformed services.

The idea of a university was first proposed in 1947 and was considered a number of times before becoming a reality. By 1975 Congress had appropriated

---

* Tables and graphs in this section are based on material presented in the studies described. References are provided in Section 13.4.

Table 13.1 Assumptions made by the Department and the Commission in their studies of the university alternative

Department of Defense	Defense Manpower Commission
Base year for cost estimates was 1981	Base year for cost estimate was 1980. Used old DOD forecasts
Productive staff years of service was the basic unit of measure	Cost per graduate was the basic unit of measure
University and scholarship graduates differ significantly in ability to meet DOD needs	Physicians from both programs would be essentially carbon copies
Expanded scholarship program would cut into civilian supply of physicians; university program would bolster it slightly	Ignored this question
Included non-Defense federal subsidies as part of scholarship program costs	Not included
New construction costs ignored	All new construction costs were counted, even previous costs

almost 80 million dollars for the construction of the medical school. More than 46 million dollars in contracts had been let by March 1976. A board of regents was appointed in May 1973 and medical faculty have since been appointed. The first class of medical students enrolled in 1975.

During congressional authorization hearings in 1975, the Defense Manpower Commission, a congressional commission, prepared a report in which it concluded that the university was an unjustifiably expensive method of procuring medical personnel for the military. In June 1975 university officials presented cost figures to the Subcommittee on Military Construction, Senate Committee on Appropriations, indicating that the university would be cost-effective. The two studies reached contrary conclusions because they were based on different assumptions, as shown in Table 13.1. In July 1975, the Surveys and Investigations Staff of the House Appropriations Committee prepared a report, based on the assumptions of the Defense Manpower Commission study, in which it concluded that the university would be too costly.

In November 1975 the GAO was asked to reexamine the question of cost-effectiveness by Senators Wendell Ford and William Proxmire.

Basic Assumptions in the GAO Study

The GAO compared the scholarship program with the Uniformed Services Medical School on physician procurement only. It assumed that the physicians trained via either program would be technically equivalent. The GAO further assumed that the only differences in the two programs would be in their costs; thus, it chose to do a cost-effectiveness analysis. This is a major weakness in the study, which the Department of Defense pointed out in its critique. The GAO should have explicitly investigated the potential uniqueness of the university concept and included benefits estimates in its analysis.* Furthermore, the two programs were intended to be complementary, not mutually exclusive. They had different purposes and could not be compared as competitors.

Because graduates of the university would be required to serve longer and, additionally, would be more likely to make military service a career, the GAO decided to use expected staff-years of service as the basic unit of analysis. Analysts used the ratio of annual equivalent cost to expected staff-years of service as the measure of effectiveness. Using 1984 (when the university was expected to be in full operation) as the standard year, they estimated the cost per expected staff-year of service (including salary and retirement costs) at about 21,500 dollars for the scholarship program and 26,200 dollars for the university program. They concluded that the university program was not cost-effective when, in fact, they could only conclude that physicians trained in the university program might cost about 400 dollars more per staff-year of service than physicians procured through the scholarship program. Whether the difference was a justifiable increment, considering the purpose of the university program and the quality of physician produced, is a question beyond the reach of their analysis.

The GAO included the following costs in its analysis:

• Capital costs not incurred as of April 1, 1976. Prior costs were appropriately considered sunk costs.
• Initial capitalized costs attributable directly to the implementation of the programs.

* Special skills training, research, patient care while in training, health care innovations, and other areas of potential uniqueness should have been considered.

Table 13.2 Scholarships program costs
988 expected graduates per year

	Annual costs, $	Cost per staff-year of expected service. $*
Educational costs		
Stipends	22,083,776	2,651
Tuition and fees	11,856,000	1,423
DOD administration	2,402,816	288
Military compensation		
Salary and bonuses	135,334,815	16,242
Retirement†	7,001,605	840
Total	178,676,197	21,444

* Assuming 8332 staff-years of service for the 988 graduates, over a 30-year period, based on DOD data.
† Estimated for the twentieth year of full operation, 2004.

Table 13.3 University program costs
175 expected graduates per year

	Annual costs, $	Cost per staff-year of expected service, $*
Educational costs		
Stipends	7,770,000	2,391
Operating costs	21,551,692	6,633
Amortization of unincurred investment costs†	3,924,786	1,208
Military compensation		
Salaries and bonuses	46,670,900	14,530
Retirement‡	4,732,660	1,474
Total	84,650,038	26,236

* Assuming 173 graduates serve in the military and 2 serve in the Public Health Service.

† Interest costs approximately 7.87 dollars for the 48.7 million dollars not expended but authorized for construction. This is probably not a legitimate cost, since the money is not really idle.

‡ Assuming 3212 staff-years of service for 173 graduates, over 30 years.

• Initial costs involved in transferring faculty from military operating units. The GAO considered these as opportunity costs *but* ignored the offsetting benefits these physicians would provide as attending physicians in their new positions, in addition to teaching.

• Program operating costs projected for the year 1984, when the university was expected to be fully operational, included: students' stipends, salaries, and fringe benefits; tuition; recruiting, advertising and other administrative costs; facilities operating costs; bonus payments for staying in the service; and military salary and retirement costs for retained physicians.

Estimated program costs are summarized in Tables 13.2 and 13.3. All costs have been discounted to 1977.

Sensitivity Analysis

The GAO conducted a limited sensitivity analysis on certain parameters of the study. For the scholarship program the parameters were program costs, student stipend costs, and retention rates. For the university program the parameters were operating budget, class size, and retention rates.

Increasing estimated scholarship program costs (tuition and fees) from 3608 dollars, which is quite small, to 8608 dollars, which is still quite small, would increase total cost per staff-year of expected service to 23,815 dollars. Increasing student stipends from 400 to 800 dollars per month, which is not unreasonable,

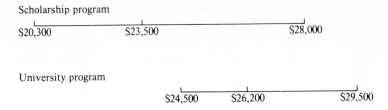

Figure 13.1 Total cost per staff-year of expected service: sensitivity ranges.

would increase total costs by about another 2000 dollars. Decreasing the retention rate (20 years of service) from 8.7 to 1.0 percent for graduates taking civilian residencies would add approximately 1000 dollars to total costs, while decreasing retention rates for those who would take military residencies would add another 1500 dollars. The total increment could easily bring total costs per staff-year of expected services to more than 28,000 dollars. The lower extreme was estimated at about 20,300 dollars.

Varying the retention rate for university graduates from 75 to 90 percent would reduce total costs per staff-year of expected service by about 1000 dollars. Increasing enrollment to 800 instead of 700 students would also reduce costs by about 1000 dollars. The total could bring costs per staff-year of expected service down to about 24,500 dollars. The upper extreme estimated by the GAO was about 29,500 dollars. Figure 13.1 summarizes the extremes. No likelihood estimates were made, so it is not clear what the extremes mean.

The immediate question is whether cost estimates for the two programs are reasonable and whether the sensitivity analyses adequately account for these uncertainties. Perhaps the two programs are really equivalent in costs. But even if GAO estimates are reasonable, the fundamental question remains: Are the higher projected costs per staff-year of expected service justified by the additional benefits which might accrue?

13.2 EARTH RESOURCES SURVEY SYSTEM (ERS)*

In 1973 the U.S. Department of the Interior, Geological Survey engaged the Earth Satellite Corporation (Earth Sat.) to develop information about the economic, social, and environmental benefits of an ERS satellite system, and to relate those benefits to the costs of the system. The system's operational period was specified as 1977 through 1986. Three system alternatives were studied:

---

* All the material in this section is extracted from the six-volume final report prepared by Earth Satellite Corporation for the U.S. Department of the Interior, *Earth Resources Survey Benefit-Cost Study*, Executive Summary, prepared by the Earth Satellite Corporation and the Booz-Allen Applied Research Corporation, November 1974.

- One satellite imaging the entire earth every 18 days in a sun synchronous orbit about 500 miles above the earth, and having the same technical capabilities as the orbiting ERTS-1 satellite (now called Landsat-1)
- Two such satellites in orbit simultaneously to provide coverage every nine days
- A fleet of high-altitude U-2 aircraft which would image the United States only from about 85,000 feet with approximately the same quality of images as produced by the satellite systems

The main objective was to estimate the relative economic benefits of the three systems in terms of improved management of the earth's natural resources. This improvement would result from the availability of electronically generated images of the surface of the earth.

Earth Satellite Corporation was constrained by the government in many ways which probably had a profound influence on the findings. Alternative system configurations were severely limited in that the unique technological capabilities of the aircraft and satellite systems and considerations of improved imaging technology during the 10-year system operating period could not be factored in. These constraints probably biased the study toward the aircraft system, on the one hand, since more detailed information can be obtained using aircraft sensing devices, and against satellites, on the other, since the then-deployed satellite sensor complements were of relatively low resolution. Moreover, cost estimates for the system alternatives were made by NASA, which, of course, had a vested interest in the satellite systems. This may have biased the study toward satellites. How these competing biases balanced out is impossible to estimate. Finally, the entire study was conducted in close liaison with a government support team and the Interagency Coordinating Committee for Earth Resources Survey Program. Even the study design was developed in cooperation with the government. Although these groups may have been biased one way or the other, they most likely encouraged accurate, realistic analysis by the contractor by requiring that all components of the study be opened to scrutiny at biweekly meetings as they developed.

Benefit Estimation

More than 10 previous benefit-cost studies based on highly speculative estimates of the value of satellite data were reevaluated initially. From this effort it became clear that there were many possible applications of ERS data across the entire spectrum of natural resources management, but most of the economic benefits would likely accrue in relatively few areas. This conclusion is demonstrated graphically in Figures 13.2 and 13.3.

Benefit estimation efforts were concentrated in those applications areas for which:

- There appeared to be the greatest potential for benefit generation.

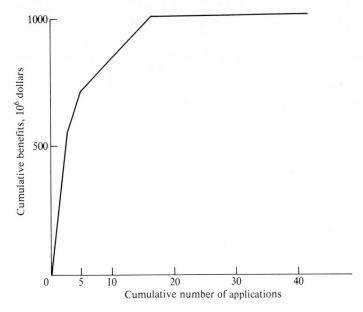

Figure 13.2 Rank order cumulation of benefits based on previous studies.

- Applicability of ERS data had been demonstrated by principal investigators on the ERTS-1 program.*
- There was sufficient information available for estimating efficiency benefits for representative applications and for extrapolation of these estimates to a broad range of similar applications with reasonable certainty.

* ERTS-1 was the first of the Landsat-type (Earth Resources Technology) satellites flown by NASA. It was launched in the early 1970s. NASA funded many research projects aimed at demonstrating the usefulness of the image data being transmitted by ERTS-1.

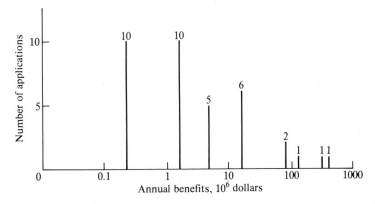

Figure 13.3 Distribution of applications by magnitude of benefits based on previous studies.

Broad application areas were selected after an extensive review of previous benefit-cost studies, tabulation of receipts from national accounts for private sector activities which might benefit, tabulation of federal budget estimates and state and local expenditures for potential application areas, a review of ERTS-1 study reports, and discussions with ERTS-1 principal investigators and government representatives of the study advisory groups. The areas were

- Agricultural production
- Water resources management
- Rangeland management
- Forestry management
- Land-use planning and management
- Environmental management
- Geologic and mineral resource management
- Marine resources management
- Disaster warning and relief
- Noneconomic and international impacts

Benefits were estimated for many activities within each application area and tallied. In forestry, agriculture, and a few other areas detailed case studies were completed, because benefits anticipated in those areas were particularly large and the quality and scope of available data were such that an especially intensive effort to estimate benefits promised to pay off.

All benefits were estimated first without regard for the influence of clouds between the satellite or aircraft imaging devices and the earth. Since the aircraft would be flown only on cloud-free days, these estimates were appropriate for that system alternative. However, experience had shown that images of the earth obscured more than 5 percent by cloud formations are practically useless, so expected benefits for the satellite systems had to be calculated using a probability model for the likelihood that an image would contain less than 5 percent cloud cover.

Economic efficiency benefits estimations were guided by the economic theory of supply and demand for management information and by the welfare economic concept of consumer surplus. The basic assumption was that the new technology would change the supply of useful data in two ways: data equivalent to those acquired by other means, such as ground surveys, would be available at lower unit cost, and entirely new and useful kinds of data would be produced. The shaded area in Figure 13.4 depicts the increase in social welfare from both impacts. Unfortunately, sufficient economic data to generate demand and supply curves were unavailable, so ad hoc projection methods to find the points $P_1 Q_1$ and $P_2 Q_2$ were necessary. Through these points demand curves could be roughed in and estimates of benefits calculated.

The value of information is a function of how it will be used; it has no implicit value. The link between the information made available and the tangible

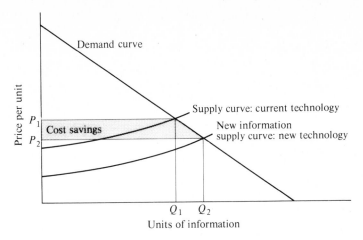

Figure 13.4 Efficiency benefits estimation.

benefits produced was conceptualized in terms of bayesian decision analysis.*
That is, it was assumed that managers could subjectively evaluate the relative
expected benefits of new and better information supplied by the specified new
technological system, in that they could weigh the utility of the information
against the inherent uncertainties and express a willingness to pay for the new
information. There were two inherent difficulties with this approach. One was
extrapolating benefits assessed in cooperation with particular managers in particular settings at a particular point in time to aggregated benefits for a broad
spectrum of situations. The second difficulty was accounting for learning and
diffusion rates among near-term and future users. Extrapolations of benefits were
done by scaling and in some cases by simulation modeling (case studies). Learning and knowledge diffusion rates were never explicitly accounted for, because
there was no model to project the change in demand for the new information as
dependency on it increased.

Cloud-Cover Analysis

The methodology employed to account for cloud cover and its impact on the
value of the ERS data was quite complex. Instead of broadly describing it, its
specific application to forest management is discussed below.

Forest Management Case Study

After reviewing all activities of the timber industry and interviewing some 200
people, including ERTS-1 principal investigators, the contractor focused this

---

* Bayes' rule was discussed in a previous chapter. Formal (bayesian) decision analysis is the topic
of Part Five of this book.

Region	Fraction per year	Completion period, yrs
South and Southeast	1/5	5
Oregon and Washington	1/7	7
Hawaii	1/15	15
Alaska	1/15	15
Others	1/10	10

Figure 13.5 Inventory rates by region.

case study on the timber inventory programs of the U.S. Forest Service (USFS) and of the Bureau of Land Management (BLM). This activity represented the major area of benefit attribution, and a great deal of high-quality data were available on the inventory process and the needs of the agencies.

Federal legislation required that all timber lands be inventoried periodically. Rates of inventory varied by region of the country as shown in Figure 13.5. Ordinarily these inventories were conducted by ground survey and low-altitude aircraft observation. Sampling sites were selected by statistical methods and detailed physical inventories of board feet of lumber were taken for those sites and extrapolated to whole forests on an equivalent site basis.

Benefits from the ERS systems proposed would result from reducing the cost of sampling and increasing the frequency with which survey data could be acquired. Principal investigators on the ERTS-1 program had demonstrated the effectiveness of a multistage sampling process which resulted in as good or better inventory estimates at much lower costs per acre over the standard procedure. Using satellite imagery taken in particular seasons of the year, they classified forests into timber volume groups, as, for example, 0 to 10,000 board feet, 10 to 100,000 board feet, and so on. Then they applied stratified random sampling techniques so that 50 percent of all samples were taken randomly from the class which composed 50 percent of the forest, 10 percent from the class which composed 10 percent of the forest, and so forth. Because each class was reasonably homogeneous, fewer samples were required to obtain estimates of a given accuracy than by purely random sampling over the entire forest. Thus, sampling costs could be reduced.

The same investigators found an even more efficient method. They used satellite images to classify forests into regions of significant change and regions of nominal change. By sampling only in regions of significant change, using the stratified sampling method previously described, they could estimate board feet in a forest by estimating the aggregated change in board feet since the previous inventory.

As it turned out, sufficient savings to acquire the workforce to complete regular periodic surveys on time, which the agencies were unable to do previously, and to reallocate funds to other project areas were anticipated. As a result of efficiencies in sampling (approximately 4 dollars per acre by the old method versus 2 dollars per acre by the new method), the forest management

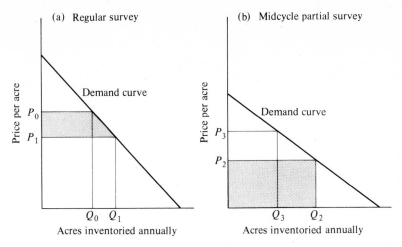

Figure 13.6 Efficiency benefits estimation—forest inventory.

agencies would have surplus funds to apply to interperiod inventory estimates in the same way that the Bureau of the Census develops intercensus population reports. The redistribution of anticipated savings was worked out with senior managers in the agencies, and, in the process, demand curves were estimated. From these, net increases in public welfare were calculated (see Figure 13.6).

Based on existing budgets $Q_0 P_0$ and the number of acres inventoried annually Q_0, the price per acre inventoried P_0 using the old sampling method was calculated. Managers were advised of the cost per acre P_1 projected for the regular inventory using the new procedure. They were also given the price P_2 for the midcycle survey using the new procedure. The managers reallocated the savings so as to complete the regular inventory of the legislatively required number of acres Q_1 and to carry out midcycle partial surveys of Q_2 acres. Remaining funds were allocated to other activities not shown here. The point $P_3 Q_3$ was obtained by quoting a different price of P_3 for the midcycle surveys and asking managers to reallocate the savings on that basis. Net benefits were then calculated as the sum of the shaded areas under the demand curves for the regular and midcycle surveys and for other activity areas to which savings were reallocated, on the assumption that the total budgets of the USFS and the BLM would remain constant.

To account for the randomness of cloud cover over regions being imaged by orbiting satellites, benefits were discounted by multiplying them by the probability that a composite image of an inventoried region, containing less than 5 percent cloud cover, could be produced in a given year. The model was based on actual ERTS data previously acquired. It accounted for the orbiting parameters of the satellite(s). The probability was estimated as:

$$P_r = \left[1 - \prod_{l=s}^{\omega} (1 - P_{r_l}) \right]^N$$

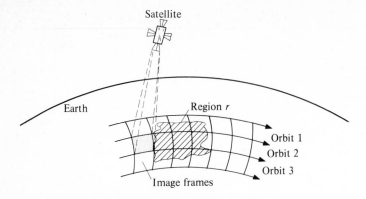

Figure 13.7 Imaging geometry.

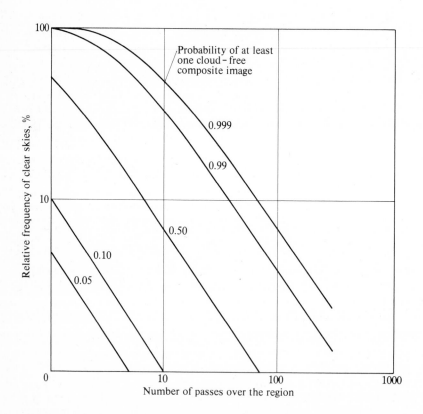

Figure 13.8 Nomogram for generalized probability of clear skies. (Produced by North American Rockwell from ERTS-1 Data Tapes.)

where N = Number of ERTS image frames required to cover the specified region

P_{r_l} = probability of getting at least one image with less than 5 percent cloud cover in region r in the lth season of the year

P_r = probability of obtaining a composite image of region r in that year, having less than 5 percent cloud cover

The symbol $\prod_{l=s}^{\omega}$ means to multiply what follows for all seasons. Reference to Figure 13.7 should help clarify the geometry of the situation.

Nomograms were constructed for each inventory region. Figure 13.8 exemplifies these. Notice that if the relative frequency of clear skies in the region were 20 percent, and the region were flown only once, there would be only a 50 percent chance of producing at least one composite image with less than 5 percent cloud cover, whereas it would have to be flown about 20 times to almost guarantee $(P_r = 0.999)$ one such composite image.

Benefits and Costs Summarized

Tables 13.4 through 13.11 summarize the efficiency (monetary) benefits estimated for each broad application area. They are given in millions of dollars per annum for all activities considered in each application area. Ranges are shown to indicate uncertainties in the estimates.

Table 13.4 Summary of benefits in agricultural production
Excluding cloud-cover impact

Application	Range of annualized benefits, $ (millions)
Crop acreage forecasting	0.70–1.40
Crop yield forecasting	0.30–0.80
Agricultural land stratification	0.10–0.50
Design and operation of irrigation projects	0.50–1.00
Pest susceptibility mapping	0.025–0.10
Flood damage assessment in agriculture	0.025–0.10

Table 13.5 Summary of benefits in water resources management
Excluding cloud-cover impact

Application	Range of annualized benefits, $ (millions)
Snow mapping and runoff engineering	9.9–27.7
Surface water extent	0.025–0.125
Location and extent of groundwater	0.100–0.500
Hydrologic response of watershed	0.025–0.100
Flood area assessment	0.125–0.600

Table 13.6 Summary of benefits in rangeland management
Excluding cloud-cover impact

Application	Range of annualized benefits, $ (millions)
Range inventory	0.5–3.5
Range monitoring	5.6–22.4
Range feed condition	2.9–4.7
Other	0.0–0.050

Table 13.7 Summary of benefits in forest management
Excluding cloud-cover impact

Application	Range of annualized benefits, $ (millions)
Forest inventory	0.8–2.7
Forest vegetation mapping	0.025–0.1
Forest fire management	0.0–0.025
Timber management	0.0–0.025
Forest pest detection and control	0.0–0.025

Table 13.8 Summary of benefits in land-use planning and management

Application	Range of annualized benefits, $ (millions)
State land-use planning	9.0–15.2
Site and route selection	0.1–0.5
Federal land planning	0.1–0.5
Cartographic mapping	0.5–1.0

Table 13.9 Summary of benefits in environmental management
Excluding cloud-cover impacts

Application	Range of annualized benefits, $ (millions)
Wetlands	0.050–0.125
Stripped lands	0.025–0.1
Water quality	0.025–0.1
Air quality	—

Table 13.10 Summary of benefits in geological exploration and mineral resources management
Excluding cloud-cover impact

Application	Range of annualized benefits, $ (millions)
Petroleum exploration	0.100–0.50
Mineral exploration	0.040–0.10

Table 13.11 Summary of benefits in marine resources management
Excluding cloud-cover impact

Application	Range of annualized benefits, $ (millions)
Living marine resources	0.0–0.025
Waterborne transportation	0.5–1.000
Ocean and coastal engineering	0.0–0.025

Table 13.12 Comparison of ERS costs and quantifiable benefits

Present value in millions of dollars at 10% discount rate

	High-altitude aircraft system*	One-satellite system	Two-satellite system
Quantifiable benefits	191–514	68–170	174–475
System costs	261–285	155–172	249–271
Quantifiable benefit-cost ratio†	0.7–2.0	0.4–1.1	0.6–1.9
Net quantifiable benefits†	−94–253	−104–15	−97–226

 * Benefits are unadjusted for cloud cover. Foreign and international benefits are excluded.
 † Ratios are computed at the highest and lowest ends of ranges.

Nonmonetary and international benefits were given very limited attention since the study was contractually focused on benefits to the United States and economic impacts were of primary concern. Even so, the contractor included in the final report a volume of about 100 pages devoted to exploring these areas.

Aggregated benefits and system costs discounted to the present (1974) were summarized as in Table 13.12.

Clearly, there is a great deal of uncertainty displayed in the presentation of the study findings. To ensure that the sponsors understood the limitations of the study, over and above estimation uncertainties, the contractors discussed the impacts of the constraints imposed by the government, by resource limitations, by the human capacity to project or imagine future events, and by the state of the art of the benefit-cost methodology on the conclusions drawn, as part of their Executive Summary, vol. 1.

13.3 POLIO VACCINE DEVELOPMENT: A CASE FOR ANALYSIS

The Secretary of HEW has begun a program of analyzing the social benefits and costs of previous medical research programs. His primary objective is to get a handle on what types of research, if funded by HEW, could be justified on a benefit-cost basis. Hopefully, retrospective analyses will provide some guidance for future funding decisions.

Assignment

As an analyst on the staff of the Secretary, you have been assigned the task of carrying out a retrospective benefit-cost analysis of the development of the poliomyelitis vaccines. The basic data for your study are included below. Treat them as your summarization of an enormous volume of raw data. You need not go beyond these data, but you may, if you wish.

You have 2 weeks to carry out your analysis and prepare a technical report

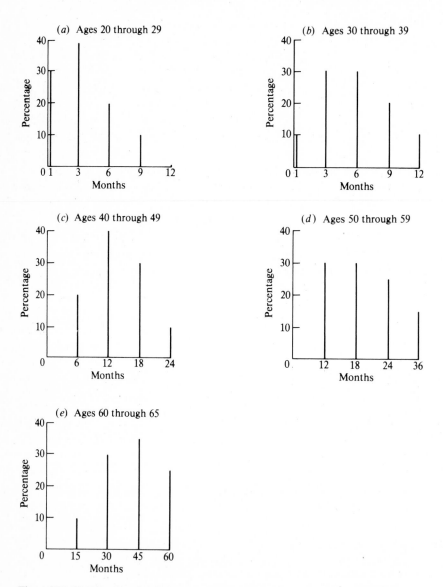

Figure 13.9 Months of work lost by age at onset.

and executive summary. Begin your report by describing the problem and how you would have liked to approach it. Describe the issues which should be considered. Describe how in fact you will proceed and rationalize your limited approach in terms of data constraints and other factors, but do not use these as excuses for not accomplishing your assignment. You must include conclusions and recommendations for the Secretary. Be factual, objective, and professional throughout.

Basic Data

Government and foundation expenditures on research for the years 1930 through 1962 are summarized in Table 13.13. Expected future annual earnings lost by age at death are given in Table 13.14. Included in these data are estimates for the value of household services of homemakers. Distributions for months of economically productive work lost due to morbidity among polio victims are given in Figure 13.9. Treatment costs are shown in Tables 13.15 and 13.16. Economic losses due to morbidity aggregate the costs for permanently paralyzed and nonparalyzed cases.

Incidence rates among age groups have remained relatively constant. About 80 percent of all new cases per year have been among persons younger than 29 years of age; 80 percent of those have been among persons between 5 and 20

Table 13.13 Estimated research awards for poliomyelitis research

Year	1970 dollars*	Year	1970 dollars
1930†	70	1944	70
1931	140	1945	70
1932	210	1946‡	242
1933	210	1947	492
1934	210	1948	746
1935	210	1949	1513
1936	210	1950	1729
1937	210	1951	2609
1938	210	1952	2744
1939	210	1953	2022
1940	210	1954	1920
1941	70	1955	2176
1942	70	1956	1962
1943	70		

   * Thousands of dollars, representing awards for research.
   † 1930–1945 rough estimates come from reviews of past funding programs.
   ‡ 1946–1956 figures are based on Science Information Exchange data.

Table 13.14 Economic losses due to death

Age at death	Expected annual earnings lost* in 1970 dollars	
	Male	Female
5–19	6,000	3,000
20–29	6,500	3,200
30–39	8,000	3,000
40–59	10,000	4,000
60–64	11,000	4,000
65–	2,000	3,000

* Earnings lost were supplemented by the economic value of household services.

Table 13.15 Treatment cost per month of morbidity

Age	Cost distribution (1970 dollars)	
Less than 40 yrs	1930	50 ± 15
40 and older		50 ± 15
Less than 40 yrs	1940	100 ± 50
40 and older		150 ± 60
Less than 40 yrs	1950	130 ± 80
40 and older		170 ± 40
Less than 40 yrs	1957	500 ± 200
40 and older		600 ± 200

Table 13.16 Treatment cost distributions per death

Year	Cost (1970 dollars)	Year	Cost (1970 dollars)
1930	30% @ $100	1950	20% @ $1000
	50% @ $200		40% @ $1500
	10% @ $300		30% @ $2000
	10% @ $500		10% @ $3000
1940	40% @ $250	1956	30% @ $1000
	50% @ $500		50% @ $2000
	10% @ $600		10% @ $4000
			10% @ $5000

Table 13.17 Incidence rates

Year	Cases	Year	Cases
1930	9,000	1944	20,000
1931	18,000	1945	14,000
1932	3,000	1946	20,000
1933	4,000	1947	13,000
1934	8,000	1948	25,000
1935	11,000	1949	42,000
1936	5,000	1950	35,000
1937	10,000	1951	30,000
1938	1,000	1952	59,000
1939	8,000	1953	36,000
1940	10,000	1954	39,000
1941	9,000	1955	30,000
1942	4,000	1956	15,000
1943	12,000		

Source: Paul Meier, "Safety Testing of Poliomyelitis Vaccine," *Science*, vol. 125, no. 3257, 1957, pp. 1067–1071

Table 13.18 U.S. population age distribution
Average 1930–1976

Age, yrs	Population, %
0–9	14
10–19	18
20–29	16
30–39	12
40–49	12
50–59	10
60–69	9
above 69	9

Table 13.19 U.S. population

Year	Millions
1930	123
1940	132
1950	151
1960	179
1970	203

years of age. The remaining 20 percent have been among persons older than 29 years; 80 percent of these were between 30 and 50 years. Statistical studies indicate that incidence rates among vaccinated persons were about 50 percent below those for unvaccinated persons.* Incidence rates are given in Table 13.17.

Population data are summarized in Tables 13.18 and 13.19. The age distribution has been fairly stable since 1930.

Some of the data given are fictitious. Treat them as though they are real.

13.4 SUPPLEMENTAL READING

Clark, Elizabeth M., and Andrew J. Van Horn: *Risk-Benefit Analysis Public Policy: A Bibliography*, informal report, Brookhaven National Laboratory, Upton, New York, Energy and Environmental Policy Center, Harvard University, November, 1976.

A Cost-Effectiveness Analysis of Selected National Family Planning Programs, Department of Economics, Pennsylvania State University, December 1969.

"Cost-Effectiveness Analysis of Two Military Physician Procurement Programs: The Scholarship Program and the University Program," Report to the Congress by the Comptroller General of the United States, U.S. Government Printing Office, Washington, D.C., May 5, 1976.

Earth Resources Survey Benefit-Cost Study, vol. I, "Executive Summary," prepared by the Earth Satellite Corporation and the Booz-Allen Applied Research Corporation, November 1974.

—— Vol. II, "Summary of Benefit Evaluation."

—— Vol. III, "Systems Effectiveness Analysis of Alternative Systems."

—— Vol. IV, "Capabilities to Derive Information of Value with ERS Data."

—— Vol. V, "Approach and Methods of Analysis."

—— Vol. VI, "Analysis of Distributional, Environmental, Social and International Impacts."

Ferreira, Joseph, Jr.: "The Long-Term Effects of Merit-Rating Plans on Individual Motorists," *Operations Research*, vol. 22, no. 5, September–October 1974.

McKean, Roland N.: *Efficiency in Government through Systems Analysis*, John Wiley & Sons, Inc., New York, 1958.

Meier, Paul: "The Biggest Public Health Experiment Ever," in Judith Tanur, et al., *Statistics, A Guide to the Unknown*, Holden-Day, Inc., San Francisco, 1972.

—— "Safety Testing of Poliomyelitis Vaccine," *Science*, vol. 125, no. 3257, 1957, pp. 1067–1071.

Merewitz, Leonard: *On the Feasibility of Benefit-Cost Analysis Applied to Remote Sensing Projects*, Special Study no. 3, Department of Business Administration, USC Berkeley, 1973.

Ray, Daryll E., and Kendall Kieth: *Crop Forecasting Benefits of Landsat: Review of Three Recent Studies*, Department of Agricultural Economics, Oklahoma State University, 1978.

Williams, A. P., et al.: *Policy Analysis for Federal Biomedical Research*, Prepared for the President's Biomedical Research Panel, Rand Corp., R-1945-PBRP/RC, March 1976.

---

* See Paul Meier, "The Biggest Public Health Experiment Ever," in Judith Tanur, *Statistics, A Guide to the Unknown*, Holden-Day, Inc., Publisher, San Francisco, 1972.

FOUR

RESOURCE ALLOCATION WITH LINEAR PROGRAMMING

Linear programming is one of the most powerful methods of analysis ever developed by applied mathematicians and management scientists. It has found extensive application in both the public and private sectors. The types of problems which have been addressed by linear programming include minimum-cost transportation, refuse collection and service delivery networks, optimal response fire and police station distributions, minimum-cost nutritious diet formulations for institutions, maximum-profit production mixes, most-efficient airline scheduling, minimum-cost air and water pollution control policy studies, energy planning, defense management, workforce planning, and many, many more.

Because it is so widely exploited and so powerful a method, linear programming algorithms are available on almost all computer systems, making it possible for analysts to solve problems with hundreds of constraints and hundreds of decision variables. The analyst's job is to understand the basics of the method and its limitations, to recognize problems which can be solved on a computer or by hand, as the case may be, and above all, to understand the results of linear programming analyses and how they relate to policy studies. Thus, the goals of the first two chapters in this part are to develop an understanding of the linear programming concept, communicate the logic of the solution process, and develop skills in casting problems in the linear programming format. The third chapter provides insight into sensitivity analysis and the interpretation of certain elements of optimal solutions in the context of additional constraints, and resources which may be recognized in the policy process only after the study is far along.

FOURTEEN

ELEMENTS OF LINEAR PROGRAMMING

Problems in which the objective is to maximize return or minimize costs within a set of constraints on available resources can often be solved using linear programming (LP). If the problem can be written in mathematical form with the objective function and the constraints expressed as linear combinations of the variables over which decision makers have control, LP algorithms can be applied to determine the optimal solution, if one exists. The solution to a LP problem is a set of numbers which specify the levels at which decision makers must set the decision variables to obtain a maximum-gain (or a minimum-loss) solution.

The purpose of this chapter is to introduce the concepts and terminology of LP, develop a feeling for the applicability of the method, and explain a basic solution algorithm. In Section 14.1 the method is explained geometrically for elementary problems in order to develop one's intuition. The corner point theorem, which is the basis of all LP solution algorithms, is explained in Section 14.2. The simplex algorithm is developed in Sections 14.3 and 14.4.

14.1 INTUITIVE OR GRAPHIC SOLUTIONS

The idea of linear programming is best conveyed by illustration. The simplest illustration is of a situation where a decision maker can distribute her or his limited labor and materials resources between two competing uses. Because there are only two uses, the problem can be portrayed geometrically and the optimal allocation of resources identified immediately.

Suppose a for-profit health clinic serves two groups of clientele. The clinic is staffed by a receptionist–records clerk, who takes all histories, pulls files, records

Table 14.1 Resource constraints and profit margins for clinic

Activity	Time consumed per patient, hours/visit		Available, hours/day
	Group P	Group G	
History, record, etc.	0.10	0.10	8
Attending nurse	0.20	0.50	32
Attending physician	0.36	0.12	24
Profit, $/visit	6	4	

symptoms, takes temperatures, and collects fees; by four attending nurses; and by three physicians. One group of patients is covered by government insurance, group G, and the other is covered by private insurance, group P. Group P contributes more to profit per visit than does group G, and consumes more physician time and less nurse time per visit. A work-time study by a consulting firm might have generated the data in Table 14.1.

Suppose the clinic wanted to maximize profits without expanding its staff and decided to do so by proportioning the distribution of patients from the two groups. Assuming, for the sake of exposition, that scheduling was not a problem, the solution could be found by a simple LP analysis. Notice that if only group P patients were accepted, the total number of patients which could be served per day would be 24 divided by 0.36, or 67. Total profits would then be 6 times 67, or 402 dollars per day, and there would be idle time for both the records clerk and nurses. On the other hand if only group G patients were accepted, patient loads could not exceed 64, because only four nurses are available. Total profits in this case would be 4 times 64, or 256 dollars per day, and physicians would have idle time.

Neither of these options is optimal. If only 60 group P patients were to be scheduled, instead of 67, there would be sufficient time to treat 20 group G patients as well. The added profit would be 4 times 20 minus 7 times 6, or 38 dollars. In fact, this is the most profitable mix of patients; it is the optimal solution.

By using the variable x to symbolize the number of group P patients scheduled, and y for group G patients, the data in Table 14.1 can be written as a set of linear inequalities in x and y, as follows:

Maximize profit P, where

$$P = 6x + 4y \quad \text{objective function}$$

such that none of these constraints are violated:

$$0.1x + 0.1y \leq 8 \quad \text{records clerk constraint}$$

$$0.2x + 0.5y \leq 32 \quad \text{nurse constraint}$$

$$0.36x + 0.12y \leq 24 \quad \text{physician constraint}$$

$$x \text{ and } y \geq 0 \quad \text{nonnegativity constraints}$$

The variables x and y are called *decision variables* because they represent the factors the managers of the clinic intended to adjust. In this problem the numbers 8, 32, and 24 represent upper limits on available professional resources, thus the inequalities in the constraint equations. The coefficients in the constraint relations represent rates of consumption of resources, and those in the objective function represent the profit margins.

All values of x and y for which none of the constraints is violated are possible solutions to this problem; they are called *feasible solutions*. A feasible solution for which the objective function is largest (maximized) is the *optimal solution*.

In Figure 14.1, the constraint lines represent all pairs of feasible solutions for which the resources would be completely used up. The areas to the left of these equalities represent the other feasible solutions. The shaded region represents all pairs of x and y values which satisfy all constraints, and the darkened boundary represents the best of all feasible solutions. The optimal solution is the point on the boundary of the feasible region first touched by the objective function of slope $-\frac{3}{2}$ as it is moved in toward the origin. It is the point $x = 60$, $y = 20$, written (60, 20), as observed previously.*

Notice that the optimal solution would be approximately $x = 27$, $y = 53$, if the profit margins in the objective function were interchanged. If they were identical, no matter what their value, the optimal solution would be all points on

---

* The solution can be calculated by observing that the optimal value (x, y) is at the intersection of the physician and records clerk constraints; that is, for the optimal solution $x + y = 80$ and $3x + y = 200$ simultaneously. According to the first equation, $x = 80 - y$. It must have the same value in the second equation, so substitute it and find $y = 20$. Now find $x = 80 - 20 = 60$.

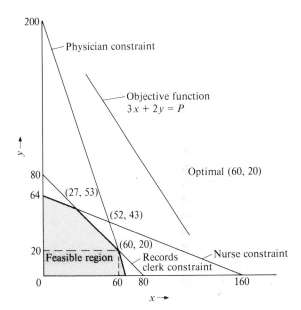

Figure 14.1 Feasible region and solution for the clinic problem.

the line of unit slope between (60, 20) and (27, 53). In other words, the optimal solution is stable as the (negative) slope (ratio of profit margins for group P to group G) becomes less steep, so long as it is greater than unity. Beyond that, a new point becomes optimal. Using this kind of sensitivity analysis, one can establish limits on the profit margins, within which each of the corners of the feasible region represents the optimal solution.

Suppose now that the clinic is operating at optimality, (60, 20), and the management is considering whether or not to hire additional staff. Assuming that the ratio of profit margins would remain the same, the solution would change as follows: Hiring only an additional records clerk would eliminate that constraint and the optimal solution would be at (52, 43). Hiring only an additional physician would eliminate that constraint, making the optimal solution (80, 0). Hiring only an additional nurse would eliminate that constraint, leaving the optimal solution unchanged at (60, 20).

In a problem as simple as the clinic problem used for illustration here, the solution can be identified rather easily and its sensitivity to variations in parameters checked just as easily. However, to solve complex problems containing many decision variables and many constraint equations one needs an efficient algorithm such as the simplex algorithm to perform these tasks. The logic and mechanics of the simplex algorithm are explained in the remaining sections of this chapter.

14.2 CORNER POINT THEOREM

The feasible region in the clinic problem (see Figure 14.1) had a very special shape. For any two points (x_1, y_1) and (x_2, y_2) in that region, every point on the straight line connecting them was also in the region. Sets of points which have this property are called *convex sets*. The idea is portrayed in Figure 14.2. It turns out that the constraint equations of LP problems define convex sets, and the

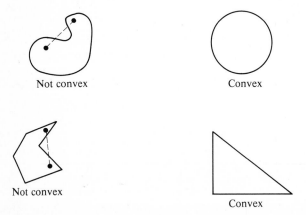

Not convex

Convex

Not convex

Convex   **Figure 14.2** Convex sets.

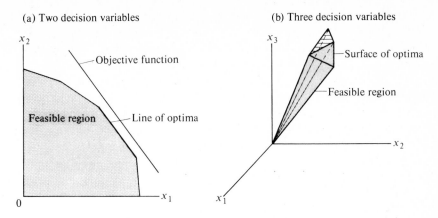

Figure 14.3 Multiple optima.

feasible region, which is actually the intersection of these convex sets, is necessarily a convex set. Mathematicians have proven that because of this characteristic of LP problems, optimal solutions, if they exist, must include at least one corner point of the convex feasible region; and if a solution is unique, it must be a corner point. Furthermore, if multiple corner points are optimal, they must be neighbors. These theoretical results are easy to visualize in two and three dimensions, as shown in Figure 14.3. These findings are formalized in the so-called *corner point theorem.*

The importance of the corner point theorem is that it proves that to find an optimal solution to a LP problem, if one exists, one need only find the coordinates of the corners of the convex feasible region and determine, by substitution into the objective function, which is optimal. Since there are only a finite number of constraints there can be only a finite number of corner points to test; thus, in principle an exhaustive check of all corner points is possible, but not necessary.

Example 14.1: Corner point theorem The five corners of the clinic problem were $(0, 0)$, $(64, 0)$, $(60, 20)$, $(26.7, 53.3)$, and $(0, 64)$. The objective function was $6x + 4y =$ profit. According to the corner point theorem, if there is an optimal solution, one or more of these corner points must be optimal. Substituting each corner point into the profit equation yields the following numbers:

Corner		Profit
$x = 0,$	$y = 0$	0.0
$x = 64,$	$y = 0$	384.0
$x = 60,$	$y = 20$	440.0
$x = 26.7,$	$y = 53.3$	373.4
$x = 0,$	$y = 64$	256.0

Clearly, the profit for the point (60, 20) is largest. There cannot be a point offering greater profit, for if there were, there would have to be at least one corner point offering the same profit. Consequently, the point (60, 20) is an optimum. Obviously, it is a unique optimum.

Unfortunately, there may be thousands of corner points in a problem, making it very difficult to determine them all and excessively time-consuming to test them in the objective function. The simplex method of searching the corners for an optimal solution was developed by George B. Dantzig in 1947 as an efficient alternative to an exhaustive search.

14.3 SIMPLEX METHOD

The idea behind the *simplex method* is that *not all corner points need to be identified and tested to find an optimal solution*. If one exists it will be found by selecting any corner point as a starting solution and then stepping to the neighboring point which offers the greatest improvement in the value of the objective function. In this way only neighboring points need be checked.

The idea is portrayed graphically in Figure 14.4. Starting at the origin the path to the optimal solution is through point 1. Points 3 and 4 never had to be considered in the process. Had point 3 been optimal, the path through point 2 would have been taken, since it necessarily has a value intermediate between those of points 1 and 3.

The trick is to find a convenient starting solution. The method for doing so is ingenious. It is founded on the theory of linear vector spaces, but operationally is quite elementary. All constraint inequalities are first converted to equalities by adding *slack variables* to those which are of the less-than-or-equal-to type and by adding *surplus variables* to those which are of the greater-than-or-equal-to type. Then to all constraints which were originally equalities and to those to which

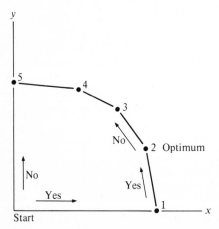

Figure 14.4 Two-dimensional protrayal of the simplex method. Assume that point 2 is optimal and that point 1 is better than point 5. Then the simplex method will move from point 0, to point 1, to point 2, and stop there.

(a) Original formulation

Maximize $P = 6x + 4y$

Subject to:
$$0.1x + 0.1y \leq 8$$
$$0.2x + 0.5y \leq 32$$
$$0.36x + 0.12y \leq 24$$
$$x, y \geq 0$$

(b) Including slack variables

Maximize $P = 6x + 4y + 0S_1 + 0S_2$

Subject to:
$$0.1x + 0.1y + S_1 \qquad\qquad = 8$$
$$0.2x + 0.5y \qquad + S_2 \qquad = 32$$
$$0.36x + 0.12y \qquad\qquad + S_3 = 24$$
$$x, y, S_1, S_2, S_3 \geq 0$$

Figure 14.5 Slack variables for the clinic problem.

surplus variables have been added, something called an *artificial variable* is added. Discussions of surplus and artificial variables are postponed to Chapter 18. The use of slack variables is illustrated in Figure 14.5, using the clinic problem of Section 14.1.

Table 14.2 Steps of the simplex algorithm

1. Add slack, surplus, and artificial variables as needed.

2. Start with the most convenient corner solution.

3. Using the objective function decide which variable to enter, that is, which corner to move to next.

4. Enter this variable at the level which provides the greatest improvement in the value of the objective function but which violates none of the constraint equations, including the non-negativity constraints.

5. Continue repeating steps 3 and 4 until no further improvements in the value of the objective function are possible. This is the optimal solution.

Notes: If at step 4 no end in improvements can be found (i.e., the variable is unconstrained) the problem is said to be *unbounded*.

If, at any step, there is ambiguity in what move to make next, the algorithm is said to be *degenerate*. Computerized versions of the algorithm have rules for overcoming degeneracy. Degeneracy also refers to optimal solutions in which one or more decision variables is zero: this situation is always preceded by the ambiguity situation.

If an artificial variable is in the solution at optimality, no feasible solution exists.

The formal algorithm of the simplex method contains five basic steps. These are listed in Table 14.2 and illustrated in Example 14.2.

Example 14.2: Logic of the simplex method demonstrated Consider again the clinic problem:

$$\text{Maximize} \quad P = 6x + 4y$$

$$\text{Subject to:} \quad 0.1x + 0.1y \leq 8 \tag{1}$$

$$0.2x + 0.5y \leq 32 \tag{2}$$

$$0.36x + 0.12y \leq 24 \tag{3}$$

$$x, y \geq 0$$

For convenience rewrite the constraints before adding slack variables by dividing the first by 0.1, the second by 0.2, and the third by 0.12. This converts the coefficients to numbers which are easier to work with. The constraints are thus converted to:

$$x + y \leq 80$$

$$x + 2.5y \leq 160$$

$$3x + y \leq 200$$

$$x, y \geq 0$$

Now add slack variables S_1, S_2, S_3.

$$\text{Maximize} \quad P = 6x + 4y + 0S_1 + 0S_2 + 0S_3$$

$$\text{Subject to:} \quad x + y + S_1 \qquad\qquad = 80$$

$$x + 2.5y \quad + S_2 \qquad = 160$$

$$3x + y \qquad\qquad + S_3 = 200$$

$$x, y, S_1, S_2, S_3 \geq 0$$

Notice that if x and y are assigned zero values, the solution of this set of constraint equations is $S_1 = 80$, $S_2 = 160$, and $S_3 = 200$, and the profit is, of course, zero. This solution is represented in the figure by the origin $(0, 0)$. It is the most convenient starting solution. To help in accounting as the analysis proceeds, denote this solution by $(0, 0, 80, 160, 200)$.

Starting solution:

$$x = y = 0$$

$$S_1 = 80 \qquad\qquad P = 6x + 4y = 0$$

$$S_2 = 160$$

$$S_3 = 200$$

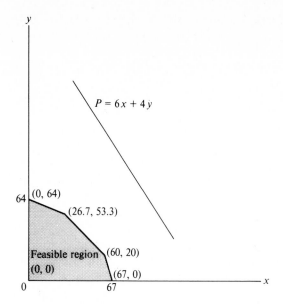

Once a starting solution has been found, the next step is to use the objective function to identify which variable, x or y, should be increased in value (brought into the solution) to increase profits. Since the coefficient of x is 6 in the objective function, while that of y is only 4, the greatest rate of increase in profit is to be gained by increasing x and leaving y at zero.

The three constraint equations place limits on how much x can be increased. According to the first equation x can be increased to 80, reducing S_1 to zero; keep in mind that S_1 may never be less than zero. The limit on x becomes obvious when the first constraint is rewritten as:

$$x = 80 - y - S_1 \tag{1'}$$

Since y is being held at zero, increases in x must be accompanied by equal decreases in S_1 in order to maintain equality, as required by the constraint. Similarly, equations 2 and 3 can be written as:

$$x = 160 - 2.5y - S_2 \tag{2'}$$

$$x = \frac{200}{3} - \frac{1}{3}y - \frac{1}{3}S_3 \tag{3'}$$

which indicate that the upper limits on x are 160 and 66.67, respectively. Clearly, equation 3' is the binding constraint and x should be increased to $x = 66.67$, or approximately 67. Thus we have the results of the first step:

$$x = \frac{200}{3} \simeq 67$$

$$y = S_3 = 0 \qquad P = 6x + 4y = 402$$

$$S_1 = 13$$

$$S_2 = 93$$

To determine whether there is a better solution, write P in terms of variables having zero values in the current solution. If any of the coefficients are positive, the corresponding variable can be increased to advantage. Since y and S_3 are zero, and x is not, substitute for x in the objective function, using equation 3', to get

$$P = \frac{6}{3}(200 - y - S_3) + 4y$$

$$= 400 + 2y - 2S_3$$

This indicates that y should be increased.

To determine the limit on increasing y, the constraints should be rewritten to indicate y as a function of S_3, which has a value of zero, and one nonzero-valued variable. That way each constraint equation will indicate what will be the effect on each nonzero variable as y is increased, and the limits will be obvious. Equations 1'', 2'', and 3'' were obtained by substituting equation 3' into equation 1' and solving for y, by substituting equation 3' into equation 2' and solving for y, and by solving equation 3' for y, respectively.

$$y = 20 - \frac{3}{2}S_1 + \frac{1}{2}S_3 \tag{1''}$$

$$y = 93 - \frac{S_2}{2.17} + \frac{1}{3(2.17)}S_3 \tag{2''}$$

$$y = 200 - 3x - S_3 \tag{3''}$$

Thus, y may not be increased beyond $y = 20$, as indicated by equation 1''; recall that $S_3 = 0$.

Once again, the question is whether profits can be further improved. Recasting the objective function in terms of zero-valued variables (S_1 and S_3), by substituting for x and y from constraint equations 1'' and 2'', gives

$$P = 440 - 3S_1 - S_3$$

Because the coefficients of the zero-valued variables S_1 and S_3 are negative, no further increases in P are possible. The optimal solution is $x = 60$, $y = 20$, $P = 440$, which, of course, is the same solution found graphically and by total enumeration of all corner points.

14.4 SIMPLEX TABLEAU

The *simplex tableau* is an efficient bookkeeping system for keeping track of the values of the variables and moving from the initial solution to the optimal solution. Instead of working with equations, one works only with the numerical coefficients of the equations of a LP problem. Rewriting equations, as in Section 14.3, is accomplished by what is called *row manipulation*. The format is demonstrated in Figure 14.6, using the clinic problem once again.

(a) Mathematical form

Maximize $P = 6x + 4y + 0S_1 + 0S_2 + 0S_3$

Subject to: $x + y + S_1 \qquad\qquad = 80$

$\qquad\qquad\quad x + 2.5y \quad + S_2 \qquad = 160$

$\qquad\qquad\quad 3x + y \qquad\qquad + S_3 = 200$

$\qquad\qquad\quad x, y, S_1, S_2, S_3 \geq 0$

(b) Tableau form

	x	y	S_1	S_2	S_3	Limit
Row 1	1	1	1	0	0	80
Row 2	1	2.5	0	1	0	160
Row 3	3	1	0	0	1	200
Objective	6	4	0	0	0	P

Figure 14.6 Simplex tableau format.

The tableau indicates immediately which variables form a solution.* In Figure 14.6, they are S_1, S_2, and S_3, because their columns consist of zeros and a single unit integer. Taking x and y to zero, the right-hand stub (limit) of the tableau indicates that $S_1 = 80$, $S_2 = 160$, and $S_3 = 200$ is the starting solution. This is the same logic employed in Section 14.3.

Notice that the objective function forms the bottom row of the tableau. In the location labeled P, the (negative) numerical values of the profit appear as a result of two manipulations performed in moving from the initial tableau to any subsequent tableau, including, or course, the optimal tableau.

Example 14.3: Tableau process for the clinic problem From Figure 14.1, the initial tableau for the clinic problem is

Initial tableau

	x	y	S_1	S_2	S_3	Limit	
Row 1	1	1	1	0	0	80	(80/1)
Row 2	1	2.5	0	1	0	160	(160/1)
Row 3	③	1	0	0	1	200	(200/3) ←
Objective	6	4	0	0	0	P	

$\qquad\quad \uparrow$

* In linear algebra these columns are called *unit basis vectors*. All vectors in a vector space of three dimensions can be generated by multiplying these by numbers and adding. In any tableau, the variables which are in the solution at that point are those which label the unit vectors in the tableau. Their values are found in the Limit column by reading across the row in which the sole 1 is located for that variable.

The starting solution is $x = 0$, $y = 0$, $S_1 = 80$, $S_2 = 160$, and $S_3 = 200$. The objective row indicates that the greatest rate of increase in P is to be obtained by increasing x, as the pointer indicates. To determine the limit on increasing x divide each number in the right-hand stub by the number in the same row of the x column, as shown. Clearly, 200/3 is the limit, so it is the third constraint which is binding, and the so-called pivot element is the circled 3.

In the next tableau the third row (row 3′) is obtained by dividing row 3 of the initial tableau by 3. The second row is obtained by subtracting row 3′ from row 2. The first row is obtained by subtracting row 3′ from row 1. The new objective row is obtained by subtracting 6 times row 3′ from the initial objective row. These are called row manipulations. They create a unit vector in the column of the variable (x, here) entering the solution, as required to determine what step, if any, to take next in the search for optimality.

Second tableau

	x	y	S_1	S_2	S_3	Limit
Row 1′	0	②/③	1	0	$-1/3$	13.3 (13.3 × 3/2) ←
Row 2′	0	2.17	0	1	$-1/3$	93.3 (93.3/2.17)
Row 3′	1	1/3	0	0	1/3	200/3 (200)
Objective	0	2	0	0	-2	$P - 400$
		↑				

The new solution consists of S_1, S_2, and x_1 at the values shown in the right-hand stub. Profit is at 400 dollars and can be improved further, as signaled by the positive 2 in the objective row. The first constraint limits increases in y to $y = 20$, the second and third constraints are less rigid, so increase y to $y = 20$.

The first row of the third tableau is simply row 1′ multiplied by $\frac{3}{2}$. The second row is row 2′ minus ($\frac{3}{2}$ multiplied by 2.17 multiplied by row 1″). The third row is row 3′ minus ($\frac{1}{3}$ multiplied by row 1″). The new objective row is the previous objective row minus row 1′ multiplied by 3.

Third tableau (optimal)

	x	y	S_1	S_2	S_3	Limit
Row 1″	0	1	3/2	0	$-1/2$	20
Row 2″	0	0	3.25	1	0.72	50
Row 3″	1	0	$-1/2$	0	1/2	60
Objective	0	0	-3	0	-1	$P - 440$

At this point, *there are no positive coefficients in the objective row.* Thus, the optimal solution, $x = 60$, $y = 20$, $P = 440$, is obtained once again; but this time the process was more efficient.

Understanding the logic and mechanics of the simplex method is essential to understanding the material in the next two chapters. To ensure that you understand what is going on, take the time to review the relation between the graphical solution, the corner point theorem, and the manipulations performed within the simplex tableau. Thoughtfully working through the problems in Section 14.5 will solidify your comprehension.

14.5 PROBLEMS

14.1 For the following LP problems, graph the constraints and identify the feasible region. If no feasible region exists, there can be no solution. You should indicate this on your graph. If the feasible region is unbounded, an optimum cannot be found. You should indicate this on your graph, if such is the case. If there is an optimal solution, indicate the values of the decision variables and objective function at optimality. Also, identify the optimal point on your graph.

(a) Maximize $P = 4.5x + 4y$

Subject to: $5x + 4y \leq 20$

$4x + 7y \leq 28$

$8x + 3y \leq 24$

$x, y \geq 0$

(b) Same as part (a) except that

$P = 4x + 4.5y$

(c) Same as part (a) except that

$P = 5x + 4y$

(d) Maximize $P = 4.5x + 4y$

Subject to: $5x + 4y \leq 40$

$4x + 7y \leq 28$

$8x + 3y \leq 24$

$x, y \geq 0$

(e) Maximize $P = 4x + 5y$

Subject to: $5x + 4y \leq 20$

$4x + 7y \leq 28$

$8x + 3y = 24$

$x, y \geq 0$

(f) Maximize $P = 4.5x + 4y$

Subject to: $5x + 4y \geq 20$

$4x + 7y \geq 28$

$8x + 3y \geq 24$

$x, y \geq 0$

(g) Maximize $P = 4.5x + 4y$

Subject to: $5x + 4y \geq 40$

$4x + 7y \leq 28$

$8x + 3y \leq 24$

$x, y \geq 0$

(h) Maximize $P = 4.5x + 4y$

Subject to: $5x + 4y \leq 20$

$4x + 7y \leq 28$

$8x + 3y \leq 24$

$x \geq 4$

$y \geq 5$

14.2 Same instructions as for Problem 14.1.

(a) Minimize $C = 4x + 8y$

Subject to: $8x + 3y \geq 24$

$40x + 70y \geq 280$

$x, y \geq 0$

(b) Minimize $C = 8x + 4y$

Subject to: $8x + 3y \geq 24$

$4x + 7y \geq 28$

$x, y \geq 0$

(c) Minimize $C = 10x + 4y$

Subject to: $8x + 3y \geq 24$

$4x + 7y \geq 28$

$x, y \geq 0$

(d) Minimize $C = 8x + 4y$

Subject to: $8x + 3y \geq 24$

$4x + 7y \leq 28$

$x, y \geq 0$

14.3 For each of the Problems 14.1a through 14.1h and 14.2a through 14.2d for which there is an optimal solution, enumerate the corner points by solving pairs of constraints for x and y at the crossover points. Evaluate P or C for each of the corner points and indicate the optimal solution. *Hint:* If two constraints are $x + 2y = 24$ and $2x + y = 36$, the corner point of the intersection of these two sets is the point (x, y) which satisfies both equalities $x + 2y = 24$ and $2x + y = 36$. To find it multiply the first equation by 2 and subtract the second equation. The result is: $0 + 3y = 12$, or $3y = 12$. Substitute $y = 4$ into the second equation to find $2x = 32$, or $x = 16$. Thus, the corner is $(16, 4)$.

14.4 Solve the following problems using the simplex tableau:

(a) Maximize $P = 10x + 8y$

Subject to:
$$8x + 3y \leq 240$$
$$3x + 7y \leq 210$$
$$5x + 3.4y \leq 170$$
$$x, y \geq 0$$

(b) Maximize $P = 10x + 6.8y$

Subject to:
$$5x + 4y \leq 170$$
$$3x + 7y \leq 210$$
$$8x + 3y \leq 240$$
$$x, y \geq 0$$

(c) Maximize $P = 9x + 7y$

Subject to:
$$8x + 3y \leq 240$$
$$0.5x + 0.4y \leq 16$$
$$0.3x + 0.7y \leq 21$$
$$x, y \geq 0$$

(d) Maximize $P = 10x + 8y$

Subject to:
$$8x + 3y \leq 240$$
$$3x + 7y \leq 210$$
$$5x + 3.4y \leq 170$$
$$x \leq 2.8, \ y \geq 0$$

Hint: $x \leq 2.8$ is a nontrivial constraint requiring a slack variable.

14.5 Use the simplex tableau to solve the following problems:

(a) Maximize $P = x + 3y + 5z$

Subject to:
$$4x + 6y + 12z \leq 23$$
$$6x + 9y + 18z \leq 34$$
$$2x + 3y + 6z \leq 12$$
$$x, y, z \geq 0$$

(b) Maximize $P = 5x + 6y + 7z$

Subject to:
$$3x + 4y + 49z \leq 60$$
$$x, y \geq 0$$

(c) Maximize $P = 5x + 6y + 7z$

Subject to:
$$3x + 4y + 50z \leq 60$$
$$4x + 3y + 10z \leq 50$$
$$x, y, z \geq 0$$

14.6 Find the least-cost diet containing the minimum weekly required elements and using two input sources. Source A costs 4 dollars per hundred pounds and source B costs 6 dollars per hundred pounds. Source A contains two pounds of element 1, four pounds of element 2, and three pounds of element 3 per hundred pounds of input. Source B contains seven pounds of element 1, three pounds of element 2, and two pounds of element 3 per hundred pounds of input. The minimum weekly requirements are 18, 15, and 12 pounds of elements 1, 2, and 3, respectively. Solve graphically.

14.7 Suppose there were two health care clinics serving three populations. Assume that the daily demand for service equaled the daily capacity of the clinics, but that different costs of delivery were incurred at the clinics, depending on the location of the individual being served. Assume, too, that it would be possible to formulate a service program which would minimize total costs. Cost and capacity data are given below:

Clinic	Population			Capacity
	P_1	P_2	P_3	
C_1	$4	$2	$1	15
C_2	$1	$3	$5	25
Demand	10	20	10	40

Let x be the number of patients from population P_1 served by clinic C_1, and y be the number from P_2 served by clinic C_1. Then the number from P_3 served by C_1 must be $15 - x - y$. Also, the number from P_1 served by C_2 must be $10 - x$, the number from P_2 served by C_2 must be $20 - y$, and the number from P_3 served by C_2 must be $25 - (10 - x) - (20 - y) = -5 + x + y$.

None of these observations can be negative and they correspond to constraints. The constraints are:

$$x + y \geq 5$$
$$x + y \leq 15$$
$$y \leq 20$$
$$x \leq 10$$
$$x, y \geq 0$$

The objective function to be minimized is

$$C = 4x + 2y + 1(15 - x - y) + 1(10 - x) + 3(20 - y) + 5(-5 + x + y)$$

(a) Draw the feasible region for this problem.

(b) Enumerate the corner points, and, by substitution into the objective function, find the optimum. (First simplify the objective function, of course.)

14.8 Solve the following problem using the simplex method:

Maximize $P = 2x + y$

Subject to: $5x + 2y \leq 10$

$3x + 5y \leq 15$

$x, y \geq 0$

14.9 A company is forced into compliance with new air pollution standards. It produces one product on two production lines. Two technologies are available to reduce emissions. The most effective technology reduces emissions by seven pollution units per production unit and adds 2 dollars to overall unit production costs. The least-effective technology reduces emissions by two units per unit of production and adds 1 dollar to overall unit production costs. The company produces four units of product per period and must reduce emissions by fourteen units per period on the combined production lines.

Write this problem as a least-cost LP problem to find the optimal production mix using the two production lines, which may have different pollution technologies applied to them.

14.6 SUPPLEMENTAL READING

Fabrycky, W. J., and G. J. Thuesen: *Economic Decision Analysis*, Prentice-Hall, Inc., Englewood Cliffs, N.J., 1974, chap. 15.

Strum, Jay E.: *Introduction to Linear Programming*, Holden-Day, Inc., San Francisco, Calif., 1972, chaps. 1, 2, 3, and 5.

Wagner, Harvey M.: *Principles of Operations Research*, 2d ed., Prentice-Hall, Inc., Englewood Cliffs, N.J., 1975, chaps. 1, 2, 3, and 4.

FIFTEEN

PROBLEM FORMULATION

Most LP problems include constraints of the greater-than-or-equal-to variety in addition to the less-than-or-equal-to variety. Minimization problems, for example, would have no bounded solution otherwise. Most LP problems also include equality constraints. To deal with these varieties using the simplex method, the concepts of surplus and artificial variables are essential. These are the subjects of Section 15.1; the revised solution method used is called the *Big M method*.

An absolute requirement in all LP problems is that all decision variables be positive for all feasible solutions. However, in many real-world problems there are free variables, meaning variables which are unconstrained in sign and can take on negative values. It turns out that problems containing free variables can be converted to problems containing only nonnegative decision variables. Thus, problems which may appear to require solution approaches other than LP, only because they contain variables which may take on negative values, can be converted to equivalent LP problems, as demonstrated in Section 15.3.

Many computerized LP algorithms are designed to solve only maximization problems. This is not a constraint, because every minimization problem can be converted to a maximization problem by changing signs of coefficients in the objective function. This procedure is demonstrated in Section 15.3. A generalized formulation of LP problems is also presented in that section.

Throughout this chapter, problem formulation is the central issue. Readers should pay special attention to how problems are cast into the LP formulation. The objectives are to provide some insight into the wide range of problems which can be analyzed using LP methods and some awareness of how they can be approached. Readers who wish to learn more about particular solution methods should refer to more advanced texts such as those included in Section 15.6.

15.1 BIG M SOLUTION METHOD

Minimization problems including only greater-than-or-equal-to constraints can be solved directly using the simplex tableau, as described in Chapter 14. Instead of adding slack variables to the constraints, one subtracts surplus variables. For example, suppose the problem were:

$$\text{Minimize} \quad C = 12x + 12y$$
$$\text{Subject to:} \quad 3x + y \geq 3$$
$$2x + 6y \geq 12$$
$$x, y \geq 0$$

Subtracting surplus variables would convert it to:

$$\text{Minimize} \quad C = 12x + 12y + 0S_1 + 0S_2$$
$$\text{Subject to:} \quad 3x + y - S_1 \quad = 3$$
$$2x + 6y \quad - S_2 = 12$$
$$x, y \geq 0$$
$$S_1, S_2 \leq 0$$

The corresponding (unadjusted) tableau would be

Unadjusted tableau

	x	y	$\bar{S}_1$	$\bar{S}_2$	Limit
Row 1	3	1	-1	0	3
Row 2	2	6	0	-1	12
Objective	12	12	0	0	C

The starting solution is not obvious, because if x and y are set equal to zero, the values of the surplus variables would be $\bar{S}_1 = -3$ and $\bar{S}_2 = -12$, which are negative numbers. Since $\bar{S}_1$ and $\bar{S}_2$ are required to be positive, this cannot be a starting solution. To find a starting solution, one must adjust the tableau to obtain two columns containing a single unit number and zeros only, while the right-hand stub contains only positive numbers. Clearly, the only possible corner is $\bar{S}_1 = 0$, $\bar{S}_2 = 0$; so adjust the tableau to contain the unit vectors in the x and y columns. When this is done, one has the initial tableau,

Initial (adjusted) tableau

	x	y	$\bar{S}_1$	$\bar{S}_2$	Limit
Row 1'	1	0	$-3/8$	$1/16$	$3/8$
Row 2'	0	1	$1/8$	$-3/16$	$15/8$
Objective	0	0	3	$3/2$	$C - 27$

In this case the adjusted initial tableau is optimal, because increasing S_1 or S_2 would *increase* costs—all numbers in the objective row of the adjusted tableau are positive.

The more efficient way to solve this kind of problem is to introduce *artificial variables* and employ the so-called Big M method of solution. Introducing artificial variables provides an obvious starting solution to an artificial problem, and the Big M method ensures that the artificial variables have zero values at the optimum so that the solution is actually the solution of the original problem.

The previous problem provides an easy illustration. Restated to include artificial variables, it would be:

$$\text{Minimize} \quad C = 12x + 12y + MA_1 + MA_2$$

$$\text{Subject to:} \quad 3x + y - S_1 \qquad\qquad + A_1 \qquad = 3$$

$$2x + 6y \qquad - S_2 \qquad\qquad + A_2 = 12$$

$$x, y \geq 0$$

Notice that the artificial variables A_1 and A_2 are included in the objective function.* The coefficient M is assumed to be so mammoth that no matter what row manipulations occur in the solution process, it dominates all numbers. Solely because M is so large the artificial variables will be "driven" to zero in the solution process.

The unadjusted tableau for this problem is

Unadjusted tableau

	x	y	$\bar{S}_1$	$\bar{S}_2$	A_1	A_2	Limit
Row 1	3	1	-1	0	1	0	3
Row 2	2	6	0	-1	0	1	12
Objective	12	12	0	0	M	M	C

---

* Henceforth neither slack nor surplus variables are shown in the objective function, because they always have coefficients of zero. For convenience the nonnegativity constraints on slack, surplus, and artificial variables are not shown; the reader should just remember that they are imposed.

The obvious starting solution for the problem is $A_1 = 3$, $A_2 = 12$, $x = y = \bar{S}_1 = \bar{S}_2 = 0$. To begin the optimization process adjust the tableau so as to replace the M's by zeros in the A_1 and A_2 columns. That is, multiply the first and second rows by M and subtract each from the objective row. The result is given below, with the terms containing M separated for convenience.

Initial (adjusted) tableau

	x	y	$\bar{S}_1$	$\bar{S}_2$	A_1	A_2	Limit
Row 1	3	1	-1	0	1	0	3 (3)
Row 2	2	⑥	0	-1	0	1	12 (12/6) ←
Objective	12	12	0	0	0	0	$C - 15M$
	$-5M$	$-7M$	$+M$	$+M$			
		↑					

Proceeding with the solution using the simplex method, the next variable to include is y, because each unit increment in y will decrease costs by $7M$ units. The binding constraint is the second one. By row manipulation, the second and third tableaus are

Second tableau

	x	y	$\bar{S}_1$	$\bar{S}_2$	A_1	A_2	Limit
Row 1'	⑧⁄③	0	-1	$1/6$	1	$-1/6$	1 (3/8) ←
Row 2'	$1/3$	1	0	$-1/6$	0	$1/6$	2 (6)
Objective	8	0	0	2	0	-2	$C - 24$
	$-8/3M$		M	$-1/6M$		$7/6M$	$-M$
	↑						

Third tableau (optimal)

	x	y	$\bar{S}_1$	$\bar{S}_2$	A_1	A_2	Limit
Row 1"	1	0	$-3/8$	$1/16$	$3/8$	$-1/16$	$3/8$
Row 2"	0	1	$1/8$	$-3/16$	$1/8$	$3/16$	$15/8$
Objective	0	0	3	$3/2$	-3	$-3/2$	$C - 27$
					M	M	

The third tableau is *optimal because there are no negative coefficients in the objective row**; M is so large that it dominates in expressions such as $-3 + M$,

---

* Whereas for maximization problems optimality was obtained when there were no more positive coefficients in the objective row, for minimization problems, optimality is obtained when there are no negative coefficients in the row, since at that point no further cost reductions are possible.

making them positive. Notice that all M terms have canceled out of the objective row in the far right column, and only C-27 appears, indicating that the minimum cost is 27 dollars. This is a requirement at optimality.

The Big M method applies whenever artificial variables are employed. The following sequence of examples demonstrate problem formulation and the use of slack, surplus, and artificial variables. Only initial and final tableaus are shown. Readers are encouraged to work through the solutions.

The first example falls into the category called *transportation problems*. Transportation problems are extremely common, and they often have nothing to do with transportation. Any minimum-cost problem which can be conceptualized as a distribution of equivalent goods or services from a variety of sources to a variety of demand centers with delivery costs that relate to the particular distribution patterns, can be cast into the form of a transportation problem. (Because it is such a common problem, special, highly efficient algorithms have been developed; these are beyond the level of this text and are not discussed.) Transportation problems are called *balanced* when supply capacity just equals demand. Unbalanced transportation problems can be balanced; when supply exceeds demand, add a fictitious demand center with demand requirements equal to the excess in supply, and when demand exceeds supply, add a fictitious supply center to balance the problem. The transportation costs associated with the new centers may be assigned values of zero, for practical purposes.

Example 15.1: Transportation problem An agency provides identical services to three population centers in a region: P_1, P_2, and P_3. All services are provided from two centers S_1 and S_2. The (average) costs of providing the services differ for each pair of service centers and population centers. Furthermore, average demand for service varies with the population center, and the service centers have unequal capacities. The data are shown in the table below.

Average costs
Dollars/unit

	P_1	P_2	P_3	Capacity
S_1	4	3	1	20
S_2	1	2	3	25
Average demand	15	20	10	45 (total)

The agency would like to direct its services distribution so as to minimize service costs. To solve this problem only two decision variables are needed, because if the agency decides to provide x units of service to population P_1 from S_1 and y units to population P_2 from S_1, it has, in fact, decided its entire services distribution scheme as follows. Since demand equals supply the problem is called *balanced*.

- If x units flow from S_1 to P_1 and y units flow from S_1 to P_2, then the remaining $20 - x - y$ flow from S_1 to P_3.
- If x units flow from S_1 to P_1, then the remaining $15 - x$ units demanded by P_1 must come from S_2.
- If y units flow from S_1 to P_2, then $20 - y$ units flow from S_2 to P_2.
- If $15 - x$ flow from S_2 to P_1 and $20 - y$ units flow from S_2 to P_2, then the remaining $25 - (15 - x) - (20 - y) = -10 + x + y$ units flow from S_2 to P_3.

Each flow must be nonnegative; therefore the constraints for the problem correspond to the flow statements above. The objective function follows from taking products of unit costs with flows.* The LP problem is

$$\text{Minimize} \quad C = 5x + 3y + 45$$
$$\text{Subject to:} \quad x + y \le 20$$
$$x \le 15$$
$$y \le 20$$
$$x + y \ge 10$$
$$x, y \ge 0$$

The initial adjusted tableau for this problem and the final tableau are shown below. The surplus variable is $\bar{S}_1$.

Initial (adjusted) tableau

	x	y	S_1	S_2	S_3	$\bar{S}_1$	A_1	Limit
Row 1	1	1	1	0	0	0	0	20
Row 2	1	0	0	1	0	0	0	15
Row 3	0	1	0	0	1	0	0	20
Row 4	1	1	0	0	0	-1	1	10
Objective	5	3	0	0	0	0	0	$C - 45 - 10M$
	$-M$	$-M$	0	0	0	M	0	

* Cost $= 4x + 3y + 1(20 - x - y) + 1(15 - x) + 2(20 - y)$

$\quad + 3(-10 + x + y)$

$= 5x + 3y + 45$

Final tableau

	x	y	S_1	S_2	S_3	$\bar{S}_1$	A_1	Limit
Row 1	0	0	1	0	0	1	-1	10
Row 2	1	0	0	1	0	0	0	15
Row 3	-1	0	0	0	1	1	-1	10
Row 4	1	1	0	0	0	-1	1	10
Objective	2	0	0	0	0	3	-3 $+M$	$C - 75$

The minimum-cost service distribution scheme is, therefore,

Optimal solution:

$x = 0$: Ship nothing from S_1 to P_1; ship 15 units from S_2 to P_1

$y = 10$: Ship 10 units from S_1 to P_2; ship 10 units from S_2 to P_2
Ship 10 units from S_1 to P_3; ship nothing from S_2 to P_3

Cost = 75

The next problem may be thought of as an optimal mix, blending, portfolio, or diet problem. All are approached the same way. In each, the objective is to minimize input costs while meeting certain requirements on outputs. The information needed is always of the form "output per unit of input" plus established output requirements.

Example 15.2: Capital budgeting The local health planning agency is studying the need for additional health delivery capacity in the area. It has divided long-term care facilities into two categories which provide service to different mixes of the same three client groups. Each category would provide additional beds at the rates shown below for each group per thousand dollars invested. The agency would like to determine the best capital investment scheme.

	New beds per $1000 invested	
Client group	Category A	Category B
1	1.00	0.80
2	0.10	0.12
3	0.40	0.16

Anticipated needs over the next 10 years are 8000 new beds for group 1, 1000 for group 2, and 2000 for group 3. Letting x represent thousands of dollars invested in category A and y represent thousands of dollars invested in category B, the LP problem has the form

$$\text{Minimize} \quad C = x + y$$

$$\text{Subject to:} \quad x + 0.8y \geq 8000$$

$$0.1x + 0.12y \geq 1000$$

$$0.4x + 0.16y \geq 2000$$

$$x, y \geq 0$$

The unadjusted tableau and final tableau are given below:

Unadjusted tableau

	x	y	$\bar{S}_1$	$\bar{S}_2$	$\bar{S}_3$	A_1	A_2	A_3	Limit
Row 1	10	8	-1	0	0	1	0	0	80,000
Row 2	1	1.2	0	-1	0	0	1	0	10,000
Row 3	4	1.6	0	0	-1	0	0	1	20,000
Objective	1	1	0	0	0	M	M	M	C

Final tableau

	x	y	$\bar{S}_1$	$\bar{S}_2$	$\bar{S}_3$	A_1	A_2	A_3	Limit
Row 1	0	0	-0.8	4	1	0.8	4	-1	4000
Row 2	0	1	0.25	-25	0	-0.25	0	0	5000
Row 3	1	0	-0.3	20	0	0.3	1	0	4000
Objective	0	0	0.05	5	0	-0.05	-1	0	$C - 9000$
						M	$11M$	M	

Thus, the optimal capital investment scheme this year is 4 million dollars by category A and 5 million dollars by category B.

The next example demonstrates that whenever a number of persons are to be distributed among the same number of job slots for which skills and job requirements are not perfectly matched, the "least-conflict" solution is to assign individuals entirely to a single job and avoid splitting their time. This result is independent of the number of job slots. Management scientists have found that the most efficient way to find the optimal solution is with network flow techniques, not discussed in this text.

Example 15.3: Personnel assignment Two assignments are to be made to two individuals. Neither individual is perfectly qualified for the job, according to their prejob tests. The administrator is considering alternative assignment strategies including assigning each person to each job for some part of the time. Because he cannot divide the jobs to match exactly the special skills of each individual, he is having trouble making the assignment. The employee ratings for the respective jobs are shown below. A score of 1 indicates a job-skill match, and a score of 10 indicates a complete mismatch.

	Job-skill scores	
Job	Employee 1	Employee 2
1	4	6
2	2	5

Letting x_{ij} represent the fraction of time employee i spends on job j, the problem is to minimize the degree of mismatch, which is characterized by the objective function (or conflict indicator) as

$$C = 4x_{11} + 2x_{12} + 6x_{21} + 5x_{22}$$

The constraints are the requirements that each person be employed 100 percent of the time and that each job be covered 100 percent of the time. Thus, the LP problem is

$$\text{Minimize} \quad C = 4x_{11} + 2x_{12} + 6x_{21} + 5x_{22}$$

$$\text{Subject to:} \quad x_{11} + x_{12} \qquad\qquad = 1$$
$$x_{21} + x_{22} = 1$$
$$x_{11} \qquad + x_{21} \qquad = 1$$
$$x_{12} \qquad + x_{22} = 1$$
$$x_{ij} \geq 0 \quad \text{all } i, j = 1, 2$$

This can be written in adjusted tableau form as below.

Adjusted tableau*

	x_{11}	x_{12}	x_{21}	x_{22}	A_1	A_2	A_3	A_4	Limit
Row 1	1	1	0	0	1	0	0	0	1
Row 2	0	0	1	1	0	1	0	0	1
Row 3	1	0	1	0	0	0	1	0	1
Row 4	0	1	0	1	0	0	0	1	1
Objective	4	2	6	5	0	0	0	0	$C - 4M$
	$-2M$	$-2M$	$-2M$	$-2M$					

* Notice that this tableau is degenerate. The variable x_{12} would be the first one to enter the solution, but constraints 1 and 2 impose the same limit on its magnitude. As a result, in the next tableau x_{12} will have a value of zero.

There is no reason to proceed with the solution by the simplex method, because the whole problem can be reduced to a single-variable problem by subtracting pairs of constraint equations. Then the solution is obvious. For example, $x_{11} = 1 - x_{12}$ and $x_{11} = 1 - x_{21}$. Thus, $x_{12} = x_{21}$. Also, $x_{22} = 1 - x_{21}$ so $x_{22} = x_{11}$. Accordingly, the objective function can be written $C = 9x_{11} + 8x_{12}$, which is minimized when $x_{11} = x_{22} = 0$ and $x_{21} = x_{12} = 1$. Assigning employee A to job 2 and employee B to job 1, each for 100 percent of the time, is optimal.

15.2 APPROXIMATELY LINEAR OBJECTIVE FUNCTIONS

Quite often problems have linear constraints and nonlinear objective functions. When nonlinear objective functions are convex downward, in minimization problems, or concave downward, in maximization problems, they can be approximated by a sequence of straight lines. Then the problems can be solved approximately by LP techniques. The next example demonstrates the process of formulating the problem only. The idea is to write the nonlinear objective function as a sum of separate linear functions, thus obtaining a linear approximation to it.

Example 15.4: Piecewise linear objective function The following problem can be rewritten as a LP problem.

$$\text{Minimize} \quad C = x + 2x^2 + 8y$$
$$\text{Subject to:} \quad x + 2y \geq 10$$
$$2x + y \geq 10$$
$$x, y \geq 0$$

The term $2x^2$ can be approximated by linear relations as shown in the graph. The points at which the linear segments touch the curve $2x^2$, above $x = 0$, $x = 2$, and $x = 4$, were selected for this exposition for simplicity. Clearly, the greater the number of linear segments, the more accurately the linear approximation matches the curve being approximated. But the more segments, the more complex the reformulation appears.*

Write $x = x_1 + x_2$, where $0 \leq x_1 \leq 2$ and $0 \leq x_2 \leq 2$. With this notation the function $2x^2$ can be written as $2x^2 \simeq 4x_1 + 12x_2$. (Notice that if $x = 3$, $x_1 = 2$, and $x_2 = 1$, so $2x^2 = 8 + 12 = 20$, which overestimates the

---

* An excellent example of the use of a piecewise linear approximation of a nonlinear objective function to a public policy issue is the article by Robert Dorfman and Henry D. Jacoby, "A Model of Public Decisions Illustrated by a Water Pollution Policy Problem," in R. Haveman and J. Margolis (eds.), *Public Expenditures and Policy Analysis*, Markham, Rand McNally, Chicago, 1970, chap. 7.

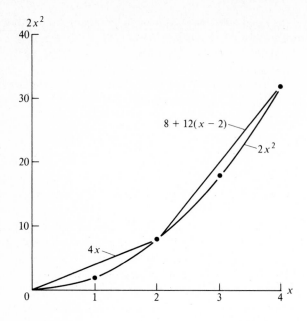

correct value of 18 by two units in this instance.) To rewrite the problem, replace x by $x_1 + x_2$ everywhere and add the two new constraints on x_1 and x_2.* Thus,

$$\text{Minimize} \quad C = 5x_1 + 13x_2 + 8y$$

$$\text{Subject to:} \quad x_1 + x_2 + 2y \geq 0$$

$$2x_1 + 2x_2 + y \geq 10$$

$$x_1 \leq 2$$

$$x_2 \leq 2$$

$$y, x_1, x_2 \geq 0$$

15.3 FREE VARIABLES

Decision problems often include variables which may be either increased or decreased. For example, workforce studies should include the possibility of discharging excess labor as well as recruiting new employees. Such important kinds of problems can be studied by LP methods by writing unconstrained or

---

* Note that x_2 will never be positive unless $x_1 = 2$, because this is a minimization problem and the increasing slope of the curve $2x^2$ forces the coefficient of x_2 to exceed that of x_1, requiring that x_1 be used up before x_2 can enter the solution.

free variables as differences in nonnegative decision variables, provided, of course, the objective function and constraints can be written as linear relations.* The solution to the so-called auxiliary problem includes the solution to the original problem.

The process of creating an auxiliary problem and obtaining the solution to the original problem is demonstrated in the following example.

Example 15.5: Free variables Suppose a clinic provided two kinds of services and was constrained in the level at which it provided those services by personnel and technology. Imagine that the opportunity to add a new staff member existed and the manager wanted to allocate that person's time in such a way that the increase in net revenue to the clinic was maximized.

Let R represent the change in net revenues, let x represent the change in service of one kind, and let y represent the change in service of the other kind. Suppose the revenue coefficients were 5 dollars per unit of x and 4 dollars per unit of y. Suppose, also, that the coefficients of technology utilization were 8 hours per unit of x and 10 hours per unit of y, while for personnel utilization rates they were 10 and 6 hours, respectively. Then, since only the labor constraint is affected (by eight hours) the LP problem would be

$$\text{Maximize} \quad R = 5x + 4y$$

$$\text{Subject to:} \quad 8x + 10y \leq 0$$

$$10x + 6y \leq 8$$

$$x, y \text{ unconstrained}$$

To solve this problem, invent new decision variables x_1, x_2, y_1, y_2 which are all nonnegative. Then the values x and y may take can be written as $x = x_1 - x_2$ and $y = y_1 - y_2$, by selecting any of an unlimited range of values for the new variables. Importantly, finding optimal values for the new variables uniquely specifies optimal values for x and y; and if there is no optimal solution to the original problem, there can be no optimal solution to the auxiliary problem. The auxiliary problem is

$$\text{Maximize} \quad R = 5x_1 - 5x_2 + 4y_1 - 4y_2$$

$$\text{Subject to:} \quad 8x_1 - 8x_2 + 10y_1 - 10y_2 \leq 0$$

$$10x_1 - 10x_2 + 6y_1 - 6y_2 \leq 8$$

$$x_1, x_2, y_1, y_2 \geq 0$$

---

* The simplex method works only because the decision variables are nonnegative. Without this constraint the pivot row in the tableau could not be determined.

The initial and final tableaus are:

Initial tableau

	x_1	x_2	y_1	y_2	S_1	S_2	Limit
Row 1	8	−8	10	−10	1	0	0
Row 2	10	−10	6	−6	0	1	8
Objective	5	−5	4	−4	0	0	R

Optimal tableau

	x_1	x_2	y_1	y_2	S_1	S_2	Limit
Row 1	1	−1	0	0	−0.112	0.18	1.54
Row 2	0	0	−1	1	−0.19	0.15	1.23
Objective	0	0	0	0	−0.205	−0.34	$R − 2.78$

The optimal solution to the original problem is $x = x_1 − x_2 = 1.54 − 0 = 1.54$, and $y = y_1 − y_2 = 0 − 1.23$; this indicates that service of the first kind should be increased by 1.54 units and service of the second kind should be reduced by 1.23 units, for a net increase in revenues of 2.78 dollars.*

The previous example could have been phrased as a situation in which the agency was operating at maximum revenue with a technology limit of 80 hours and a personnel limit of 60 hours. If eight additional personnel hours become available, the question would be how to move to a new optimum within the new labor constraint. These sorts of questions fall into the category called *sensitivity or postoptimality analysis*, which is the subject of Chapter 16.

15.4 COMPUTERIZATION

Linear programming algorithms are available for most computer systems. Even some hand-held calculators have **LP** packages, for small problems. Different systems have unique characteristics which must be mastered before one can effectively use them. However, the processes of analysis are basically the same, even though newer systems use algorithms other than the simplex method.

Most systems will automatically assign a very large value to M for use in the

---

* When working through the tableau manipulations, keep in mind that the first row may not be multiplied by −1, because that would change the sense of the inequality. Thus, in the intermediate tableau, it is the second constraint which is binding.

Big M solution method when the need for it has been signaled by the entry of artificial variables. When the user is required to select the value of M, he or she simply multiplies the largest coefficient in the original problem by 1000 or more and enters that as the value of M.

Many systems require that all problems be entered as maximization problems. This is reasonable, since all minimization problems can be converted to maximization problems by multiplying the objective function by -1. After all, reducing the value of C from, say, 10 to 5 (optimal) is equivalent to increasing $-C(=P)$ from -10 to -5 (optimal). The only difference in the two solutions is in the sign of the optimal objective value; the values of the decision variables are the same in both cases. To convince yourself that this is true, solve the following two problems:

Minimize	$C = 5x + 4y$	Maximize	$P = -5x - 4y$
Subject to:	$8x + 10y \geq 80$	Subject to:	$8x + 10y \geq 80$
	$10x + 6y \geq 60$		$10x + 6y \geq 60$
	$x, y \geq 0$		$x, y \geq 0$

15.5 PROBLEMS

15.1 Use the Big M method to solve the following problems:

(a) Minimize $C = 5x + 4y$

Subject to: $8x + 10y \geq 800$

$10x + 6y \geq 600$

$x, y \geq 0$

(b) Minimize $C = 5x + 4y$

Subject to: $8.8x + 8.8y \geq 770$

$7.8x + 10y \geq 800$

$10.4x + 6.2y \geq 600$

$x, y \geq 0$

(c) Maximize $P = 10x + 10y$

Subject to: $8x + 10y \leq 80$

$8.4x + 8y = 67$

$10x + 6y \leq 60$

15.2 Imagine that you are in the process of assigning two people to two job slots. Your personnel psychologist assessed the incompatibility of each individual to each job and forwarded the following table of incompatibilities to you with a recommendation that you either assign individuals according to their lowest incompatibility score or that you divide their time among jobs so as to minimize average job dissatisfaction.

Incompatibility scores

Job	Employee 1	2
1	2	6
2	5	2

(a) Formulate the problem as a LP problem.

(b) Find the optimal solution.

(c) What do you think of your psychologist's recommendation?

15.3 Transportation problems can be solved using LP techniques, if the demand and supply levels are balanced. Suppose you had a problem in which demand was for only 40 units of a good or service distributed among four consumer units, while supply available from the three sources was 45 units. How might you amend the problem to balance it? *Hint*: You want some way of accounting for those units which will not be transported. The cost of not transporting or delivering something is zero.

15.4 Solve the following transportation problem.

Table of costs

Sources	Destination D_1	D_2	D_3	D_4	Supply
S_1	3	4	1	0	22
S_2	1	3	3	0	25
Demand	15	20	10	2	47

Interpret destination D_4 in the context of Problem 15.3, above.

15.5 Find an approximate solution to the following problem by rewriting the objective function as a piecewise linear approximation.

Minimize $C = x^2 + 0.8y$

Subject to: $8x + 10y \geq 80$

$10x + 7y \geq 70$

$x, y \geq 0$

15.6 Find the optimal solution to the following problem. Then use the method for dealing with free variables to find the new optimal solution, if the second constraint is tightened by one unit, from 120 to 119.

Maximize $P = 20x + 24y$

Subject to: $3x + y \leq 48$

$3x + 4y \leq 120$

$x, y \geq 0$

15.7 There are two incinerators in St. Louis. The ash from these incinerators is transported to one of three landfill sites. The city has decided to dump a total of 1500 tons at site 1, 2000 tons at site 2, and 1200 tons at site 3. The total number of tons available from incinerator 1 is 2200. Incinerator 2 generates 2500 tons of ash. Trucking costs per ton are:

$	From incinerator	To site
3	1	1
1	2	1
4	1	2
3	2	2
1	1	3
3	2	3

Write the LP formulation (constraints and objective function) for this problem, if the objective is to minimize costs.

15.8 A company is forced into compliance with new air pollution standards. It produces one product on two production lines. Two technologies are available to reduce emissions. The most effective technology reduces emissions by seven pollution units per production unit and adds 2 dollars to overall unit production costs. The least effective technology reduces emissions by two units per unit of production and adds 1 dollar to overall unit production costs. The company produces four units of product per period and must reduce emissions by fourteen units per period on the combined production lines.

Write this problem as a least-cost LP problem to find the optimal production mix using the two production lines, which may have different pollution technologies applied to them.

15.6 SUPPLEMENTAL READING

Dorfman, Robert, and Henry D. Jacoby: "A Model of Public Decisions Illustrated by a Water Pollution Policy Problem," in R. Haveman, and J. Margolis (eds.), *Public Expenditures and Policy Analysis*, Markham, Rand McNally, Chicago, 1970, chap. 7.

Fabrycky, W. J., and G. J. Thuesen: *Economic Decision Analysis*, Prentice-Hall, Inc., Englewood Cliffs, N.J., 1974, chap. 15.

Patten, Thomas H., Jr.: *Manpower Planning and the Development of Human Resources*, Wiley-Interscience, New York, 1971, chap. 3.

Smith, A. R.: *Models of Manpower Systems*, proceedings of a conference, NATO Scientific-Affairs Division, The English Universities Press Limited, London EC4, 1969.

Strum, Jay E.: *Introduction to Linear Programming*, Holden-Day, Inc., Publisher, San Francisco, 1972, chaps. 1, 2, 3, and 5.

Wagner, Harvey M.: *Principles of Operations Research*, 2d ed., Prentice-Hall, Inc., Englewood Cliffs, N.J., 1975, chaps. 1, 2, 3, and 4.

SIXTEEN

SENSITIVITY ANALYSIS

The profit margins or cost coefficients which multiply decision variables in objective functions are seldom known with certainty. They must be estimated from available data and often in large, important studies they represent averages or crude approximations. Similarly, coefficients in constraint relations are uncertain. Even the upper or lower limits on available resources are generally only approximately known. Since solutions to LP problems obviously depend on these parametric values, one should test the sensitivity of solutions by varying them over appropriate ranges, just as the parameters of analysis were varied in benefit-cost studies, and present the results to decision makers. Efficient techniques for conducting sensitivity analyses using the simplex tableau are the topics of this chapter.

Another use of sensitivity analysis is called postoptimality analysis. After an optimal solution for a problem has been found, managers often ask "what if" questions: What would be the marginal gain if a few more units of a resource could be obtained? Suppose costs or prices were revised by such and such amounts, how would that change the optimal allocation of resources? Or, what is the second-best solution if exogenous factors reduce the utility of the obtained solution? Postoptimality analysis also permits one to impose new constraints and evaluate their impacts without reworking the entire problem. Finally, one may have the option of adding a new service or product and want to determine at what level to include it so as to make optimal use of available resources. Methods for investigating all but one of these questions are discussed in this chapter also. Only the question of the second-best solution is deferred to other books or to computerized algorithms.

Row vector Column vector Matrix

$$[2, 1, 0] \qquad \begin{bmatrix} 2 \\ 1 \\ 0 \end{bmatrix} \qquad \begin{bmatrix} 3 & 4 & 2 & 1 \\ 5 & 1 & 8 & 7 \\ 20 & 0 & 4 & 6 \end{bmatrix}$$

3-dimensional 3-dimensional 3- by 4-dimensional

Figure 16.1 Vector and matrix notation.

Uncertainty and postoptimality issues are handled by the same analytical procedures. This chapter is divided into sections which address these general methods. All the methods employ elementary vector operations which are explained in Section 16.1.

16.1 VECTOR AND MATRIX OPERATIONS

Mathematicians have developed efficient ways to manipulate arrays of numbers. These arrays are called *vectors* if they contain only one row or one column of numbers. The number of elements (numbers) in a vector is called its *dimension*. Arrays containing more than one row and more than one column are called *matrices*. Matrices may be square or rectangular in shape. If a matrix has five rows and six columns, it is said to be of dimension 5 by 6 or (5×6); the first number refers to the number of rows while the second refers to the number of columns.* Examples are provided in Figure 16.1. Matrices can be multiplied, if the multiplying matrix has the same number of columns as the multiplied matrix has rows. The rules for matrix multiplication are illustrated in Example 16.1.

Vectors and matrices may be multiplied by a number; simply multiply every element by the number. Division by a number works the same way.

Example 16.1: Multiplication by a constant

Specific	*General*

$$2[1, 3, 4] = [2, 6, 8] \qquad C[a_1, a_2, a_3] = [Ca_1, Ca_2, Ca_3]$$

$$3\begin{bmatrix} 1 \\ 3 \\ 4 \end{bmatrix} = \begin{bmatrix} 3 \\ 9 \\ 12 \end{bmatrix} \qquad C\begin{bmatrix} b_1 \\ b_2 \\ b_3 \end{bmatrix} = \begin{bmatrix} Cb_1 \\ Cb_2 \\ Cb_3 \end{bmatrix}$$

$$2\begin{bmatrix} 2 & 1 & 5 \\ 3 & 0 & 6 \end{bmatrix} = \begin{bmatrix} 4 & 2 & 10 \\ 6 & 0 & 12 \end{bmatrix} \qquad C\begin{bmatrix} a_{11} & a_{12} & a_{13} \\ a_{21} & a_{22} & a_{23} \end{bmatrix} = \begin{bmatrix} Ca_{11} & Ca_{12} & Ca_{13} \\ Ca_{21} & Ca_{22} & Ca_{23} \end{bmatrix}$$

* Vectors are matrices of dimension 1 by n or n by 1.

Vectors of dimension $1 \times n$ can multiply vectors of dimension $n \times 1$; that is, row vectors multiply column vectors. The product is a single number.* Matrices of dimension n by m can multiply matrices of dimension m by n. The result is a matrix of dimension n by n.

Example 16.2: Vector and matrix multiplication

Vector × vector

$$[2, 1, 3] \begin{bmatrix} 1 \\ 2 \\ 3 \end{bmatrix} = 2 + 2 + 9 = 13$$

$\quad$ 1 by 3 $\qquad$ 3 by 1

Matrix × matrix

$$\begin{bmatrix} 2 & 3 & 1 \\ 4 & 5 & -2 \end{bmatrix} \begin{bmatrix} 1 & 2 \\ 2 & 1 \\ 1 & 2 \end{bmatrix} = \begin{bmatrix} (2 + 6 + 1) & (4 + 3 + 2) \\ (4 + 10 - 2) & (8 + 5 - 4) \end{bmatrix}$$

$\quad$ 2 by 3 $\qquad$ 3 by 2 $\qquad\qquad$ 2 by 2

Matrix × vector

$$\begin{bmatrix} 2 & 3 & -1 \\ 4 & 5 & 2 \end{bmatrix} \begin{bmatrix} 3 \\ 2 \\ 2 \end{bmatrix} = \begin{bmatrix} (6 + 6 - 2) \\ (12 + 10 + 4) \end{bmatrix}$$

$\quad$ 2 by 3 $\quad$ 3 by 1 $\qquad$ 2 by 1

and

$$\begin{bmatrix} 2 & 3 & -1 \\ 4 & 5 & 2 \end{bmatrix} \begin{bmatrix} x_1 \\ x_2 \\ x_1 \end{bmatrix} = \begin{bmatrix} (2x_1 + 3x_2 - x_3) \\ (4x_1 + 5x_2 + 2x_3) \end{bmatrix}$$

$\quad$ 2 by 3 $\quad$ 3 by 1 $\qquad$ 2 by 1

Adding and subtracting vectors or matrices is done element by element. Therefore, only vectors or matrices of the same dimension may be added or subtracted.

---

* This is called the *dot or scalar product*, because it produces a number. There is another kind of vector multiplication called the *cross or vector product*, because it produces a vector; it is not needed here.

Example 16.3: Vector and matrix addition

Vector + vector

$$[2, 7, 10] + [1, 2, -3] = [3, 9, 7]$$

$$\begin{bmatrix} 2 \\ 7 \\ 10 \end{bmatrix} + \begin{bmatrix} 0 \\ 1 \\ 0 \end{bmatrix} = \begin{bmatrix} 2 \\ 8 \\ 10 \end{bmatrix}$$

Matrix + matrix

$$\begin{bmatrix} 2 & 5 & -6 \\ 3 & -10 & 4 \end{bmatrix} + \begin{bmatrix} 2 & 0 & 6 \\ -3 & -10 & 1 \end{bmatrix} = \begin{bmatrix} 4 & 5 & 0 \\ 0 & -20 & 5 \end{bmatrix}$$

You may have noticed that the simplex tableau contains matrices and vectors. In fact all LP problems can be written in matrix notation. For example, consider this problem:

$$\text{Maximize} \quad P = 4x + 5y$$

$$\text{Subject to:} \quad 10x + 8y \leq 40$$

$$6x + 10y \leq 30$$

$$x, y \geq 0$$

Including slack variables it can be written as follows (where the nonnegativity constraints are not shown):

$$10x + 8y + S_1 + 0S_2 = 40$$

$$6x + 10y + 0S_1 + S_2 = 30$$

$$4x + 5y + 0S_1 + 0S_2 = P$$

This same set of equations is obtained by multiplying the matrix of coefficients and the column vector of decision and slack variables as follows:

$$\begin{bmatrix} 10 & 8 & 1 & 0 \\ 6 & 10 & 0 & 1 \\ 4 & 5 & 0 & 0 \end{bmatrix} \begin{bmatrix} x \\ y \\ S_1 \\ S_2 \end{bmatrix} = \begin{bmatrix} 40 \\ 30 \\ P \end{bmatrix}$$

Notice also that the initial tableau for the problem is equivalent to the matrix formulation, but in the simplex format the column vector of variables is moved to the position of column headings for convenience, thus:

	x	y	S_1	S_2	Limit
Row 1	10	8	1	0	40
Row 2	6	10	0	1	30
Objective	4	5	0	0	P

Each column and each row in the tableau represents a vector which may be manipulated. The process of moving from one tableau to the next is actually row vector manipulation.

There is a standard shorthand notation for vectors and matrices which, once mastered, makes sensitivity analysis quite easy. Vectors are written as bold lowercase letters, as in $\mathbf{x}$ or $\mathbf{b}$. Matrices are labeled with bold uppercase letters, as in $\mathbf{A}$. In this notation, matrix and vector products are written as follows:

$$\mathbf{Ax} = \mathbf{b} \quad \text{and} \quad \mathbf{ab} = \mathbf{c}$$

Sums of vectors are written simply as

$$\mathbf{a} + \mathbf{b} = \mathbf{d}$$

16.2 SIMPLEX TABLEAU VECTORS

Certain vectors and a very special matrix formed from entries in specific columns and rows of the simplex tableau (slightly revised as shown below) are the keys to sensitivity analysis. These are defined in this section.

Consider again the following LP problem:

$$\text{Maximize} \quad P = 4x_1 + 5x_2$$
$$\text{Subject to:} \quad 10x_1 + 8x_2 \leq 40$$
$$6x_1 + 10x_2 \leq 30$$
$$x_1, x_2 \geq 0$$

The familiar initial tableau for this problem is

Adjusted initial tableau
Old format

	x_1	x_2	S_1	S_2	Limit
Row 1	10	8	1	0	40
Row 2	6	10	0	1	30
Objective	4	5	0	0	P

For sensitivity analysis it is convenient to revise the tableau format by adding a special column to account for P in lieu of leaving it in the limits column of the objective row. Thus the revised initial format is

Adjusted initial tableau
Revised format

	x_1	x_2	S_1	S_2	$-P$	Limit
Row 1	10	8	1	0	0	40
Row 2	6	10	0	1	0	30
Objective	4	5	0	0	1	0

The two formats are equivalent, as can be seen by multiplying the corresponding matrix products below to obtain the same set of equations.

$$\begin{bmatrix} 10 & 8 & 1 & 0 \\ 6 & 10 & 0 & 1 \\ 4 & 5 & 0 & 0 \end{bmatrix} \begin{bmatrix} x_1 \\ x_2 \\ S_1 \\ S_2 \end{bmatrix} = \begin{bmatrix} 40 \\ 30 \\ P \end{bmatrix}$$

$$\begin{bmatrix} 10 & 8 & 1 & 0 & 0 \\ 6 & 10 & 0 & 1 & 0 \\ 4 & 5 & 0 & 0 & 1 \end{bmatrix} \begin{bmatrix} x_1 \\ x_2 \\ S_1 \\ S_2 \\ P \end{bmatrix} = \begin{bmatrix} 40 \\ 30 \\ 0 \end{bmatrix}$$

The special vectors required for sensitivity analysis are extracted from specific locations in the revised format of the simplex tableau.

For the problem at hand, there are five column vectors in the interior of the revised tableau, namely, x_1, x_2, S_1, S_2, and $-P$. As one progresses from the initial to the optimal tableau, the numerical entries in each column vector, with the exception of $-P$, change due to row manipulation, as follows:

Initial tableau

	x_1	x_2	S_1	S_2	$-P$	Limit
Row 1	10	8	1	0	0	40
Row 2	6	⑩	0	1	0	30
Objective	4	5	0	0	1	0

Second tableau

	x_1	x_2	S_1	S_2	$-P$	Limit
Row 1'	⑤.2	0	1	-0.8	0	16
Row 2'	0.6	1	0	0.1	0	3
Objective	1	0	0	-0.5	0	-15

Optimal tableau

	x_1	x_2	S_1	S_2	$-P$	Limit
Row 1″	1	0	0.192	−0.154	0	3.06
Row 2″	0	1	−0.115	0.192	0	1.16
Objective	0	0	−0.192	−0.346	1	−18.05

The unit vectors which composed the S_1 and S_2 columns of the initial tableau, there indicating the starting solution, appear in the x_1 and x_2 columns of the final tableau, there indicating the optimal solution. The numbers which replaced them in the S_1 and S_2 columns of the optimal tableau have special significance. They provide a record of the entire process of row manipulation which led from the initial tableau to the optimal tableau. As such they contain the key to inquiries into how the optimal solution would differ if new constraints or decision variables were added to the original problem, if resource constraints were relaxed, or if coefficients of the decision variables were modified in the constraints or the objective function.

The matrix composed of all but the objective row of the S_1 and S_2 columns at optimality is given a special label, namely, $\mathbf{B}^{-1}$; for the problem just solved

$$\mathbf{B}^{-1} = \begin{bmatrix} 0.192 & -0.154 \\ -0.115 & 0.192 \end{bmatrix}$$

The matrix $\mathbf{B}^{-1}$ is called the *basis inverse*.* It has the unique property that the product $\mathbf{B}^{-1}$ with the original resource limits vector gives the optimal solution to the LP problem; thus $\mathbf{B}^{-1}\mathbf{b} = \mathbf{x}$, or

$$\begin{bmatrix} 0.192 & -0.154 \\ -0.115 & 0.192 \end{bmatrix} \begin{bmatrix} 40 \\ 30 \end{bmatrix} = \begin{bmatrix} 3.06 \\ 1.16 \end{bmatrix} = \begin{bmatrix} x_1 \\ x_2 \end{bmatrix}$$

To sum up, the following definitions apply to any LP problem:

• The matrix composed of the numbers appearing at optimality in the columns corresponding to the unit vectors of the initial tableau (objective row and $-P$ column excluded) is called the basis inverse. It is given the symbol $\mathbf{B}^{-1}$.

• The limits column of the initial tableau (objective row excluded) is labeled $\mathbf{b}$. The symbol $\mathbf{b}$ applies only to the numbers of the initial limits column, it is never applied to the vector of numbers which appear in the limits column of intermediate or optimal tableaus.

• The other column vectors of the tableau corresponding to individual decision variables are labeled x_1, x_2, etc., as appropriate to the problem.

• The column vector composed of the solution to the LP problem (values of the decision variables at optimality) is abbreviated $\mathbf{x}$, so that $\mathbf{B}^{-1}\mathbf{b} = \mathbf{x}$.

The tools are now at hand, so on with postoptimal sensitivity analysis.

---

* The name is founded in the theory of linear algebra. We shall not discuss its theoretical basis nor the full extent of its meaning. We shall simply use its name for convenience and illustrate its use in sensitivity analysis in Section 16.3.

16.3 REVISING RESOURCE LIMITS

Resource limits are seldom rigidly fixed in real problems. Consequently, LP problems are typically solved using reasonable estimates for the limits. And investigations into how the solution would differ if the actual limits were somewhat larger or smaller are accomplished postoptimally. Impacts are easily estimated so long as the change in resource limits from the original problem is within certain bounds; the bounds are easily calculated.

Example 16.4: Impact of changed resources

$$\text{Maximize} \quad P = 4x_1 + 5x_2$$

$$\text{Subject to:} \quad 10x_1 + 8x_2 \le 40$$

$$6x_1 + 10x_2 \le 30$$

$$x_1, x_2 \ge 0$$

Adjusted initial tableau*

x_1	x_2	S_1	S_2	$-P$	Limit
10	8	1	0	0	40
6	10	0	1	0	30
4	5	0	0	1	0

Final tableau

x_1	x_2	S_1	S_2	$-P$	Limit
1	0	0.192	−0.154	0	3.06
0	1	−0.115	−0.192	0	1.16
0	0	−0.192	−0.346	1	−18.05

Recall that the product of the basis inverse matrix $\mathbf{B}^{-1}$ and the original resource limits vector $\mathbf{b}$ produces the optimal solution to the original problem $\mathbf{B}^{-1}\mathbf{b} = \mathbf{x}$, or:

$$\begin{bmatrix} 0.192 & -0.154 \\ -0.115 & 0.192 \end{bmatrix} \begin{bmatrix} 40 \\ 30 \end{bmatrix} = \begin{bmatrix} 3.06 \\ 1.16 \end{bmatrix} = \begin{bmatrix} x_1 \\ x_2 \end{bmatrix}$$

$$P = 4(3.06) + 5(1.16) \simeq 18.05$$

---

* Henceforth the left-hand "Row" labeling is discontinued.

Now suppose that the resource constraints vector were modified slightly from $\mathbf{b}$ to $\mathbf{b^*}$, where

$$\mathbf{b^*} = \begin{bmatrix} 40 \\ 31 \end{bmatrix} = \begin{bmatrix} 40 \\ 30 \end{bmatrix} + \begin{bmatrix} 0 \\ 1 \end{bmatrix} = \mathbf{b} + \Delta\mathbf{b}$$

Since the modification is small, the same sequence of row modifications which solved the original problem will solve the modified problem, and the basis inverse matrix for the two problems will be identical. Thus, to calculate the impact of the modification on the optimal solution simply take the product $\mathbf{B}^{-1}\mathbf{b^*} = \mathbf{x^*}$; that is:

$$\begin{bmatrix} 0.192 & -0.154 \\ -0.115 & 0.192 \end{bmatrix} \begin{bmatrix} 40 \\ 31 \end{bmatrix} = \begin{bmatrix} 2.91 \\ 1.35 \end{bmatrix} = \begin{bmatrix} x_1^* \\ x_2^* \end{bmatrix}$$

$$P = 4(2.91) + 5(1.35) \simeq 18.40$$

Comparing the solution of the modified problem to the solution of the original problem, one sees that relaxing the second constraint by one unit of resource produced a profit increase of about $\Delta P = 18.40 - 18.05 = 0.35$ and changed the optimal solution, $x_1 = 3.06$, $x_2 = 1.16$ to $x_1^* = 2.91$, $x_2^* = 1.35$.

There is an even simpler way to observe these impacts. Instead of working with the basis inverse matrix consider the more comprehensive matrix $\mathbf{M}$, which includes $\mathbf{B}^{-1}$ as well as the $-P$ column and the appropriate elements of the objective row. Also extend the resource limits vector and the solution vector to include the elements in the objective row; label these $\mathbf{b}_m$ and $\mathbf{x}_m$. For the original problem:

$$\mathbf{M} = \begin{bmatrix} 0.192 & -0.154 & 0 \\ -0.115 & 0.192 & 0 \\ -0.192 & -0.346 & 1 \end{bmatrix} \quad \mathbf{b}_m = \begin{bmatrix} 40 \\ 30 \\ 0 \end{bmatrix} \quad \mathbf{x}_m = \begin{bmatrix} x_1 \\ x_2 \\ -P \end{bmatrix}$$

Notice that at optimality $\mathbf{M}\mathbf{b}_m = \mathbf{x}_m$;

$$\begin{bmatrix} 0.192 & -0.154 & 0 \\ -0.115 & 0.192 & 0 \\ -0.192 & -0.346 & 1 \end{bmatrix} \begin{bmatrix} 40 \\ 30 \\ 0 \end{bmatrix} = \begin{bmatrix} 3.06 \\ 1.16 \\ -18.05 \end{bmatrix}$$

The same is true for the modified problem;

$$\begin{bmatrix} 0.192 & -0.154 & 0 \\ -0.115 & 0.192 & 0 \\ -0.192 & -0.346 & 1 \end{bmatrix} \begin{bmatrix} 40 \\ 31 \\ 0 \end{bmatrix} = \begin{bmatrix} 2.91 \\ 1.35 \\ -18.40 \end{bmatrix}$$

Furthermore, since $\mathbf{b}_m^* = \mathbf{b}_m + \Delta\mathbf{b}_m$, the differences in solutions and profits are immediately observed by taking the product $\mathbf{M}\,\Delta\mathbf{b}_m = \Delta\mathbf{x}_m$;

$$\begin{bmatrix} 0.192 & -0.154 & 0 \\ -0.115 & 0.192 & 0 \\ -0.192 & -0.346 & 1 \end{bmatrix} \begin{bmatrix} 0 \\ 1 \\ 0 \end{bmatrix} = \begin{bmatrix} -0.154 \\ 0.192 \\ -0.346 \end{bmatrix}$$

This formulation makes it obvious that the addition of one unit of resource in the second constraint is worth 0.346 units in additional profit, and indicates the numerical adjustments required in operating level (x_1 and x_2) to move from the original optimum to the new optimum.

The number -0.346 in the second column of the last row of the M matrix is called the shadow price for the second constraint. It tells one instantly the rate at which profits (or costs) will change as the resource limit for the second constraint is modified; in particular each unit increase will yield 0.346 additional units of profit, while each unit decrease will reduce profits by 0.346 units.

The shadow price for the first constraint is -0.192.

The technique illustrated in Example 16.4 works only when the resource modifications are within certain bounds. For example, if the first resource were increased by more than 10 units, the sequence of row manipulations would not have remained the same and the entire problem would have had to be worked through from the beginning to find the new solution; this situation is illustrated in Figure 16.2.

One can easily determine the bounds of resource modification. Observe that the nonnegativity constraints on the decision variables, namely, x_1, x_2, etc., ≥ 0,

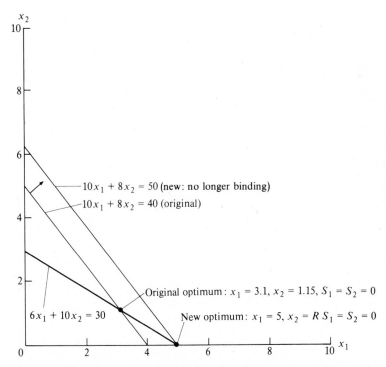

Figure 16.2 Relaxing resource constraint $\mathbf{b}^* = \mathbf{b} + \Delta\mathbf{b}$.

mean that the solution vector $\mathbf{x}$ must be nonnegative; $\mathbf{x} \geq 0$. Thus a solution is feasible if $\mathbf{B}^{-1}\mathbf{b} \geq 0$. Since $\mathbf{B}^{-1}$ encapsulates the sequence of row manipulations which led to optimality of the original problem, modifications of the resource limits vector to $\mathbf{b}^*$ for which $\mathbf{B}^{-1}\mathbf{b}^* \geq 0$ yield an optimal solution upon the same sequence of manipulations, namely, $\mathbf{B}^{-1}\mathbf{b}^* = \mathbf{x}^*$. The bounds are therefore determined by solving the inequality $\mathbf{B}^{-1}\mathbf{b} \geq 0$ for the elements of $\mathbf{b}$. For example, consider the problem of Example 16.3 with the second constraint modified. The inequality takes the form

$$\begin{bmatrix} 0.192 & -0.154 \\ -0.115 & 0.192 \end{bmatrix} \begin{bmatrix} 40 \\ b_2 \end{bmatrix} \geq \begin{bmatrix} 0 \\ 0 \end{bmatrix}$$

Taking the product of the first row of $\mathbf{B}^{-1}$ with $\mathbf{b}$ yields the inequality $b_2 \leq 49.87$. The product of the second row of $\mathbf{B}^{-1}$ with $\mathbf{b}$ yields $b_2 \geq 23.96$. Thus, so long as $23.96 \leq b_2 \leq 49.87$, the technique illustrated in Example 16.4 is applicable.

Solving the inequality

$$\begin{bmatrix} 0.192 & -0.154 \\ -0.115 & 0.192 \end{bmatrix} \begin{bmatrix} b_1 \\ 30 \end{bmatrix} \geq \begin{bmatrix} 0 \\ 0 \end{bmatrix}$$

establishes the bounds on b_1 as $24.06 \leq b_1 \leq 50.08$.

Computerized algorithms determine the bound on the resource limits for this kind of postoptimal analysis. Within these limits, one can calculate the impacts of resource modifications on profits by multiplying the appropriate shadow price times the change in the resource level; for example, relaxing the first constraint of the problem of Example 16.4 by two units ($\Delta b_1 = 2$) would increase profits by $0.192 \times 2 = 0.384$ units.

16.4 REVISING RATES OF PROFIT (COST) OR RESOURCE CONSUMPTION

Any revision in one coefficient in the objective function or a constraint represents an increment in a column vector of the unadjusted initial tableau of the original problem. All such revisions take the form $\mathbf{x}^* = \mathbf{x} + \Delta\mathbf{x}$, where $\Delta\mathbf{x}$ is a multiple of a unit vector. For example, if the x_1 column of the original problem were

$$\mathbf{x}_1 = \begin{bmatrix} 10 \\ 6 \\ 4 \end{bmatrix}$$

and the first element were revised upward by two units, the x_1 column of the modified problem would be

$$\mathbf{x}_1^* = \mathbf{x}_1 + \Delta\mathbf{x}_1$$

$$= \begin{bmatrix} 10 \\ 6 \\ 4 \end{bmatrix} + 2 \begin{bmatrix} 1 \\ 0 \\ 0 \end{bmatrix}$$

Therefore, the impact of any sequence of row manipulations on Δx could be determined by observing the impact on the unit vector shown. If the sequence of row manipulations which led to the optimal solution of the original problem were applied to the modified problem, the x_1 column would contain the numbers in the x_1 column of the optimal tableau plus 2 times the numbers in the S_1 column of the optimal tableau.

The next example demonstrates the process of assessing the impact of revisions in coefficients on the optimal solution, when the sequence of row manipulations for the original and modified problems can be assumed to be the same.

Example 16.5: Revised profit and resource rates Imagine that the limits of estimation error were ± 0.5 for the price factor of x_1 in Example 16.3. The initial and optimal tableaus of the original problem were

Initial tableau

x_1	x_2	S_1	S_2	$-P$	Limit
10	8	1	0	0	40
6	10	0	1	0	30
4	5	0	0	1	0

Optimal tableau

x_1	x_2	S_1	S_2	$-P$	Limit
1	0	0.192	-0.154	0	3.06
0	1	-0.115	0.192	0	1.16
0	0	-0.192	-0.346	1	-18.05

The x_1 column of the modified problem containing the upper limit of error $(+0.5)$ would be

$$\begin{bmatrix} 10 \\ 6 \\ 4.5 \end{bmatrix} = \begin{bmatrix} 10 \\ 6 \\ 4 \end{bmatrix} + 0.5 \begin{bmatrix} 0 \\ 0 \\ 1 \end{bmatrix}$$

After the same sequence of row manipulations the x_1 column of the modified problem would be

$$\begin{bmatrix} 1 \\ 0 \\ 0 \end{bmatrix} + 0.5 \begin{bmatrix} 0 \\ 0 \\ 1 \end{bmatrix} = \begin{bmatrix} 1 \\ 0 \\ 0.5 \end{bmatrix}$$

Therefore, instead of an optimal tableau for the modified problem the same sequence of row manipulations would result in the following unadjusted tableau:

Unadjusted preoptimal tableau

x_1	x_2	S_1	S_2	$-P$	Limit
1	0	0.192	-0.154	0	3.06
0	1	-0.115	0.192	0	1.16
0.5	0	-0.192	-0.346	1	-18.05

Adjusting—subtracting 0.5 times row 1 from the objective row—yields the new optimal solution.

Optimal tableau for the modified problem

x_1	x_2	S_1	S_2	$-P$	Limit
1	0	0.192	-0.154	0	3.06
0	1	-0.115	0.192	0	1.16
0	0	-0.287	-0.271	1	-19.58

Most computerized LP algorithms determine the ranges of variability in coefficients beyond which the original sequence of row manipulations would lead to infeasible solutions for the modified problem. Within these ranges, the analyst may investigate the impact of variations in price and resource consumption rates and the implications of uncertainty regarding their values on the optimal solution, by the method demonstrated above.

16.5 NEW CONSTRAINTS OR DECISION VARIABLES

Adding a new constraint can, at most, reduce the range of feasible solutions. If the new constraint is not binding, it cannot alter the problem. Then the optimal solution of the original problem is also the optimal solution of the modified problem. Obviously, when a new constraint is imposed, the first action must be to substitute the original optimal values of the decision variables into the new constraint to see if the original optimal solution satisfies it. If it does, the solution is still optimal. If it does not, simply add the new constraint to the optimal tableau of the original problem as a new row and adjust. If the adjusted tableau indicates nonfeasibility, the new constraint has removed the original optimal corner from the feasible region. Unless the feasible region has been entirely eliminated, pivoting on some negative element in the adjusted tableau will surmount this difficulty. The process is demonstrated in the next example.

Example 16.6: Adding a new constraint Imagine that management imposed a new constraint $3.4x_1 + 4.4x_2 \leq 15$ on the problem in Example 16.5. The original optimal solution, $x_1 = 3.06$, $x_2 = 1.16$ does not satisfy the new constraint, so the old optimal corner has been cut away. The starting tableau for the new problem is the original optimal tableau with the new constraint included as a new first row with a new slack variable, as shown.

Starting tableau for modified problem

x_1	x_2	S_1	S_2	S_3	$-P$	Limit	
3.4	4.4	0	0	1	0	15	} New constraint
1	0	0.192	−0.154	0	0	3.06	} Final tableau or original problem
0	1	−0.115	0.192	0	0	1.16	
0	0	−0.192	−0.346	0	1	−18.05	

The adjusted and optimal tableaus are given below; the pivot element was selected because the objective element of that column −0.192 is less negative than −0.346.

Adjusted tableau
Unit vectors in the x_1 and x_2 columns

x_1	x_2	S_1	S_2	S_3	$-P$	Limit
0	0	−0.14	−0.33	1	0	−0.6
1	0	0.192	0.154	0	0	3.06
0	1	−0.115	0.192	0	0	1.16
0	0	−0.192	−0.346	0	1	−18.05

Intermediate tableau

x_1	x_2	S_1	S_2	S_3	$-P$	Limit
0	0	1	2.40	−7.10	0	4.30
1	0	0	−0.61	1.30	0	2.30
0	1	0	0.47	−0.82	0	1.65
0	0	0	0.14	−1.42	1	−17.25

Optimal tableau

x_1	x_2	S_1	S_2	S_3	$-P$	Limit
0	0	0.15	1	-2.96	0	1.79
1	0	0.09	0	-0.51	0	3.39
0	1	-0.07	0	0.57	0	0.80
0	0	-0.02	0	-0.01	1	-17.50

Thus, the new optimal corner is $x_1 = 3.39$, $x_2 = 0.80$, and $S_2 = 1.79$, and the overall profit is reduced to 17.5 units. The new constraint definitely had an impact.

If one were to question the desirability of adding a new product or service, a new capital investment, or a new source of nutrients in a diet problem, one would have to investigate the impact of adding a new decision variable to the problem. This adds a column vector which is included in the initial tableau as the first column. The next example demonstrates a process for incorporating a new decision variable into the problem of Example 16.6.

Example 16.7: Adding a decision variable Suppose that in Example 16.6 a new service were contemplated which would consume resources at the rates of one unit per unit of output for the first resource and two units for the second resource. The net revenue margin for the new service might be 2 dollars per unit of service. The problem is to determine the impact this addition would have on P.

Call the new decision variable x_0. Then the initial tableau for this modified problem differs from that for the original problem by the inclusion of the new column:

$$x_0 = \begin{bmatrix} 1 \\ 2 \\ 2 \end{bmatrix} = \begin{bmatrix} 1 \\ 0 \\ 0 \end{bmatrix} + 2 \begin{bmatrix} 0 \\ 1 \\ 0 \end{bmatrix} + 2 \begin{bmatrix} 0 \\ 0 \\ 1 \end{bmatrix}$$

Clearly, the optimal tableau of the original problem must now be modified by the addition of a new first column equal to the S_1 column plus 2 times the S_2 column plus 2 times the $-P$ column, if it is to represent the modified problem after the same set of row manipulations have been performed. Thus, the preoptimal and optimal tableaus are

Preoptimal tableau for the modified problem

x_0	x_1	x_2	S_1	S_2	$-P$	Limit
-0.115	1	0	0.192	-0.154	0	3.06
0.269	0	1	-0.115	0.192	0	1.16
1.116	0	0	-0.192	-0.346	1	-18.05

Optimal tableau

x_0	x_1	x_2	S_1	S_2	$-P$	Limit
0	6.99	2.99	1	-0.503	0	24.90
1	2.99	4.99	0	0.499	0	14.97
0	-1.86	-4.94	0	-1.28	1	-29.48

The new solution indicates that the new service should replace the x_2 service and the x_1 service for a net increase in revenue of about $11\frac{1}{2}$ units.

16.6 ROLE OF THE COMPUTER

Standardized computer algorithms for solving LP problems greatly facilitate the conduct of sensitivity analysis to investigate the importance of uncertainties as well as to inquire into the impacts of adding new decision variables and new constraints. Once analysts have run a LP problem on a computer, they should extract as much additional information as possible by conducting extensive sensitivity analyses. The information so gained is likely to be of greater value to decision-making bodies than the optimal solution itself. To cite an example not discussed in this chapter, political or other considerations exogenous to a LP problem being analyzed could prohibit implementing the optimal solution. Thus, knowing the second-best or even third-best solution, which might actually be implementable, may be more important than knowing the optimal solution. By methods similar to those discussed in this chapter, these solutions may be extracted from the computer analysis also. (Output formats and even capabilities vary with computer systems. Your computer facilities staff will gladly demonstrate their system and explain the output for you.)

16.7 PROBLEMS

16.1 Multiply the following:

(a) 2 times $\begin{bmatrix} 1 \\ 0 \\ 1 \end{bmatrix}$

(b) 10 times $\begin{bmatrix} -2 \\ 6 \\ 3 \end{bmatrix}$

(c) $[8, 7, 3, 2, 5] \begin{bmatrix} 2 \\ 4 \\ 3 \\ 6 \\ 1 \end{bmatrix}$

(d) $[8, 7, 6, 4] \begin{bmatrix} 2 \\ 1 \\ 3 \end{bmatrix}$

(e) $[1, 0, 0] \begin{bmatrix} 0 \\ 1 \\ 0 \end{bmatrix}$

(f) $\begin{bmatrix} 2 & 4 & 3 \\ 1 & 7 & 2 \end{bmatrix} \begin{bmatrix} x_1 \\ x_2 \\ x_3 \end{bmatrix}$

(g)
$$\begin{bmatrix} 1 & 0 & 0 \\ 0 & 1 & 0 \\ 0 & 0 & 1 \end{bmatrix} \begin{bmatrix} x_1 \\ x_2 \\ x_3 \end{bmatrix}$$
(h)
$$\begin{bmatrix} 0 \\ 1 \\ 0 \end{bmatrix} \begin{bmatrix} 0, & 1, & 0 \end{bmatrix}$$

16.2 Find the optimal solutions for the original problem and for the variations shown. Use the methods of this chapter. Draw the diagram for the original problem, first.

(a) Original problem:

$$\text{Maximize} \quad P = 9x_1 + 10x_2$$

$$\text{Subject to:} \qquad 6x_1 + 9x_2 \leq 54$$

$$10x_1 + 4x_2 \leq 40$$

$$x_1, x_2 \geq 0$$

(b) Add the new constraint:

$$7x_1 + 6x_2 \leq 42$$

(c) New objective function:

$$P = 9x_1 + 12x_2$$

(d) New objective function:

$$P = 9x_1 + 15x_2$$

(e) New decision variable, x_0, where

$$\mathbf{x}_0 = \begin{bmatrix} 4 \\ 2 \\ 7 \end{bmatrix} \text{first} \quad \text{and} \quad \mathbf{x}_0 = \begin{bmatrix} 1 \\ 1 \\ 1 \end{bmatrix} \text{next}$$

(f) New constraint coefficients:

$$6x_1 + 8x_2 \leq 54$$
$$10x_1 + 5x_2 \leq 40$$

(g) New resource constraints:

$$\mathbf{b} = \begin{bmatrix} 5 \\ 2 \\ 0 \end{bmatrix}$$

16.3 You are an analyst for an agency. You have formulated an important decision problem as a LP problem. You have estimated coefficients for the objective function and constraint relations by analyzing agency data and by talking to agency personnel. The general form of your problem and probability distributions for the coefficients are given below. Explain the components of an extensive sensitivity analysis for this problem. Find the optimal solution using first the mean value and then the mean plus one standard deviation for b_2, a_{11}, and c_1.

(a) General problem:

$$\text{Maximize} \quad P = c_1 x_1 + c_2 x_2$$

$$\text{Subject to:} \qquad a_{11} x_1 + a_{12} x_2 \leq b_1$$

$$a_{21} x_1 + a_{22} x_2 \leq b_2$$

$$x_1, x_2 \geq 0$$

(b)	Coefficient	Mean	Standard deviation
	a_{11}	7	1.5
	a_{12}	6	0.5
	a_{21}	4	1.0
	a_{22}	10	2.0
	c_1	9	0.5
	c_2	10	0.5
	b_1	42	5.0
	b_2	40	2.0

16.4 A farmer is faced with the following situation. He has 1200 acres of land on which he can grow some mixture of corn and soybeans. He also has the opportunity of putting all but 500 acres in the land bank. He intends to decide how to use his land on a purely economic basis, using the information below. He will act to maximize his profit.

(*a*) He has a supply contract with a fertilizer company, which will guarantee him 3000 units of fertilizer if he wants it; but the price will fluctuate with the price and supply of natural gas, since it is essential to fertilizer production. At the same time he might be able to purchase a few hundred units more at the going price, if he works it right with another farmer.

(*b*) He has a supply contract with the seed company also. This would guarantee him enough corn seed to plant 1400 acres in corn and enough soybean seed to plant 800 acres, should he wish to. However, he will have to pay the market price at the time of purchase. Furthermore, the company can guarantee only that the quantity of seed purchased will plant the specified acreage within ± 10 percent, since seed is sold by the pound, not the grain.

(*c*) He is *nearly* certain that he can get a total of 8000 units of labor distributed as needed throughout the cropping year. However, the economic situation is such that he cannot disregard the possibility that only 6000 units will be available.

(*d*) To produce optimal yields he must use half again as much fertilizer per acre on the corn as he must on the soybeans. Corn requires three units per acre.

(*e*) The production of soybeans is more labor-intensive than corn. On the average, corn requires five units of labor per acre and soybeans require eight units per acre.

(*f*) Because of all the uncertainties in the economic situation of next year (consumer demand; foreign export; seed, fertilizer, and labor supplies; and, of course, aggregate United States production) he cannot accurately forecast his profit per acre for either crop. In order to make some reasonable estimates of the profit picture for the next year, he has gathered data on the prices (in constant dollars) farmers in his region have received in the past few years, as well as data on production changes from year to year. He has analyzed these data and added many other considerations to generate judgmental probability distributions for his profit per acre for corn and for soybeans:

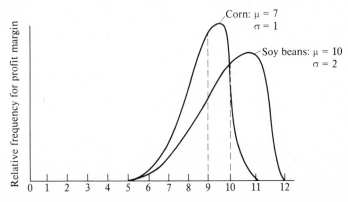

Questions:

1. Write this LP problem in mathematical form, with constraints and profit factors reflecting the farmer's expectations.

2. Solve the LP problem to determine the farmer's optimal planting scheme and his expected profit.

3. Based on this solution, what unit fee per acre would the government have to exceed to induce the farmer to retire his land (i.e., put it in the land bank)?

4. Analyze the sensitivity of this unit fee per acre (see 3) to the uncertainties (described above) in the seed, land, labor, and fertilizer constraints. To which of the constraints are these shadow prices (fees) sensitive, and how? If a possible change in a constraint would affect the shadow prices, find the new shadow prices and the new optimal solution.

5. Are the uncertainties in the profit factors of the objective function likely to affect the optimal allocation of land (optimal solution of the LP problem)? Answer this by determining whether adjustments of the profit factors used in the problem (mean values) by plus or minus one standard deviation would change the optimal solution. Indicate what the new optimal solutions would be, if any exist.

6. Finally, suppose at the last minute the farmer found he could get no corn seed, but everything else worked out exactly as expected. How should the farmer allocate his land between raising soybeans and retiring it if the government has decided to pay him 12 units per acre for the first 100 acres he retires, 11 units for the next 200 acres he retires, 10.5 units for the next 200 acres, 9 units for the next 200 acres, and, of course, nothing at all for land retired in excess of 700 acres. What will be his profit if he has no cost for retiring land? (Assume he has *exactly* 800 acres of soybean seed, 8000 units of labor, 3000 units of fertilizer, and the profit coefficient for soybeans is 10 units.)

16.5 The LP problem below has the optimal tableau shown. Calculate the additional profit to be realized for an increase of two resource units in the first constraint.

$$\text{Maximize} \quad P = 2x + y$$

$$\text{Subject to:} \quad 5x + 2y \leq 10$$

$$3x + 4y \leq 12$$

$$x, y \geq 0$$

Optimal tableau

x	y	S_1	S_2	b
1	0	8/35	−1/7	16/14
0	1	−3/14	5/14	30/14
0	1	−5/14	−1/14	$P - 62/14$

16.6 Solve the following problem using the simplex method. Find the limits on b_1 and b_2

$$\text{Maximize} \quad P = 2x + y$$

$$\text{Subject to:} \quad 5x + 2y \leq 10$$

$$3x + 5y \leq 15$$

$$x, y \geq 0$$

16.7 A company is forced into compliance with new air pollution standards. It produces one product on two production lines. Two technologies are available to reduce emissions. The most effective technology reduces emissions by seven pollution units per production unit and adds 2 dollars to overall unit production costs. The least effective technology reduces emissions by two units per unit of production and adds 1 dollar to overall unit production costs. The company produces four units of product per period and must reduce emissions by fourteen units per period on the combined production lines.

Write this problem as a least-cost LP problem to find the optimal production mix using the two production lines, which may have different pollution technologies applied to them. Carry out a postoptimality analysis on the resource limits.

16.8 SUPPLEMENTAL READING

Dorfman, Robert, and Henry D. Jacoby: "A Model of Public Decisions Illustrated by a Water Pollution Policy Problem," in R. Haveman, and J. Margolis Rand McNally, Chicago, 1970, chap. 7.

Hadley, G.: *Linear Programming*, Addison-Wesley Publishing Company, Inc., Reading, Mass., 1962, chap. 11.

Strum, Jay E.: *Introduction to Linear Programming*, Holden-Day, Inc., Publisher, San Francisco 1972, chap. 8.

Wagner, Harvey M.: *Principles of Operations Research*, 2d ed., Prentice-Hall, Inc., Englewood Cliffs, N.J., 1975, chap. 5.

SEVENTEEN

AN APPLICATION OF LINEAR PROGRAMMING AND A CASE FOR ANALYSIS

Linear programming has found numerous applications in a great variety of environmental policy studies. An application to air pollution regulatory policy is discussed in Section 17.1. Section 17.2 is a case for analysis. Data appropriate for an evaluation of alternative sewage management policies are included along with a description of the decision problem to be addressed.

17.1 AIR POLLUTION CONTROL FOR ST. LOUIS

Since 1969 Dr. Robert E. Kohn has been using LP methods to investigate policy options and contrast approaches to controlling air pollution in the St. Louis airshed. He has published many papers on his findings as well as two books, in which he summarizes his work. One piece of his work is described in this section. The LP problem he formulated to analyze the question of the economic feasibility of governmentally imposed regulations on the sulfur content of coal burned in the region is described and his findings are summarized. References to Dr. Kohn's work are provided in Section 17.3.

Dr. Kohn formulated a least-cost linear program for meeting imposed pollution limitations. The objective function is a linear cost function and the constraints are legislatively imposed pollution reduction requirements and projected production levels. Dr. Kohn considers a region in which the only polluting activity is cement production and where the technology, costs, and production levels are given in the following LP problem:

Minimize $C = \$0.14x_1 + 0.18x_2$

Subject to: $x_1 + x_2 = 2{,}500{,}000$ production constraint

$1.5x_1 + 1.8x_2 \geq 4{,}200{,}000$ pollution reduction requirement

$x_1, x_2 \geq 0$

where

- x_i is the production level to which abatement technology i is applied.
- $\$0.14$ and $\$0.18$ are the annual equivalent capital and operating costs for the first and second control technology alternatives.
- 2,500,000 pounds of cement are produced annually.
- The first technological alternative would reduce dust particulates by 1.5 pounds per barrel, from 2 pounds per barrel, and the second would reduce particulates by 1.8 pounds per barrel of cement produced.
- The industry has been required to reduce emissions by 4,200,000 pounds annually.

The basic issue investigated in the study was a regulation adopted by the Missouri Air Conservation Commission on March 14, 1967, prohibiting the burning of coal with greater than 2 percent sulfur content. The technological options for large power plants were to burn high-sulfur coal and desulfurize stack emissions, burn higher-priced sulfur coal, or switch to natural gas. For smaller power plants and industrial, commercial, and residential users, the first option was considered too expensive. Because coal burning also produces nitrogen oxides, particulates, hydrocarbons, and carbon monoxide, Dr. Kohn included them in his analysis on the assumption that the optimal utilization of technological alternatives would be influenced by emission regulations for these pollutants also.

The St. Louis airshed includes the city of St. Louis, St. Louis County, St. Charles County and Jefferson County in Missouri and Madison, St. Clair and Monroe Counties in Illinois. The total area is approximately 3600 square miles and the population in the region is about 2.5 million. Within the region, Dr. Kohn identified 94 emission sources for particulates and sulfur dioxide. Only 1965 emission data were available, so he projected emission rates (millions of pounds per year) to 1975. At the regional air quality monitoring station, he then related average annual concentrations q_i for pollutant i to emission flow rates by the linear equation

$$q_i = m_i e_i + b_i$$

where b_i represented background pollution (i.e., not generated in the region), e_i represented the emission rate, and m_i was a constant derived from observed data for 1963. He used this simple relation because adequate meteorological transport models were not available. He carefully explained how the inadequacies of this approach might have influenced this finding.

Table 17.1 Emission reductions required

Pollutant (i)	Concentration goal* (q_i')	Emission reduction in 1975 (millions of pounds) (r_i)
Carbon monoxide	5 ppm	1865
Hydrocarbons	3.1 ppm	525
Nitrogen oxides	0.069 ppm	110
Sulfur dioxide	0.02 ppm	990
Particulates	75 μg/m^3	165

Source: Kohn, 1971, "Application of Linear Programming to a Controversy on Air Pollution Control," *Management Science*, vol. 17, no. 10, June 1971, Table 2.
* ppm = parts per million.
μg/m^3 = micrograms per cubic meter.

Using his relation, Dr. Kohn converted air quality goals into required emission reduction rates as follows: If the air quality goal for pollution i were q_i', then $e_i' = (q_i' - b_i)/m_i$ is the maximum allowable emission flow and the required reduction in emission rates is:

$$r_i = e_i - e_i'$$

These numerical values for q_i' and r_i are summarized in Table 17.1, for major pollutants.

The generalized LP problem for the analysis contained nine technological decision variables and took the form

$$\text{Minimize} \quad \mathbf{C} = \mathbf{cx}$$

$$\text{Subject to:} \quad \mathbf{Ax} \leq \mathbf{s}$$

$$\mathbf{Bx} \geq \mathbf{r}$$

$$\mathbf{x} \geq 0$$

where $\mathbf{c}$ = nine-dimensional cost vector associated with the nine-dimensional decision vector $\mathbf{x}$
$\mathbf{A}$ = production rate matrix for the 94 polluting sources
$\mathbf{S}$ = vector of production requirements for the polluting sources
$\mathbf{B}$ = abatement matrix for each technology for each source
$\mathbf{r}$ = vector of required abatement levels for each pollution type

The general categories of pollution sources are listed in Table 17.2. Each category includes numerous specific activities, which total 94.

After finding the least-cost technology mix, which could reduce the emissions in Table 17.2 by 990 million pounds of sulfur dioxide, Dr. Kohn extracted from it that portion of the solution which was relevant to coal-burning in the region. Coal consumption projected for 1975 technologies employed to meet

Table 17.2 Sulfur emissions in the absence of abatement in St. Louis region in 1975

Category	Emission rate (10^6 lbs/yr)
Transportation	13.6
Combustion of fuel oil	55.1
Combustion of coal by industry	176.5
Combustion of coal by commercial and institutional users	15.1
Combustion of coal by utilities	755.2
Combustion of natural and by-product gases	44.7
Combustion of refuse	2.2
Industrial processes	273.4
Combustion of coal in residences	54.2
Total	1390.0

sulfur concentration regulations and annual costs of the control methods are summarized in Table 17.3.

Dr. Kohn's complete solution was so detailed that he was able to identify optimal technologies for specific activities on the basis of combustion efficiencies and the characteristics of the technological solutions available. He concluded that less low-sulfur coal should be consumed than others had forecast. This he attributed to the combustion efficiency data and to the impact of abatement regulations for other pollutants, especially nitrogen oxides.*

Dr. Kohn tested the sensitivity of his findings to variations in prices of natural gas and low-sulfur coal, to variations in Public Health Service estimates of nitrogen oxide emission rates for different boiler types, and to variations in efficiencies for desulfurization technology. Dr. Kohn's work provided much insight into costs, the interrelationships between concentrations of different pollu-

---

* His linear program accounted for five pollutants. Technologies employed to control one pollutant also affect emissions of other pollutants and the LP method takes advantage of these interactions in arriving at a least-cost solution. Thus, although low-sulfur coal controls sulfur emissions, the additional requirement to control nitrogen oxides simultaneously makes switching to gas the least-cost decision for some pollution sources.

Table 17.3 Least-cost solution for coal-burning in St. Louis

Technology	10^6 tons of coal	Annual costs of control (10^6)
Desulfurization of stack gases	8.27	$8.31
Conversion to low-sulfur coal	0.49	1.30
Conversion to natural gas	0.75	3.35
Conversion to coke oven gas	0.03	−0.01*
Remote electricity generation	2.07	2.07

* Indicates a cost saving due to more efficient utilization of resources.

tants and various control technologies, and other pollution control options. It provided an analytical basis for discussion and a framework within which various policy questions could be investigated. Because questions on pollution control continue to arise, his approach may well find continued use.

17.2 SLUDGE MANAGEMENT OPTIMIZATION FOR THE BISTATE REGION: A CASE FOR ANALYSIS

The Bistate Sewage Management Agency was chartered this year after many years of interstate and intrastate political struggle. The purpose of the agency is to develop and manage an efficient sewage system for the bistate metropolitan area. The agency receives tax monies from both states and charges users' fees. It has the authority to invest capital as it sees fit, although it must have the approval of the Bistate Utilities Commission in order to issue bonds to raise capital.

There is widespread agreement among politicians on both sides of the river that sludge management programs in each state are wholly inadequate even when viewed in the perspective of single-state operations. Politicians and citizens alike have made it clear to the agency that a coordinated, economically efficient, environmentally sound system using existing facilities and deposition sites is to be the highest-priority objective of the new agency. Once that is achieved, longer-range issues may be addressed.

Assignment ·

You are an analyst working for the agency. You are part of a study team formed to address the immediate problem of optimizing the existing system and consider subsequently long-term issues such as preparing to meet future needs. With others of your team you have inventoried sewage treatment facilities, incinerators, and landfill sites currently in use on both sides of the river. You have estimated capacities for these and conducted site visits to observe their operation. You have marked out roads linking treatment plants to landfill sites and incinerators and measured the mileage along each route. Existing and possible future sites and routes are illustrated in Figure 17.1.

In addition, your engineer economist and accounting people have compiled cost data. They have data on transportation costs composed of labor, operations, and amortization charges per ton of sludge moved per mile (cost per ton-mile). They have calculated fixed and variable costs per ton processed at treatment and incineration facilities. They know the rated processing capacity at these facilities and how capacity varies as a function of air and water pollution regulations. They also have projections for sludge disposal capacity requirements over the next decade and capital requirements over the next decade should new incinerator or landfill sites be acquired; processing facilities are considered adequate. Some of these data are summarized in the next subsection.

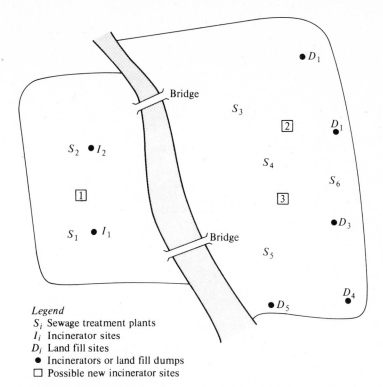

Legend
S_i Sewage treatment plants
I_i Incinerator sites
D_i Land fill sites
● Incinerators or land fill dumps
□ Possible new incinerator sites

Figure 17.1 Sewage treatment system for the Bistate region.

Your immediate job is to find the most efficient sludge disposal system for the current situation within environmental constraints. Ignore possible political considerations until you have found an optimal routing network. Then postulate political constraints which might affect your solution and investigate their impacts. (Think transportation problems.)

Next, consider the possible future need for another incinerator. Find the optimal location for it, again ignoring political considerations initially. Be sure to account for uncertain environmental constraints. Finally, postulate typical realistic political constraints on siting and on timing of capital investments and investigate the economic and other impacts these could have.

Prepare an efficiently written, thorough technical memorandum for your agency covering your work. Include a carefully written executive summary and table of contents. Remember, executives often read only executive summaries, introductions, conclusions, and recommendations; yet they must understand the options and their relative merits.

Data Summaries

All sewage must be preprocessed to remove water. Sludge with 70 percent of the water removed may be deposited at landfill sites. However, 80 percent of the

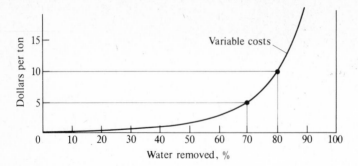

Figure 17.2 Costs of water removal (per ton of dewatered sludge produced). Fixed costs: 60,000 dollars per annum.

water must be removed prior to incineration. Eighty percent of grit containing heavy metals must be removed prior to landfill. Grit represents 1 percent of 70 percent dewatered sludge by weight. There is a linear relation between percent of grit removed and percent of heavy metals removed. Grit is always incinerated. The residue from incineration is an inert, odorless ash. The heavy metals go off as stack emissions. The ash is chlorinated and dried prior to disposal as clean landfill.

Variable and fixed costs of water and grit removal are given in Figures 17.2 and 17.3. Incineration costs for existing facilities are given in Table 17.4. Fixed costs include amortized capital, interest, insurance, and all operating costs which are incurred just to keep the facilities operational. Variable costs are the marginal costs of operation including maintenance, energy, workforce, and the like per ton processed. Total annual preprocessing costs per ton of dewatered sludge produced are the sums of fixed and variable costs for each operation.

Disposal costs for a ton of sludge at a landfill site is a function of the number of tons disposed of per period. Landfill sites are leased at fixed annual

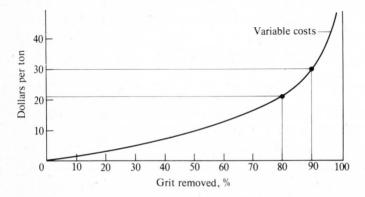

Figure 17.3 Costs of grit removal (per ton of grit produced). Fixed costs: 20,000 dollars per annum.

Table 17.4 Incineration costs of 80% dewatered sludge for existing operating constraints

Fixed costs per incinerator = $100,000
Variable costs per ton incinerated = $30.00

rates and maintained at constant levels. The variable portion of the annual cost of a landfill site is the per-ton dumping charge. Fixed and variable costs for existing sites are given in Table 17.5.

Dry ash from incinerators is disposed of at no cost to the agency.

Trucking costs for one-way hauls from sewage treatment sites to incinerators or landfill sites may be calculated from the mileage data in Table 17.6.

Table 17.5 Disposal costs and annual sludge capacity for landfill sites
Each site has a 20-yr lifetime from the present

Site no.	Fixed cost ($/yr)	Dump charge ($/ton)	Annual capacity (tons/yr)
D_1	1000	0.50	50,000
D_2	1000	0.60	50,000
D_3	1200	0.90	75,000
D_4	4000	1.00	125,000
D_5	5000	1.50	100,000

Table 17.6 Shortest one-way haul mileage

Disposal sites	Sewage treatment sites					
	S_1	S_2	S_3	S_4	S_5	S_6
I_1	2	8	12	12	13	16
I_2	7	2	17	19	20	35
D_1	30	22	5	8	12	9
D_2	25	24	9	7	9	2
D_3	24	34	12	8	7	4
D_4	27	39	20	12	9	9
D_5	26	43	20	10	8	11
	Distances for proposed sites					
1	4	4	20	22	16	19
2	28	19	3	3	15	6
3	21	30	10	5	8	5

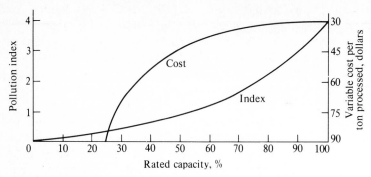

Figure 17.4 Pollution and processing cost by incinerator capacity.

Trucking costs including all operating costs are given in Table 17.7 for two levels of dewatering. These data assume truck leasing. Thus there are no capital costs to account for and trucks are fully utilized. Fixed costs not included are negligible.

Air pollution standards are about to be revised. Currently an index of 4 is

Table 17.7 Trucking costs for one-way haul mileage

Dollars per ton of material

% dewatered	Mileage				
	5	10	20	40	80
70	13	14	16	20	27
80	16	20	24	36	50

Table 17.8 Sludge volume and disposal capacities

Sewage treatment site	Volume* of sludge (tons/yr)		Sludge disposal site	Rated capacity (tons/yr)	Cost per ton processed
	Now	In 10 yrs			
S_1	70,000	135,000	I_1	90,000	—
S_2	85,000	155,000	I_2	90,000	—
S_3	70,000	90,000	D_1	50,000	—
S_4	80,000	100,000	D_2	50,000	—
S_5	70,000	100,000	D_3	75,000	—
S_6	25,000	45,000	D_4	125,000	—
			D_5	100,000	—
			I_2 (new)	120,000	50–80†

* Average volume for 70 and 80 percent dewatered sludge.

† Assuming a pollution index level of 2. These costs include fixed and variable costs. The range accounts for uncertainty in level of use.

applied to incineration. The operating costs of Table 17.4 are appropriate for this index level. In 10 years the required index level will probably be 2. To improve air quality (decreasing index level) incinerators must be operated at reduced capacity, as shown in Figure 17.4. This in turn changes the cost per ton processed.

Currently the combined capacity of the incinerators and dumps exceeds the demand placed upon the system by existing industrial and residential communities. However, in 10 years demand is expected to exceed capacity. An additional incinerator with the capacity indicated in Table 17.8 will meet the new demand easily. If an incinerator is to be built, design and financing efforts must be initiated within the next 2 years. Using the data provided determine which site shall be selected.

17.3 SUPPLEMENTAL READING

Atkinson, Arthur, and Richard S. Gaines, (eds.): "Development of Air Quality Standards," Environmental Resources, Inc., Charles E. Merrill Publishing Company, Riverside, Calif.

Darby, William P., Paul J. Ossenbruggen, and Constantine J. Gregory: "Optimization of Urban Air Monitoring Networks," *Journal of the Environmental Engineering Division, Proceedings of the American Society of Civil Engineers*, vol. 100, no. EE3, June 1974, pp. 577–591.

Kohn, Robert E.: "Air Quality, the Cost of Capital, and the Multiproduct Production Function," *Southern Economic Journal*, vol. XXXVIII, no. 2, October 1971, pp. 156–160.

————: "Application of Linear Programming to a Controversy on Air Pollution Control," *Management Science*, vol. 17, no. 10, June 1971, pp. B-609 to B-621.

————: "A Cost-Effectiveness Model for Air Pollution Control with a Single Stochastic Variable," *Journal of the American Statistical Association*, vol. 67, no. 337, March 1972, pp. 19–22.

————: "Industrial Location and Air Pollution Abatement," *Journal of Regional Science*, vol. 14, no. 1, 1974, pp. 55–63.

————: "Input-Output Analysis and Air Pollution Control," in Edwin S. Mills (ed.), *Economic Analysis of Environmental Problems*, Columbia University Press, New York, 1975.

————: "Joint-Outputs of Land and Water Wastes in a Linear Programming Model for Air Pollution Control," *1970 Social Statistics Section Proceedings of the American Statistical Association*.

————: "Labor Displacement and Air-Pollution Control," *Operations Research*, vol. 21, no. 5, September–October, 1973, pp. 1063–1070.

————: *A Linear Programming Model for Air Pollution Control*, MIT Press, Cambridge, Mass., 1978.

————: "Linear Programming Model for Air Pollution Control: A Pilot Study of the St. Louis Airshed," *Journal of the Air Pollution Control Association*, vol. 20, no. 2, February 1970, pp. 78–82.

————: "A Mathematical Programming Model for Air Pollution Control," *School Science and Mathematics*, June 1969, pp. 487–494.

————: "Optimal Air Quality Standards," *Econometrica*, vol. 39, no. 6, November 1971, pp. 983–995.

————: "Price Elasticities of Demand and Air Pollution Control," *The Review of Economics and Statistics*, vol. LIV, no. 4, November 1972, pp. 392–400.

———— and Donald E. Burlingame: "Air Quality Control Model Combining Data on Morbidity and Pollution Abatement," *Decision Sciences*, July 1971, pp. 300–310.

———— and Eric Weger: "Pollution Control by Locational Change and Technological Abatement," *Journal of the Air Pollution Control Association*, vol. 23, no. 12, December 1973, pp. 1045–1047.

Russell, Clifford S., and William J. Vaughan: *A Linear Programming Model of Residuals Management for Integrated Iron and Steel Production*, Resources for the Future, Inc., Washington D.C., July 1974.

FIVE

FORMAL DECISION ANALYSIS

Formal decision analysis is a *normative paradigm* or methodology for evaluating the relative merits of alternative courses of action when the consequences are uncertain. It is *dynamic*. It entails early analysis as a guide to initial decision making, observation of short-term impacts, and changing situations, and further analysis as a guide to revising decisions. The basis of decision analysis is the seminal work, *Theory of Games and Economic Behavior*, by John Von Neumann, a mathematician, and Oscar Morgenstern, an economist, published in 1947. The formal methodology, as it exists today, is due primarily to Howard Raiffa and his colleagues, who have developed it into a practical and useful tool.

Perhaps the greatest benefit to be gained from studying formal decision analysis is skill in structuring decision problems and selectively applying analytical techniques to evaluate alternatives. Decision making under uncertainty is central to human existence, and formal decision analysis provides a rational framework for analyzing alternatives. Whether one is considering daily personal choices, alternative public programs, public policy issues, or capital investment options, the analytical process is the same and the types of information needed are similar.

Formal decision analysis consists of six major steps. These are:

- *Establish goals and objectives.* Without these it is impossible to evaluate the relative merits of alternative courses of action.
- *Develop measurable criteria of effectiveness.* These are the criteria by which decision makers will judge the potential success of alternative courses of action, and against which progress toward goal achievement will be measured.
- *Identify alternative courses of action* by which the goals and objectives may be achieved. These may be chronological sequences of decisions and information acquisition.
- *Identify possible indirect impacts of each alternative* and develop measurable indicators for these if possible.

- *Probabilistically link alternative courses of action with potential outcomes.* These probabilistic links measure the strength of the relation between actions and consequences and may be purely judgmental, purely experimental (statistical), or mixed.
- *Select the course of action for which the expected utility is largest.* When the utility function varies linearly with payoffs, the rule is to select the alternative offering the largest expected payoff.

Chapters 18, 19, and 20 develop the analytical facets of decision analysis. Throughout, goals and objectives are assumed. This may rankle experienced managers, because they know full well that establishing goals and setting objectives is priority business. In the absence of these, no amount of analysis will lead to good decisions. However, the purpose of this text is to communicate methods of analysis. The process of setting objectives is familiar to most managers and is covered extensively in the literature on general management.

There are two distinct formats for structuring decision problems. One is called *extensive* and the other *normal*. Both are illustrated in Chapter 18. The preferability for each format in specific situations is discussed, and calculational algorithms are illustrated for each.

In real decision situations managers usually have opportunities to delay action until additional information is acquired. This is sometimes called *reducing uncertainty*. Of course, delays consume resources; these are indirect costs of acquiring more information. Furthermore, information acquisition can have direct costs. Consequently, managers are faced with the question of whether or not additional information is worth its costs. Formal decision analysis includes a procedure for calculating the expected value of additional information. This issue is discussed at length in Chapter 19.

The concept of utility is introduced only in Chapter 20. This is done for two reasons. First, all the material in Chapters 18 and 19 is operational whether consequences are measured in monetary, utility, or other units. This is important, because utility functions for complex multiattribute decision problems may be impossible to assess. If one could not utilize the concepts of maximum expected return and the marginal value of additional information without including explicit utility functions, decision analysis would be of limited practical value. As it is, the very process of structuring the problem and collecting and evaluating the data according to the decision analysis algorithm can often provide sufficient insight to rationalize a course of action, without reference to explicit utility functions.

The second reason for postponing an introduction of utility until the end is that it is a difficult concept to grasp and operationalize even for simple situations. If it were introduced too soon, it would probably interfere with the learning of other material.*

---

* The definitive work on multi-attribute utility theory and applications was written by Ralph Keeny and Howard Raiffa. Their book is entitled, *Decisions with Multiple Objectives; Preferences and Value Tradeoffs* (John Wiley & Sons, New York, 1976).

EIGHTEEN

DECISION TREES AND ALGORITHMS

Many rules for decision making in the face of uncertainty have been advanced. There is, in fact, a whole field of endeavor called the mathematical theory of decisions. To illustrate a few of the many rules consider a problem where three choices are possible, D_1, D_2, D_3, and the payoffs for each choice vary according to which of four possible scenarios are eventually realized. In Table 18.1, all the payoffs are positive, so even if one knew nothing of the uncertainties, one could make a rational choice. In other words, one could choose that decision which maximizes the minimum gain, namely, D_3; this is called the *maximin rule*.

Consider the same table of payoffs, with the exception that all payoffs are negative. A rational decision is easy here, also, even with no knowledge of the likelihood values for the payoffs. Simply choose that decision which minimizes the maximum loss, namely, D_1; this is called the *minimax rule* and is commonly employed in games against an opponent.

Neither of these rules makes very much sense when some payoffs are positive and others are negative. In that case one needs some estimate of the likelihoods of the various outcomes. Table 18.2 illustrates this point. According to the maximin rule, D_2 and D_3 are equally good choices. However, suppose the likelihood of outcome S_1 is 1.0. In that case D_3 is clearly optimal, for D_2 would net a loss of 10 units whereas D_3 would net a gain of 18 units; D_2 and D_3 are not equally good choices. Similarly, according to the minimax rule, D_1 is the best decision; yet if the likelihood of outcomes S_4 is very large, D_3 is clearly preferred.

The fact is that the minimax and maximin rules are reasonable only when one has no reason to believe that the likelihoods of the various outcomes are unequal. Then, the minimax rule is reasonable if all payoffs are negative and the maximin rule is reasonable if all payoffs are positive. If the payoffs are of mixed sign or likelihoods are not uniform, these rules lose their credibility.

Whereas the minimax and maximin decision rules focus on the magnitudes of payoffs without attention to the probabilistic nature of decision making, there

Table 18.1 Payoffs

Decision	Possible outcome or scenario			
	S_1	S_2	S_3	S_4
D_1	10	3	15	5
D_2	10	5	16	3
D_3	18	6	4	5

is a rule which borders on the other extreme. This might be called the *most likely outcome rule*. Here one chooses the course of action which has the maximum payoff under the assumption that the most likely outcome in fact obtains. Consider the data in Table 18.3. According to the most likely outcome rule, D_1 is the best decision if $P(S_1)$ is largest, D_2 is the best decision if $P(S_2)$ is largest, and D_1 and D_2 are equally good if $P(S_3)$ is largest.

The difficulty with this rule is that it ignores relatively large likelihoods of substantial negative outcomes. In this example, if $P(S_1) = 0.4$, $P(S_2) = 0.5$, and $P(S_3) = 0.1$, the likelihood rule would select D_2, with a payoff of 9 units, if b were true. Yet, there is a 40 percent chance that S_1 is true, with a payoff of -10. Decision D_2 does not seem to be the unequivocally best choice in this situation. Clearly, the maximum-likelihood rule has greater credibility when one possible outcome is very nearly certain. However, when the likelihoods of outcomes are relatively uniform, its credibility is minimized.

If a decision rule is to be generally credible, it must somehow balance the possible payoffs in a decision situation against the likelihoods of their occurring. A widely accepted decision rule which does this is the *maximum expected value rule*, first introduced in Chapter 13. Suppose decision D can produce three possible outcomes S_1, S_2, and S_3 which are exhaustive and mutually exclusive, and the likelihoods are $P(S_1)$, $P(S_2)$, and $P(S_3)$. Let the payoff values be $V(S_1)$, $V(S_2)$, and $V(S_3)$; since individuals may differ on the value of an outcome, the V's here represent either the individual or group values appropriate to the problem. Then the expected value of this decision is:

$$E(D) = P(S_1) \times V(S_1) + P(S_2) \times V(S_2) + P(S_3) \times V(S_3)$$

Table 18.2 Payoffs

Decision	Possible outcome or scenario			
	S_1	S_2	S_3	S_4
D_1	10	3	15	-5
D_2	-10	5	16	-3
D_3	18	-6	-4	5

Table 18.3 Payoffs

Decision	Possible outcome or scenario		
	S_1	S_2	S_3
D_1	10	−8	5
D_2	−10	9	5

With this rule a balance between payoff values and likelihoods is struck by taking a weighted average of the payoff values. A low payoff with a high likelihood may count as much as a large payoff with a small likelihood. Notice, too, that this rule accounts for all payoffs and all likelihoods whereas the other decision rules do not.

The decision rule for formal decision analysis is an extension of the maximum expected value rule. The concept "value" is generalized to utility. *Utility functions* are analytical expressions of relative preferences for measurable consequences and of attitudes toward risk. In relatively simple decision situations for individual decision makers at particular points in time, utility functions can be assessed. However, in complex decision situations, which include many outcome variables and/or more than one decision maker, assessing utility functions can be impractical or impossible. Then the maximum expected utility decision rule is inoperative. One must then resort to maximizing expected gains or minimizing expected losses in other units of measurement, such as lives saved or monetary gain.

18.1 DECISION TREES

Conceptually, it is convenient to view decision problems as branching trees with decision nodes for alternative choices and chance nodes for uncertain outcomes. There are two forms of decision tree. One is simply a chronological ordering of choices and possible outcomes. This is called the *extensive form*, because every decision point and every array of chance outcomes is explicitly indicated. The other is called the *normal form*. It is more compact and explicitly indicates only abbreviated strategies of action and the possible scenarios which describe the major contingencies. Since analysis of extensive form trees is the easiest to follow, the normal form is introduced last even though it offers some important analytical advantages.

Extensive Form Trees

Consider the elementary decision problem summarized in Table 18.4. There are two competing projects A and B, which have uncertain NPE payoffs. The uncer-

Table 18.4 Likelihoods and payoffs

State	Likelihood	Net present equivalents Project A	Net present equivalents Project B
S_1	0.2	10,000	20,000
S_2	0.6	8,000	−2,000
S_3	0.2	8,000	10,000

tainties arise from the fact that the payoffs are associated with scenarios or events over which the manager has incomplete control; these are sometimes called *states of nature*. The scenarios are mutually exclusive and exhaustive. This is a requirement for every chance node in a decision tree, because the calculational algorithm is based on the assumption that chance nodes represent probability distributions.

The extended form decision tree for this decision problem is shown in Figure 18.1. Notice that the same set of uncertain events is associated with each alternative. These are the so-called states of nature which influence payoffs and are generally beyond the control of the decision maker(s). Obviously, uncertain events which do not influence payoffs are not incorporated into decision trees, because outcomes are independent of such events. If the payoffs for any alternative are identical for each state of nature, that chance node need not be included in the tree.

To evaluate this problem just calculate the expected payoff for each decision option and choose the one with the greatest expected return, namely,

$$E(A) = -0.2 \times 10,000 + 0.6 \times 8000 + 0.2 \times 8000 = 4400$$

$$E(B) = 0.2 \times 20,000 - 0.6 \times 2000 + 0.2 \times 10,000 = 4800$$

Since $E(B) > E(A)$, decision B is the best choice, based on the information provided.

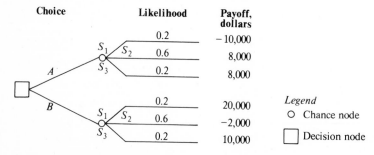

Figure 18.1 Extensive form decision tree.

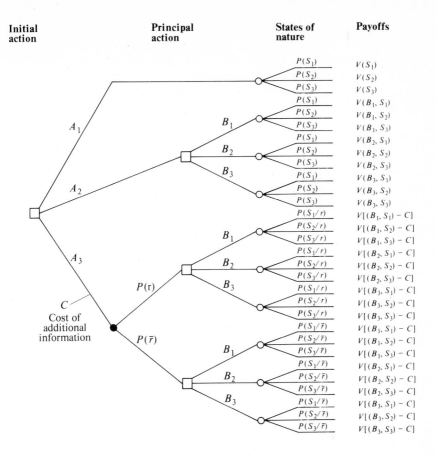

Figure 18.2 Decision tree including one intermediate observation: extensive form.

A more complex decision problem is diagramed in Figure 18.2. Included are conditioning events for one course of action which are inappropriate for the other. Probabilities, costs, and payoffs are indicated symbolically; they are not assigned specific values. The payoffs are functions of the decisions made along each branch and of the state of nature. They may be in the form of NPEs, annual or future equivalents, or the decision makers' utility payoffs.

Take particular note of the fact that probabilities are conditioned only on previous chance outcomes, not on previous decisions, whereas payoffs are functions of both the actions taken and the state of nature, which is a chance outcome.

The branch labeled A_1 represents no action taken and the acceptance of whatever chance offers. Call it the status quo. It may be excluded only if the payoffs for the other branches represent increments over the status quo. Branch A_2 represents the option to elect to do B_1, B_2, or B_3 on the basis of existing information alone. Branch A_3 represents the option to purchase additional information, at a cost of C dollars, to better understand which state of nature is more likely, before choosing among B_1, B_2, and B_3. The actions B_1, B_2, and B_3 are the principal choices in the decision problem. They could represent project alternatives, capital investments of other types, political alternatives, actions a physician might take on behalf of a patient, or any number of other types of alternative choices.

The algorithm for evaluating this complex tree is precisely the one already illustrated, namely, calculate the expected net payoff for each action and choose the one with the greatest expected return. The analysis of complex trees begins at the extreme tips and progresses backward through the tree. For example, evaluate the A_2 branch in Figure 18.2 as follows.

Step 1:

$$E(A_2, B_1) = P(S_1)V(B_1, S_1) + P(S_2)V(B_1, S_2) + P(S_3)V(B_1, S_3)$$

$$E(A_2, B_2) = P(S_1)V(B_2, S_1) + P(S_2)V(B_2, S_2) + P(S_3)V(B_2, S_3)$$

$$E(A_3, B_3) = P(S_1)V(B_3, S_1) + P(S_2)V(B_3, S_2) + P(S_3)V(B_3, S_3)$$

Step 2: Keep only the choice with the largest expected payoff in step 1. Assign that value to the A_2 branch. It will be used to compare A_2 with A_1 and A_3.

The final comparison of the A branches establishes the best course of action which (assuming A_3 is best) is of the type:

Solution: Choose A_3. If result r is observed, choose action B_1. If result $\bar{r}$ is observed, choose B_3.

Normal Form Trees

The decision problem described in Table 18.4 can also be diagramed in terms of strategies. A *strategy* specifies which action to take in the event of specific contingencies. The solution of extensive form analysis is, in fact, a strategy. This particular decision problem is quite elementary in that the strategies are single actions. The normal form tree depicts each strategy conditionally on each state likelihood, as shown in Figure 18.3. The payoffs shown are expected payoffs conditioned on the state of nature.

To evaluate a normal form tree use s to symbolize a strategy and calculate the products

$$E(s) = -0.2 \times 10,000 + 0.6 \times 8000 + 0.2 \times 8000 = 4400; s = A$$

$$E(s) = 0.2 \times 20,000 - 0.6 \times 2000 + 0.2 \times 10,000 = 4800; s = B$$

Since the expected return for strategy B is greatest, select B as the course of action.

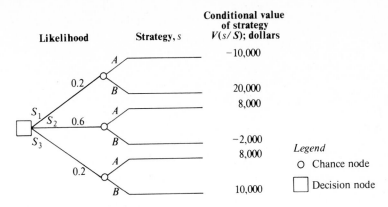

Figure 18.3 Normal form tree.

The more complex problem symbolized in Figure 18.2 is a bit more interesting to analyze in normal form. There are 19 strategies for action and observation for this decision problem. These are listed in Table 18.5. Again the symbol s_i is a shorthand notation for strategy i. The 10 unique and contending strategies are diagramed as a normal form tree in Figure 18.4.

Decision problems of this complexity are quite common. Numerical examples are evaluated in subsequent sections.

Table 18.5 Strategies for the general problem

A_1
  s_1: Status quo

A_2
  s_2: Choose B_1
  s_3: Choose B_2
  s_4: Choose B_3

A_3
  s_5: If r, choose B_1. Otherwise choose B_2
  s_6: If r, choose B_1. Otherwise choose B_3
  s_7: If r, choose B_2. Otherwise choose B_1
  s_8: If r, choose B_2. Otherwise choose B_3
  s_9: If r, choose B_3. Otherwise choose B_1
  s_{10}: If r, choose B_3. Otherwise choose B_2
  s_{11}: If $\bar{r}$, choose B_1. Otherwise choose B_2 (same as s_7)
  s_{12}: If $\bar{r}$, choose B_1. Otherwise choose B_3 (same as s_9)
  s_{13}: If $\bar{r}$, choose B_2. Otherwise choose B_1 (same as s_5)
  s_{14}: If $\bar{r}$, choose B_2. Otherwise choose B_3 (same as s_{10})
  s_{15}: If $\bar{r}$, choose B_3. Otherwise choose B_1 (same as s_6)
  s_{16}: If $\bar{r}$, choose B_3. Otherwise choose B_2 (same as s_8)
  s_{17}: Choose B_1 (no advantage over s_1)
  s_{18}: Choose B_2 (no advantage over s_2)
  s_{19}: Choose B_3 (no advantage over s_3)

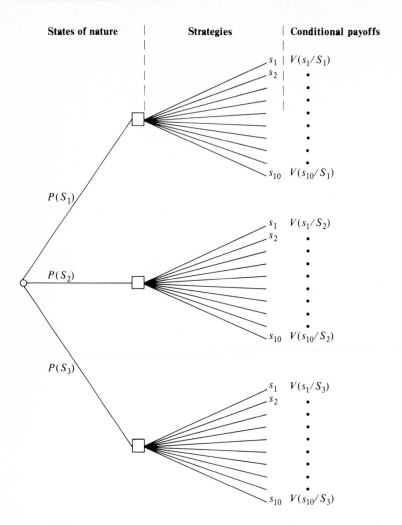

States of nature | Strategies | Conditional payoffs

s_1 | $V(s_1/S_1)$
s_2
s_{10} $V(s_{10}/S_1)$

$P(S_1)$

s_1 $V(s_1/S_2)$
s_2
s_{10} $V(s_{10}/S_2)$

$P(S_2)$

$P(S_3)$

s_1 $V(s_1/S_3)$
s_2
s_{10} $V(s_{10}/S_3)$

Figure 18.4 Decision tree including one intermediate observation; inverted tree for normal form.

18.2 EQUIVALENCE OF EXTENSIVE AND NORMAL FORMS

When the maximum expected utility decision rule is employed, the extensive and normal forms of analysis lead to the same choice of action, provided that probabilities and payoffs have been consistently assigned in the analyses.

Since the two forms of analysis are equivalent, the choice of which to use for a particular problem depends on convenience and available information. This topic is the subject of the next section.

The following discussion demonstrates the conversion of an extensive form tree to a normal form tree. Both trees are then evaluated and the best option is indicated. Monetary payoffs are used as proxies for utility values. This is legitimate so long as the utility function is linear, as explained in Chapter 20.

Imagine that two projects have been proposed, B_1 and B_2. Sufficient funds to implement only one project are available. It is not clear which project offers the largest NPE, because many future uncertainties will influence final outcomes. Working with the responsible public manager, analysts have developed scenarios which represent the two possible states of nature, S_1 and S_2. Estimated incremental payoffs for choosing B_1 or B_2 as opposed to holding the status quo, under assumptions that S_1 or S_2 will actually be the state of nature, are shown in Table 18.6. Likelihood estimates for S_1 and S_2 are also shown.

Analysts have devised a scheme for obtaining additional information regarding the likelihoods of S_1 and S_2. Using historical evidence they have estimated frequencies with which results r or $\bar{r}$ from their scheme would predict states S_1 and S_2. Their estimates are:

$$P(S_1/r) = 0.8 \qquad P(S_2/r) = 0.2$$
$$P(S_1/\bar{r}) = 0.4 \qquad P(S_2/\bar{r}) = 0.6$$

Using these estimates, they calculated $P(r) = P(S_1)P(r/S_1) + P(S_2)P(r/S_2) = 0.5$.

On the basis of these likelihoods their scheme seems valuable, but it would cost 4000 dollars. The question they must answer before recommending this course of action is whether or not the additional information is actually worth that much. The answer to this question is at the heart of subsequent calculations.

The extensive form decision tree for the problem is shown in Figure 18.5, with all estimates, including the 4000 dollars, indicated on appropriate branches. The expected net present benefits for each decision sequence, shown at the base of the tree, were calculated starting at the extreme right and moving back along the branches calculating expected values for each chance node and exercising the maximum expected value (utility) rule at each decision node. The steps in the extensive form evaluation process are:

• Without the benefit of additional information, the expected return for project B_1 would be $0.6 \times 16{,}000 + 0.4 \times -6000 = 7200$. The value of project B_2 would be only 2000 dollars. Thus B_1 would be the wisest economic choice. The expected value of action A_1 is, therefore, $E(A_1) = \$7200$.

• If management were to opt for the analysts' scheme for gathering additional information and the results were r, project B_1 would be indicated. Project

Table 18.6 Estimates of payoffs and state likelihoods

Scenario or state	Project payoff (10^3 dollars)		State likelihood
	B_1	B_2	
S_1	16	−10	0.6
S_2	−6	20	0.4

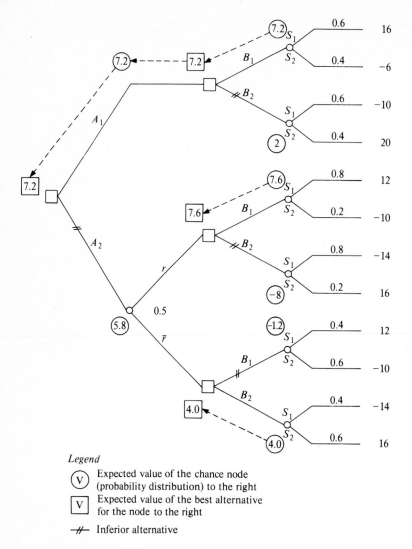

Legend

  ⓥ Expected value of the chance node
      (probability distribution) to the right

  ⬛ⱽ Expected value of the best alternative
      for the node to the right

  ⧸⧸⧸ Inferior alternative

Figure 18.5 Extensive form tree analysis.

B_2 would be indicated otherwise. The respective payoffs would be 7600 and 4000 dollars. Since results r and $\bar{r}$ are considered equally likely, the expected return for the best options on branch A_2 is $0.5 \times 7600 + 0.5 \times 4000 = \$5800$.

  ● Clearly, the analysts' scheme is not worth the cost in this case. However, had it cost only 2000 dollars, it would have provided a 600 dollar margin. Even at the estimated 4000 dollar cost, if the conditional likelihoods had been $P(S_1/r) = 0.9$ and $P(S_1/\bar{r}) = 0.2$, the expected NPE of the A_2 branch would have been approximately $E(A_2) \simeq \$9400$ for a marginal value of $E(A_2) - E(A_1) = \$2200$.

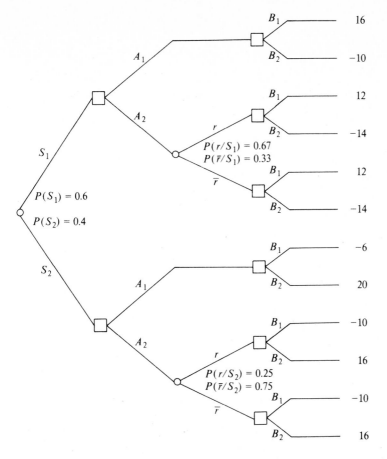

Figure 18.6 Inverted decision tree for normal form analysis.

To analyze this problem in normal form, it is convenient to invert the tree, as shown in Figure 18.6. From this tree the strategies, relevant payoffs, and likelihoods are easily identified. The strategies are listed in Table 18.7.

Table 18.7 Strategies

A_1
    s_1: Choose B_1
    s_2: Choose B_2

A_2
    s_3: Choose B_1 only if r is observed
    s_4: Choose B_2 only if r is observed
    s_5: Choose B_1
    s_6: Choose B_2

Table 18.8 Conditional values of strategies
Thousands of dollars

s_1:

$$V(s_1/S_1) = 16 \qquad V(s_1/S_2) = -6$$

s_2:

$$V(s_2/S_1) = -10 \qquad V(s_2/S_2) = 20$$

s_3:

$$V(s_3/S_1) = 0.67 \times 12 - 0.33 \times 14 = 3.4$$
$$V(s_3/S_2) = -0.25 \times 10 + 0.75 \times 16 = 9.5$$

s_4:

$$V(s_4/S_1) = -0.67 \times 14 + 0.33 \times 12 = -5.4$$
$$V(s_4/S_1) = 0.25 \times 16 - 0.75 \times 10 = -3.5$$

The conditional value of each strategy is calculated in Table 18.8. Since s_5 and s_6 are obviously inferior to s_1 and s_2, respectively, their values are not calculated. The NPE values may be taken from either tree. The probability values are calculated from those in the extensive form tree, using Bayes' rule as follows:

$$P(r/S_1) = \frac{P(S_1/r)P(r)}{P(S_1)} = \frac{0.8 \times 0.5}{0.6} = 0.67 \qquad P(\bar{r}/S_1) = 0.33$$

$$P(r/S_2) = \frac{P(S_2/r)P(r)}{P(S_2)} = \frac{0.2 \times 0.5}{0.4} = 0.25 \qquad P(\bar{r}/S_2) = 0.75$$

Strategies and their conditional values are diagramed in Figure 18.7. To complete the analysis calculate the expected values for each strategy from $E(s) = E(s/S_1)P(S_1) + E(s/S_2)P(S_2)$. The results are given in Table 18.9. Clearly, strategy s_1 is the best course of action. The optimal decision using the extensive

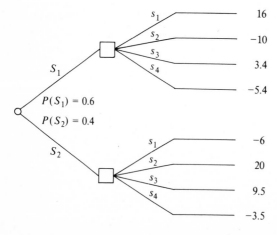

Figure 18.7 Strategies and conditional values.

Table 18.9 Strategy values

$E(s_1) = \$7200$	$E(s_3) = \$5800$
$E(s_2) = 2000$	$E(s_4) = -4600$

form of analysis also was to choose project B_1 without investigating further. The next-best strategy is s_3. The extensive form of analysis provided precisely the same result. The other two strategies were found to be dominated early in the extensive form analysis. The value $E(s_2)$ actually appears in the extensive form decision tree. The value $E(s_4)$ does not appear in the tree because it was never explicitly evaluated, although it could be extracted easily.

18.3 SITUATIONAL ADVANTAGES

The normal form of analysis can be more efficient than the extensive form. Often possible strategies can be seen to be inferior for all states of nature, even before numerical evaluations have begun, as in the last section. This reduces the calculational effort and can be particularly important when there are a large number of strategies. Although the same kind of reduction in problem size can be accomplished in extended form analysis, the process is more efficient in the normal form. Whereas decision trees must be constructed in the extended form prior to thinning, strategy elimination is accomplished readily, without the need for construction of a decision tree in normal form analysis.

The only problem with early strategy domination is that in the process of reducing the magnitude of the problem, one might inadvertently overlook important strategies or mistakenly discard a viable one. When the data for the problem are very uncertain, this is a particularly serious problem, especially since these early decisions are not likely to be subjected to sensitivity analyses which are conducted at the conclusion of the evaluations. This is a very common problem and often arises, as in the following example on medical decision analysis.

Example 18.1: Viable strategies omitted Physicians, like most of us, often ignore viable strategies in the process of diagnosis and treatment. They tend to focus on a few elementary strategies, namely, do nothing or apply a diagnostic test and proceed as if the test were infallible. Since tests are seldom infallible this narrow range of options may not include the optimal strategy. The situation is typified by a decision-analytic evaluation of diagnostic and treatment strategies for cancer of the pancreas, as follows.

Assuming the availability of an early-warning test for cancer of the pancreas, diagnosticians might consider the options indicated in the decision tree below. Clearly there are other strategies. For example, one could operate on the basis or pretest evidence and bypass the testing option.

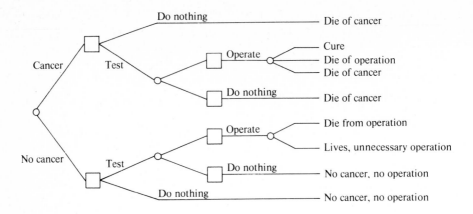

Table 18.9 Outcomes per thousand patients

Strategy	Cure	Die from		Unnecessary operation	No cancer No operation
		Operation	Cancer		
Do nothing	0	0	12	0	988
Test: if positive, operate; if negative do not	4.32	5.9(4.94)*	6.72	44.46	938.6
Operate; no test	5.4	100(98.8)*	5.4	889.2	0

* Deaths to noncancerous patients are indicated in parentheses.

Table 18.9 was constructed from an analysis of this problem in the medical literature. The authors of the article considered only two strategies: (1) Do nothing. (2) Test; operate only if positive. That is, they considered the consequence indicated in the first two rows. Another writer added the third row to illustrate precisely the point of this example. Notice that the additional strategy is optimal if the objective is to minimize deaths from pancreatic cancer, or to maximize the cure rate. By ignoring this strategy, the original authors ignored perhaps the most important reason for early diagnosis. The only conclusions they could draw from their limited analysis were: If minimizing deaths due to the operation is the goal, it is best to act consistently with the test results. If minimizing total deaths to the population is the goal, it is best to do nothing.

Another advantage to the normal form of analysis stems from the fact that the most uncertain information in a decision problem is usually the future state of nature. Decision makers can have but limited influence over the state of nature; they must speculate on what it will be and act according to their judg-mental estimates of relative likelihoods. For this reason, it is often desirable to carry the analysis as far as possible without incorporating this uncertainty into it. In normal form analysis, strategies are evaluated conditionally on the state of nature, and only in the last analytical step must the very uncertain state likeli-hoods be introduced; whereas in the extended form of analysis, state likelihoods are introduced in the initial evaluation step. The sensitivity of strategy selection is then particularly easy to assess.

The problem discussed in Section 18.2 is a case in point. The objective function was the linear relation

$$E(s) = P(S_1)V(s/S_1) + P(S_2)V(s/S_2)$$

The probabilities $P(S_1)$ and $P(S_2)$ are merely weights which may take any values between zero and unity, so long as $P(S_1) + P(S_2) = 1$. The particular values assigned to these weights determine which strategy is best. As indicated in Figure 18.8, the optimal strategy was s_1 because it was the first point contacted as the indicated objective function was moved in toward the origin. So long as the ratio $P(S_1)/P(S_2)$, which is the slope of the objective function, is more negative than the slope of the line connecting s_3 and s_1, namely, -1.23, strategy s_1 remains optimal. For probability values in the ratio of -1.23, both strategies s_1 and s_2 are equally desirable on the basis of the economic analysis and the choice must be made using other criteria. If the state likelihood ratio is between -1.23 and -0.86, strategy s_3 is optimal. Finally, if the ratio is less negative than 0.86, strategy s_2 is best.

Clearly, inaccurate estimates of state likelihoods can lead to adopting infer-

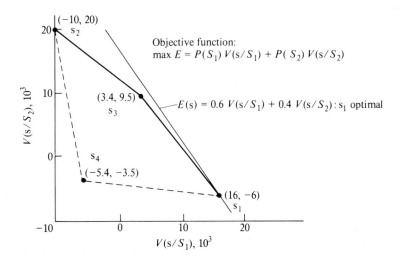

Figure 18.8 Sensitivity of strategy selection to state likelihoods.

ior strategies. Since state likelihoods are often quite uncertain, sensitivity analyses on them provide important decision-relevant information. The normal form of analysis obviously facilitates this kind of sensitivity analysis, and for that reason may be preferred when the most important uncertainties are state likelihoods.

There is one other advantageous aspect to the normal form of analysis; it is especially easy to observe in medical decision problems where the states of nature are disease states. Physicians assess state likelihoods initially on the basis of the patient's medical history and subjective information called symptoms. To refine their assessments they may opt to request laboratory analyses of different sorts before prescribing treatment. Laboratory results r are statistically calibrated against known cases where specific disease states were known to exist. Thus, physicians are left with calibration likelihood $P(r/S_i)$ and state likelihood $P(S_i)$. These are precisely the likelihoods needed for normal form analysis. Conversely, the posterior state likelihoods $P(S_i/r)$ and $P(S_i/\bar{r})$ would have to be calculated, using Bayes' rule, if the extended form of analysis were to be pursued.

One important advantage to the extensive form of analysis is that computer algorithms of the dynamic programming type (not discussed in this text) have been written to carry out decision analyses. One powerful, interactive version may be purchased from the University of Iowa, Iowa Testing Programs. It was developed under a National Science Foundation grant and is quite inexpensive to purchase. It was written for very large computer systems. Of course, many prefer to write their own programs for personal use. The relative advantages of the two forms are summarized in Table 18.10.

Table 18.10 Relative advantages of extensive and normal forms of analysis

Extensive form	Normal form
Requires explicit considerations of all options, thereby reducing chance of inadvertently overlooking important strategies. In policy analysis this can be critical advantage.	Reduces analytical burden when many possible strategies. Dominated strategies are easily identified.
No need to count and enumerate large number of strategies.	Need not construct large decision tree.
Computer algorithms of dynamic programming type have been developed for extensive form analysis.	Probability data are commonly available in the form $P(S_i)$ and $P(r_j/S_i)$. These are directly applicable—no need to apply Bayes' rule to calculate $P(S_i/r_j)$.
	Most of the analysis can be completed conditionally on states of nature. This can be very important since great uncertainty usually prevails regarding likelihoods $P(S_i)$. Sensitivity analysis to determine probability ranges within which alternative decisions are preferable is facilitated.

18.4 PROBLEMS

18.1 Imagine that you are in the process of deciding whether to replace your old car this year or have some mechanical and body work done on the old buggy and try to hold off for another year. The estimated base costs for each alternative are given below.

Net costs for one year

New car	Fix old car
$2000	$1000

The likelihoods that additional costs will be incurred to keep each car in "good" operating condition throughout the year are given in the next table.

Additional costs and judgmental likelihoods

States of nature	New car		Old car	
	Cost	Likelihood	Cost	Likelihood
S_1	1000	0.001	1000	0.5
S_2	500	0.01	500	0.45
S_3	100	0.989	100	0.05

(a) Draw the decision tree in extended form showing all costs and likelihoods in the proper branches.

(b) Use the maximum expected value rule to identify the best course of action.

(c) Check the sensitivity of your selection to the probability estimates. Adjust the state probability distributions until a different decision emerges as best.

18.2 Suppose the additional cost figures were related to the kind of service the automobiles would be expected to provide. For example, the likelihood that a 1000 dollar additional cost would be encountered for the new car might be 0.001 if service level a were required, but 0.009 if level b were required.

(a) Modify the first problem by using the data given in the table below and answer the same three questions.

(b) Solve this modified problem in normal form as well. Identify all strategies and select the best.

Additional costs and conditional likelihoods by service level

Alternative service level additional cost	New car		Old car	
	a	b	a	b
1000	0.001	0.009	0.5	0.8
500	0.01	0.09	0.45	0.2
100	0.989	0.901	0.05	0

18.3 Draw the decision tree and select the best project using the information given. Carry out the evaluation using the maximum expected payoff rule, the minimax rule, the maximin rule, and the maximum payoff under most likely conditions rule. For what likelihood distribution do the other rules select out the same course of action as the maximum expected payoff rule? The payoffs for D_1, D_2, and D_3 are, respectively $(100, 25, -60)$, $(90, -700, 50)$ and $(-70, 1000, -20)$. The probabilities are $(0.2, 0.3, 0.5)$.

18.4 A utilization review team is charged with reviewing hospital admissions within 24 hours to determine the appropriateness of the admission and then of the treatment program, for confirmed admissions. This team has a set of data to guide its decisions. For patients with disease d_1, there are two treatment paths A and B from which they may select. The outcomes of treatment have been categorized as:

$$d = \text{death}$$

$$c = \text{total cure}$$

$$f = \text{fractional cure}$$

The likelihoods and payoff values for each treatment path are given in the table.

Outcome likelihoods

Outcome	If treatment A	If treatment B	Value of outcome
d	0.2	0.01	-10
c	0.8	0.79	10
f	0	0.2	0

(a) Calculate the expected values for each possible treatment.

(b) Which choice is best according to the maximum expected utility rule (assume the value numbers reflect utility values accurately)?

(c) Which choice is best for those who value *life* very highly?

(d) Which choice is preferable according to the most likely outcome rule if:

- Total cure is the objective?
- At least partial cure is the objective?

18.5 Congress is considering a funding bill to replace the Mississippi River Locks and Dam No. 26. There is much controversy regarding the bill, with the Mississippi River Valley Railroad Consortium and environmental societies fighting the bill. The U.S. Army Corps of Engineers submitted data on estimated benefits and costs to Congress which were based on statistical analyses of tonnage throughput under existing conditions and proposed conditions. Congress is deliberating three options. On the basis of the data submitted it may authorize funding or not. The third option is to require the Corps to hire an independent consultant to study operating practices of the towboat companies and the locks and try to identify alternative operating procedures which would result in essentially the same annual net benefits the Corps attributed to the proposed new locks and dam.

Politically, Congress would like to avoid funding the new construction. Congressional leaders have met with experts on the issue and have concluded that there is a reasonable likelihood that a more beneficial alternative to the new locks and dam could be worked out. Using the information given, diagram this decision problem and determine how large the likelihood of the existence of an alternative must be to make the search economically worthwhile. Assume that the consulting contract and necessary delay in action would reduce annual benefits of the new locks and dam by about $\frac{1}{2}$ percent, if they should eventually be authorized.

Data submitted by the Corps

	Marginal annual equivalent benefits to perpetuity	
	Mean	Standard deviation
New locks and dam	$10,000,000	$2,000,000
Old locks and dam	0	
Alternative operating procedures	10,000,000 (1.01)	2,000,000

The distribution of marginal benefits is assumed to be approximately normal.

18.6 The city is considering building a new fire station to replace an old station. The old site is insufficient for the new facility, so two new sites are being considered. Consultants have estimated the travel times to all zones in which the station has first-priority response obligations. The data are in the form of travel time distributions for an arbitrary call and are based on the actual frequencies of calls into each zone in the past 5 years. For the two alternative sites the travel time distributions are stated below:

Travel time in minutes

Site	Distribution	Mean	Variance
A	Normal	12	4
B	Normal	10	6

So far as the city is concerned the only decision variable is response time, because all cost elements except those related to response travel are the same for both sites and the public has expressed no preferences for either site.

(a) Draw the decision tree for this problem.
(b) Which is the best site, if the maximum expected benefit rule is used?
(c) Critique your conclusion on the basis of the distributional variances.
(d) Which site is preferable when the distribution for B has a variance of 9, while that for A is 1? Assume that the consequences become severely negative as travel times increase beyond 15 minutes.

18.7 A construction firm recently received a government contract to build a highway. The job will be completed in 2 years. The firm won the contract with cost estimates based on using present construction equipment. With new equipment, profit per unit would be increased. However, this would be true only if the contract were for more than 2 years. Management feels there is a 45 percent chance for an additional 2-year contract, a 35 percent chance for an additional 4-year contract, and a 20 percent chance of no contract renewal. Using only present equipment, total profit for a 2-, 4-, and 6-year contract is estimated as 45,000, 85,000, and 125,000 dollars respectively. If the new equipment is purchased, total profit would change to $-100,000$, 90,000, and 180,000 dollars, respectively. If the expected-value rule is used, should the new equipment be purchased? Does the most likely future criterion seem reasonable in this decision? Why? Which choice would a minimaxer make?

18.8 The city has just purchased a new machine and is deciding whether to retire the old one or keep it for standby service. Annual costs to retain it are an estimated 1800 dollars per year. The primary advantage of retaining it is that production can be continued when the new machine is "down" for repairs; this advantage is valued at 40 dollars per hour of use. Repairs average about eight hours. There are about 2000 working hours per year. The estimated repair frequency is:

Number of times "down" for repair per year	Probability
0–2	0.10
3–5	0.20
6–8	0.40
9–11	0.20
12–14	0.10

Should the old machine be retained if repairs follow the frequency distribution given?

18.9 Alternatives A and B have identical lives and first costs; each has zero salvage. Several sets of outcome possibilities exist for the cash flow patterns of alternatives A and B. You have analyzed each and produced the following comparison:

Outcome	NPE_A	p_A	NPE_B	p_B
Pessimistic	− $200,000	0.2	− $400,000	0.3
Most likely	200,000	0.7	300,000	0.4
Optimistic	600,000	0.1	700,000	0.3

(a) Find the expected value of NPE for each alternative.

(b) State the circumstances under which you might choose alternative A.

(c) Sketch a decision tree for this choice; show $E(NPE)$ for each alternative.

18.5 SUPPLEMENTAL READING

Behn, Robert D., and James W. Vaupel: "Teaching Analytical Thinking," *Policy Analysis*, vol. 2, no. 4, Fall 1976, p. 663.

Brown, Rex V., Andrew S. Kahr, and Cameron Peterson: *Decision Analysis for the Manager*, Holt, Rinehart and Winston, New York, 1974, chaps. 1–8.

Fabrycky, W. J., and G. J. Thuesen: *Economic Decision Analysis*, Prentice-Hall, Inc., Englewood Cliffs, N.J., chap. 10.

Gohagan, John K.: "Decision Analysis in Medicine," to the editor, *JAMA*, vol. 237, no. 7, Feb. 14, 1977.

Kassouf, Sheen: *Normative Decision Making*, Prentice-Hall, Inc., Englewood Cliffs, N.J., 1970, chaps. 1–5.

Moore, P. G., and H. Thomas: *The Anatomy of Decisions*, Penguin Books, Inc., Baltimore, 1976, chaps. 1–8.

Raiffa, Howard: *Decision Analysis: Introductory Lectures on Choices under Uncertainty*, Addison-Wesley Publishing Company, Inc., Reading, Mass., 1968, chaps. 1, 2, 5, and 6.

Von Neumann, John and Oscar Morgenstern: *Theory of Games and Economic Behavior*, 2d ed., Princeton University Press, Princeton, N.J., 1947.

NINETEEN

VALUE OF ADDITIONAL INFORMATION

In most decision situations one has the option of deciding an issue on the basis of existing information or postponing the final decision until additional information has been collected and processed. The big question is always: Will it be cost-beneficial to delay a decision? In this chapter a decision-analytic procedure for investigating the possible worth of collecting *additional information regarding the states of nature* is explained. The idea is that one would like to compare the marginal cost of electing to acquire the additional information, to the incremental gain in benefits to be expected as a result of it. The entire procedure is prospective, therefore probabilistic. Its value is in the guidance it offers at the time when the choice must be made, which is, obviously, prior to actually obtaining the additional information.

The costs of postponing a decision may be of many types. Some options might by foreclosed as a result of the delay. There may be a charge for the information in the form of consulting fees or experimental costs. There may be operating costs in terms of labor and interest on debt. There may be intangible costs such as impairment of one's public image or psychological or physical stress. Throughout the discussions of this chapter, however, costs will be characterized as economic costs. The extension of the idea to other types of costs is immediate in terms of utility functions, which are discussed in Chapter 20.

The value of additional information regarding the likelihoods of outcomes is a function of the decision rule employed. Neither the minimax nor the maximin rules account for likelihoods, so refined likelihood estimates contribute nothing to decision making when they are used. The only impact would be an additional cost. On the other hand the most likely outcome rule turns on the likelihood estimates, and additional information which modified them could cause a public manager to discard what initially seemed to be the best choice in favor of one of its competitors. Similarly the maximum expected utility (value) rule compounds payoffs and likelihoods, so revised likelihood estimates could lead to revised decisions.

The expected value of perfect information and expected value of imperfect information concepts discussed in this chapter relate specifically to the maximum expected utility (value) rule of formal decision analysis. The concept of opportunity loss, which is briefly discussed, relates to all decision rules.

19.1 OPPORTUNITY LOSS

Opportunity loss is the difference between the return you would receive if you could perfectly forecast the future state of nature and make your decision accordingly, and the return you would actually receive if your decision were based on imperfect information. Opportunity loss depends on the decision made and the actual state of nature.

For example, consider the simple decision problem represented in Table 19.1. If a person were to choose decision D_1 on the basis of perfect information that the actual state was S_1, he would receive 10 units of return. He would have made the same decision if he had known that the true state was S_1, so he lost no opportunity as a result of having imperfect information. On the other hand, had his decision been D_2 he would have lost 5 units. Had he known that the true state was S_1, he would have chosen D_1 instead. Then, as opposed to

Table 19.1 Payoffs

Decision	State of nature S_1	S_2
D_1	10	-7
D_2	-5	14

Table 19.2 Opportunity losses (OL)

Decision	State of nature S_1	S_2
D_1	0	21
D_2	15	0

losing 5 units, he would have gained 10 units. In other words, for state S_1, the decision D_2 represents an opportunity loss (OL) of 15 units; OL $=$ $10 - (-5) = 15$.

The opportunity loss for each decision and outcome state for this problem are given in Table 19.2.

19.2 EXPECTED VALUE OF ADDITIONAL INFORMATION

Because the worth of additional information must be estimated prior to obtaining the information, it must be estimated probabilistically. The decision rule for decision analysis is an expected-value rule. Therefore, the estimated worth of additional information must be in the form of an expected value. If one could purchase perfect information on the true state of nature, its estimated worth would be called the expected value of perfect information. If one could purchase only imperfect information, its worth would be called the expected value of sample (imperfect) information.

Consider the problem data in Table 19.3. Decision D_2 is the best decision because its expected value is largest; $E(D_2) = 4.5$ whereas $E(D_1) = 1.5$. Now suppose that information on the true state of nature could be acquired, at a cost, of course. Suppose the information would result from a scientific study such that if the results (information) were in category r, the state probabilities would be known to be $P(S_1/r) = 0.99$ and $P(S_2/r) = 0.01$; whereas if they were in $\bar{r}$, the state probabilities would be $P(S_1/\bar{r}) = 0.09$ and $P(S_2/\bar{r}) = 0.91$.* Then, one could prospectively calculate the expected values for each decision for each possible information category. These are:

$$E(D_1/r) = 0.99 \times 10 - 0.01 \times 7 = 9.8$$

$$E(D_2/r) = -0.99 \times 5 + 0.01 \times 14 = -4.8$$

$$E(D_1/\bar{r}) = 0.09 \times 10 - 0.91 \times 7 = 5.4$$

$$E(D_2/\bar{r}) = 0.09 \times 5 + 0.91 \times 14 = 12.3$$

* This would imply that $P(r) = 0.455$ and $P(\bar{r}) = 0.545$ since
$$P(S_1) = P(S_1/r)P(r) + P(S_1/\bar{r})P(\bar{r}) \text{ and } P(S_1) = 0.5.$$

Table 19.3 Payoffs

Decision	State S_1	S_2	Estimated state probabilities
D_1	10	−7	$P(S_1) = 0.5$
D_2	−5	14	$P(S_2) = 0.5$

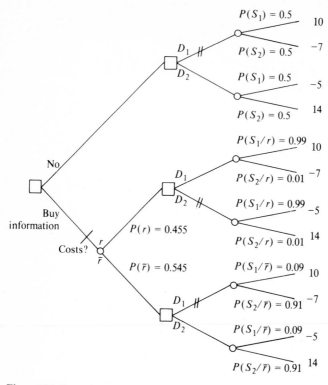

Figure 19.1 Extensive form decision tree.

The extensive form decision tree is shown in Figure 19.1. The projected expected gross return on the additional information is

$$E(\text{buy}) = 0.455 \times 9.8 + 0.545 \times 12.3 = 11.2 \text{ units}$$

Therefore, the *expected value of the sample information* (EVSI) for the problem is

Extended form analysis

$$\text{EVSI} = \left(\begin{array}{l} \text{Expected} \\ \text{value of the} \\ \text{decision to} \\ \text{acquire } sample \\ \text{information} \end{array} \right) - \left(\begin{array}{l} \text{Expected value of} \\ \text{the best decision} \\ \text{on the basis of} \\ \text{the } original \\ \text{information} \end{array} \right)$$

$$= 11.2 - 4.5$$
$$= 6.7 \text{ units}$$

This means that the sample information has a prospective worth of 6.7 units and if it can be purchased for less than that, it should be.

Now consider the case where the additional information available is so good that if r results from the study, the state probabilities would be $P(S_1/r) = 1$, $P(S_1/\bar{r}) = 0$, $P(S_2/r) = 0$, and $P(S_2/\bar{r}) = 1$. This is perfect information in that the scientific study would remove all uncertainty about the true state of nature (a rare situation). The best choice if the study produced result r would be D_1, for a return of 10 units. Otherwise D_2 would be best, with a return of 14 units. Since the likelihoods of r and $\bar{r}$ are $P(r) = P(\bar{r}) = 0.5$, the expected value of the decision problem when the perfect information is purchased is $E(\text{buy}) = 0.5 \times 10 + 0.5 \times 14 = 12$ units. Therefore, the *expected value of the perfect information* (EVPI) is

Extensive form analysis

$$\text{EVPI} = \begin{pmatrix} \text{Expected} \\ \text{value of the} \\ \text{decision to} \\ \text{acquire the } perfect \\ \text{information} \end{pmatrix} - \begin{pmatrix} \text{Expected value of} \\ \text{the best decision} \\ \text{on the basis of} \\ \text{the } original \\ \text{information} \end{pmatrix}$$

$$= 12 - 4.5$$
$$= 7.5 \text{ units}$$

Perfect information has a prospective worth of 7.5 units in this problem. This is the upper limit on the worth of additional information, since imperfect information must be worth less. If additional information can be acquired, but only at a cost greater than 7.5 units, the best course of action is to proceed without it. If the information would cost less than 7.5 units but would be imperfect, it could be a good buy; but that would have to be determined by estimating the EVSI on the basis of the probabilities $P(S_1/r)$ and $P(S_2/r)$.

The EVSI and EVPI can be estimated when the analysis is done in normal form, as well. (The idea was introduced in the previous chapter.) For the problem at hand the strategies are given in Table 19.4.

The revised state probabilities for imperfect (sample) information given

Table 19.4 Strategies and conditional values

Strategy	$V(s/S_1)$	$V(s/S_2)$
s_1: Choose D_1	10	-7
s_2: Choose D_2	-5	14
s_3: Choose D_1 only if r is the result	$[10\ P(r/S_1) - 5\ P(\bar{r}/S_1)]$	$[-7\ P(r/S_2) + 14\ P(\bar{r}/S_2)]$
s_4: Choose D_2 only if r is the result	$[-5\ P(r/S_1) + 10\ P(\bar{r}/S_1)]$	$[14\ P(r/S_2) - 7\ P(r/S_2)]$
s_5: Choose D_1 if either r or $\bar{r}$	—Dominated by s_1—	
s_6: Choose D_1 if either r or $\bar{r}$	—Dominated by s_2—	

previously were $P(S_1/r) = 0.99$, $P(S_1/\bar{r}) = 0.09$, $P(S_2/r) = 0.01$, and $P(S_2/\bar{r}) = 0.91$. These imply that

$$P(r) = 0.455 \qquad P(r/S_1) = 0.9 \qquad P(r/S_2) = 0.01$$

$$P(\bar{r}) = 0.545 \qquad P(\bar{r}/S_1) = 0.1 \qquad P(\bar{r}/S_2) = 0.99$$

Sample information of this quality implies that $V(s_3/S_1) = 8.5$, $V(s_3/S_2) = 13.8$, $V(s_4/S_1) = -3.5$, and $V(s_4/S_2) = -6.8$, by substitution into the expressions of Table 19.4. Clearly, in this particular case, strategy s_4 is not a contender.

The strategies remaining for comparison are: s_1—choose D_1 without collecting sample information; s_2— choose D_2 without collecting sample information; and s_3—collect the sample information and if the results are r, choose D_1; otherwise choose D_2. The expected values for each strategy are found from

$$E(s) = V(s/S_1)P(S_1) + V(s/S_2)P(S_2)$$

The expected values are summarized in Table 19.5.

Strategy s_3 is preferable if the information is not too costly. As before, the expected value of sample information is

Normal form analysis

$$\text{EVSI} = \left(\begin{array}{l} \text{Expected} \\ \text{value of the} \\ \text{best strategy} \\ \text{including } \textit{sample} \\ \text{information} \end{array} \right) - \left(\begin{array}{l} \text{Expected} \\ \text{value of the} \\ \text{best strategy} \\ \text{based on } \textit{initial} \\ \text{information only} \end{array} \right)$$

$$= E(s_3) - E(s_2)$$

$$= 11.2 - 4.5$$

$$= 6.7 \text{ units}$$

Had perfect information been available, $P(S_1/r) = 1$ and $P(S_2/\bar{r}) = 1$, the probabilities needed for Table 19.4 would have been $P(r/S_1) = 1$ and $P(\bar{r}/S_2) = 1$, as one might suspect. Then the conditional values for strategies s_3 and s_4 would have been $V(s_3/S_1) = 10$, $V(s_3/S_2) = 14$, $V(s_4/S_1) = -5$, and $V(s_4/S_2) = -7$. Again, strategy s_4 would not be a contender. The expected

Table 19.5 Expected values of strategies

Strategy	$E(s)$
Initial information only	
s_1	1.5
s_2	4.5
With sample information	
s_3	11.2

Table 19.6 Expected values of strategies

Strategy	$E(s)$
Initial information only	
s_1	1.5
s_2	4.5
With perfect information	
s_3	12.0

values of the other strategies are given in Table 19.6. Again, strategy s_3 is desirable if the perfect information is not too costly. Just as in extensive form analysis, the expected value of perfect information, which represents the upper limit on the worth of additional information, is

Normal form analysis

$$\text{EVPI} = \begin{pmatrix} \text{Expected} \\ \text{value of the} \\ \text{best strategy} \\ \text{including } perfect \\ \text{information} \end{pmatrix} - \begin{pmatrix} \text{Expected} \\ \text{value of the} \\ \text{best strategy} \\ \text{based on } initial \\ \text{information only} \end{pmatrix}$$

$$= E(s_3) - E(s_2)$$
$$= 12 - 4.5$$
$$= 7.5 \text{ units}$$

One final note regarding EVPI: The opportunity losses for all decision-state pairs in this problem are given in Table 19.7. The *expected opportunity loss* (EOL) for a decision D is calculated from

$$\text{EOL}(D) = P(S_1)\,\text{OL}(D/S_1) + P(S_2)\,\text{OL}(D/S_2)$$

For this problem, $\text{EOL}(D_1) = 10.5$ units and $\text{EOL}(D_2) = 7.5$ units. Notice that the expected opportunity loss for decision D_2 is equal to the expected value of perfect information. It is always the case that the minimum of all expected

Table 19.7 Opportunity losses (OL)

Decision	S_1	S_2
D_1	0	21
D_2	15	0

opportunity losses for a given problem is also the expected value of perfect information;

$$\min \text{EOL}(D) = \text{EVPI} \qquad \text{for all decisions } D$$

This fact offers an opportunity to quickly estimate the upper limit on the worth of additional information in a decision problem.

To repeat a statement made early in this chapter, one usually has an opportunity to delay a decision on a course of action for the purpose of collecting additional information on the states of nature. This would be a reasonable thing to do if the expected benefits attributable to the additional information exceeded the expected costs of acquiring the information. The methods of this chapter permit one to estimate an upper limit on the potential value of additional information as well as the expected value of imperfect information which may actually be available, whether extended form or normal form analysis is being used.

As a general guide, one should quickly estimate the EVPI, perhaps using the EOL procedure. Then, if additional information of a specific quality (that is, the probabilities $P(S_i/r)$ and $P(S_i/\bar{r})$ or, instead, $P(r/S_i)$ for all states S_i are known) is available at a cost lower than the EVPI, calculate the EVSI. If the EVSI is positive, it would be wise to postpone choosing a course of action until the additional information had been interpreted, provided, of course, that its expected cost were less than the projected EVSI. Otherwise, postponing the decision would not be cost-beneficial.

As always, the sensitivity of choices (optimal decisions) to the parameters of analysis, especially the likelihoods and payoffs, must be checked and the results reported clearly.

19.3 PROBLEMS

19.1 For the following decision problems calculate the EOL, EVPI, and EVSI. Assume that additional information could fall into three categories, r_1, r_2, and r_3. Do each problem by first assuming $P(S_2/r_1) = 0.5$, $P(S_2/r_2) = 0.2$, $P(S_1/r_1) = 0.2$, $P(S_1/r_2) = 0.5$, $P(S_3/r_1) = 0.3$, and $P(S_3/r_2) = 0.3$. What happens if $P(S_i/r_1) = P(S_i/r_2) = P(S_i/r_3) = \frac{1}{3}$ for $i = 1, 2, 3$?

Payoff table A

Decision	States		
	S_1	S_2	S_3
D_1	−2	10	4
D_2	20	−20	3
Probabilities $P(S)$	1/3	1/3	1/3

Payoff table B

Decision	States		
	S_1	S_2	S_3
D_1	50	70	−10
D_2	20	−5	40
D_3	−60	200	50
Probabilities $P(S)$	1/3	1/3	1/3

19.2 Is it true that if all states are equally likely and all the conditional probabilities of the form $P(r_i/S_j)$ are equal, there is no benefit in delaying a decision for the sake of acquiring additional information? Explain.

19.3 Is it true that if the states are equally likely and all the posterior probabilities of the form $P(S_i/r_j)$ are equally likely, then there is no benefit in delaying a decision for the sake of acquiring additional information? Explain.

19.4 Is it true that if the states are not equally likely but the posterior probabilities $P(S_i/r_j)$ are equally likely, then delaying a decision for the sake of additional information is not beneficial? Explain.

19.5 Suppose you were a public manager and a consulting company was analyzing alternative road surfaces for you. The company presented you with information regarding the expected (repair-free) lifetime for alternative surfaces based on the best professional judgment of its engineers. You are planning a multimillion dollar road resurfacing program and want to "have the facts" about the alternative surfaces.

(a) What questions would you ask the consultant in order to get a better grip on the state likelihoods?

(b) Draw a decision tree roughly representing this decision problem. Explain your assumptions. You have no numbers to assign to branches, so your job is to conceptualize the problem.

19.6 The Medical Services Corporation is considering opening a medical screening unit in a certain city. They have estimated revenues and cost for two contingencies, namely, local unions will sign a 5-year preemployment physical examination contract with them next month, or they will not.

The corporation feels that until the Bistate Health Systems Planning Agency (HSA) decides on the legitimacy of a similar screening clinic in the neighboring state, the odds that the unions will sign are 50-50. On the other hand, if the other clinic is approved, the corporation feels that the unions will sign the contract, with a likelihood of about 75 percent.

The economic data for two design options are given below:

Annual costs and revenues

	Design A		Design B	
Unions sign?	Yes	No	Yes	No
Probabilities				
If HSA says "yes"	0.75	0.25	0.75	0.25
If HSA says "no"	0.5	0.5	0.5	0.5
Annual equivalent capital required ($, $\times 10^3$)	30	30	20	20
Annual equivalent operating costs				
Fixed (10^3)	300	300	200	200
Variable (per patient)	2	2	1	1
Revenue	600	400	400	360
Patient load ($\times 10^3$)	20	10	10	9

(a) The corporation is uncertain about opening the clinic now. The directors feel they have no way of estimating the likelihood of the agency approving the clinic in the other state. What do you think?

(b) What strategies are open to the corporation?

(c) Which is the best strategy?

(d) What are the EVPI and EVSI in this case?

19.7 For the following decision problem calculate the minimum expected opportunity loss. What is the *maximum* amount the decision maker should be willing to pay for additional information?

Payoff table

	S_1	S_2
D_1	20	-8
D_2	10	-2
D_3	-10	30
State probabilities	0.7	0.3

19.4 SUPPLEMENTAL READING

Brown, Rex V., Andrew S. Kahr, and Cameron Peterson: *Decision Analysis for the Manager*, Holt, Rinehart and Winston, New York, 1974, chap. 3.

Kassouf, Sheen: *Normative Decision Making*, Prentice-Hall, Inc., Englewood Cliffs, N.J., 1970, chaps. 1–5.

Raiffa, Howard: *Decision Analysis: Introductory Lectures on Choices under Uncertainty*, Addison-Wesley Publishing Company, Inc., Reading, Mass, 1968, chaps. 0, 1, 2, 5, and 6.

TWENTY

UTILITY AND MULTIATTRIBUTE DECISIONS

Utility is a complex concept. It is personal and it is situational. It reflects one's current status, one's preferences for certain kinds of consequences, and one's attitude toward uncertainty. After all, taking a chance is not very risky if one has nothing to lose, if anticipated losses are insignificant, if one cares little for the kinds of consequences anticipated, or if the chance of a loss is very small. But taking a chance is serious business for one who stands to lose an entire fortune or even one's life.

Economists, psychologists, and statisticians have developed a theory of utility which incorporates all these elements. The theory provides a useful framework for evaluating decision alternatives from the perspective of an individual decision maker. Operationally, however, it is limited, for one must generate a numerical utility function which captures one's preference for outcomes and attitude toward risk in a particular situation. This is seldom possible in fast-moving, complex real-world decision situations. Still the theory is very insightful and well worth studying, for multiattribute decision making is standard practice in the public sector, and the concept of utility underlies all techniques for comparing alternatives.

The fundamental concepts of utility theory are developed in Section 20.1. Procedures for operationalizing the theory in very simple situations characterized by a single payoff factor, such as money, are developed in Section 20.2. Operationalization in slightly more complex situations is the subject of Section 20.3.

20.1 RISK PROFILES

A dollar is not a dollar. A wealthy person may view the loss or gain of 10 or even 100 dollars as being of little consequence. On the other hand, gains or losses of tens of dollars mean much more to a poor person, and a loss of 100 dollars might even be devastating. Economists would describe these different perspectives in terms of utility. They would say that the marginal utility of a dollar more or less is negligible for wealthy individuals. For the less wealthy, the marginal utility of an extra dollar is relatively larger and the marginal utility of a dollar lost may be even greater. Similar contrasts are applicable to other factors having nothing to do with wealth, such as lives saved or personal prestige. The idea is illustrated in Figure 20.1.

Furthermore, individuals hold different views regarding the relative worth of multiple payoff factors. Some value human life and human comfort above all else, while others think the natural environment and animal life should have greater weight in certain decision situations. Some value the life of a child more highly than the life of an adult, while others take the opposite view. Some consider personal economics to be the dominant factor in decision making, while others care more about their personal image than about anything else. Here, too, one's perspective depends on one's current status. The wealthy person with little need for additional money may be quite willing to expend much money and time for a small increase in his nonprofessional image. Someone of enormous ethical stature may forego great personal economic gain to preserve or slightly enhance her stature. On the other hand, the parents of starving children may care little for their image in contrast to additional food and clothing for their family.

The relative worth of a pair of decision factors can often be depicted in terms of trade-off curves, as in Figure 20.2. They are referred to as *curves of equal or isopreference*, meaning that every point on a curve represents a situation which has equal value to every other point on the curve. Such curves are generated by characterizing a situation and trading off the competing factors, while holding

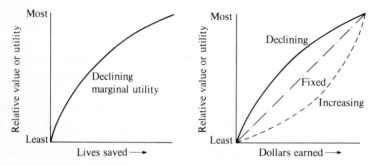

Figure 20.1 Relative value of more or less.

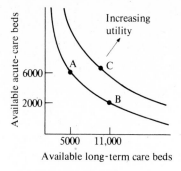
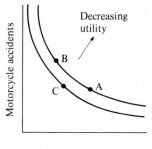

Figure 20.2 Isopreference curves.

the total value or utility constant. Individual curves in an isopreference diagram represent unique levels of utility, as for example, in Figure 20.2, where the points A and B represent equally desirable situations which, in turn, are less desirable than point C.

Uncertainty is the third component in the concept of utility. Some people will sacrifice a great deal to avoid even a very small chance of a loss. Some have great difficulty choosing a course of action when uncertainty is present. Others will go so far as to pay for an opportunity to take a gamble, to which the casinos of Las Vegas testify lavishly. For the most part, the different perspectives seem to reflect different situations in life, as discussed earlier. But it is also the case that individuals subjectively evaluate the odds differently, even in apparently identical situations where the odds have been statistically established.

Decision making under uncertainty appears to represent a balance of the perceived likelihoods of possible multifaceted consequences or scenarios against a subjective evaluation of the relative merits of each. The idea is captured in the expected utility decision rule, where utility values for complex, mutually exclusive outcomes multiplied by the likelihoods for these outcomes, are summed for each decision option to provide a single measure of worth for comparison. In Table 20.1, for example, decision D_1 is best on a measure of expected utility.

Table 20.1 Relative worth (utility) of payoffs

	Outcome	Scenarios	Expected utility
Decision options	S_1	S_2	
D_1	100	-50	70
D_2	-300	700	-100
Probabilities	0.8	0.2	

20.2 ASSESSING A SIMPLE UTILITY CURVE

Consider the decision situation in Table 20.2. Which decision you would choose depends on the value k. Select a value of k for which you consider the two decision options equally meritorious. If you selected a number smaller than 10, you are risk-averse in this situation. If you selected a number larger than 10, you are risk-prone. If you selected exactly 10, you are risk-neutral.

For every person there is a specific value of k for which the two decision options are equivalent. This is called one's *certainty-equivalent payoff*. It is a numerical measure of one's attitude toward risk in this situation. A risk-neutral person plays the averages and accepts payoffs at their face values, as insurance companies do. To this person the two options are equivalent only if they have the same expected payoff. Neither risk-averse nor risk-prone individuals play the averages. A risk-averse person, being fearful of the potential for a substantial loss, values D_1 somewhat less than its expected payoff. To this person the two options are equivalent when k is less than $E(D_1)$. The risk-prone individual, on the other hand, sees in D_1 an opportunity for a substantial gain. In this person's view, decision D_1 is worth more than its average value, and the two options are equivalent only if k is larger than $E(D_1)$.

Figure 20.3 illustrates these observations for particular values of k. For the risk-averse person, the difference between the expected payoff $E(D_1)$ and the certainty equivalent k is called the *insurance premium*. It represents this person's willingness to sacrifice in order to avoid the risk of greater loss. For the risk-prone person the difference between the certainty equivalent and the expected payoff is called the *risk premium*. It represents the minimum price this person would charge someone for the rights to the decision option D_1.

The objective of assessing a person's utility function for a decision situation is to develop a numerical scale of relative worth. This can be done for any individual payoff factor such as dollars or lives saved, so long as the factor itself is measurable.

The first step in the assessment process is to establish the best and worst payoffs. To these assign maximum and minimum utility values. These may be any numbers selected for convenience, since only their relative values, not their absolute values, have any influence on decision making. The most convenient numbers are 0 for the worst possible payoff and 1 for the best.

Table 20.2 Certainty-equivalent decision problem
Dollar payoffs

Decisions	S_1	S_2	$E(D)$
D_1	100	-80	10
D_2	For certain payoff of k		k
State probabilities	0.5	0.5	

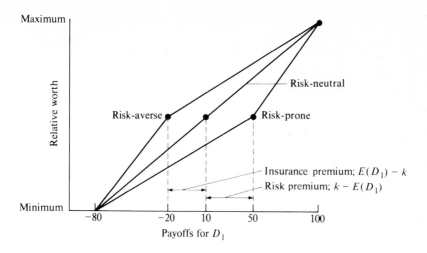

Figure 20.3 Relative worth or utility.

Next calculate the expected payoff and determine the certainty equivalent for the decision situation. Also calculate the expected utility for the extreme payoffs, 0.5. This represents the relative worth of the decision and must therefore correspond to the utility of the certainty equivalent.

To determine the third and fourth points on the utility curve, first substitute the value k just obtained and its utility for the best outcome in the decision problem. This creates a new problem and a corresponding new certainty equivalent, the utility of which is halfway between 0.5 and 0. Then repeat this step, but this time substitute k for the worst payoff. The certainty equivalent this time will necessarily have a utility value midway between 0.5 and 1.0.

This process may be continued indefinitely, but the five points so far determined will ordinarily suffice. Plot all points on a graph of utility versus payoff. This graph provides the approximate utility values to substitute for payoffs in the evaluation of decision alternatives associated with the real decision problem.

Example 20.1 illustrates assessment of the process when money is the payoff variable.

Example 20.1: Utility for money Assume the decision maker is risk-averse. Imagine that this person is faced with a decision for which the payoff extremes are 100,000 dollars and $-75{,}000$ dollars. The decision maker's utility assessment is illustrated with a decision tree and a corresponding utility curve.

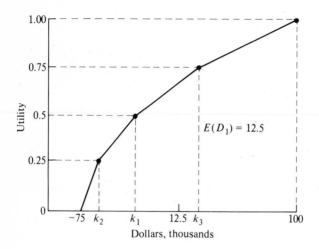

		Step 1		Step 2		Step 3	
		Payoff	Utility	Payoff	Utility	Payoff	Utility
		100	1	k_1	0.5	100	1
		-75	0	-75	0	k_1	0.5
		k_1	0.5	k_2	0.25	k_3	0.75

20.3 MULTIATTRIBUTE UTILITY FUNCTIONS

In complex decision problems, outcomes tend to be multifaceted. In some cases, trade-off analysis can be used to establish substitution rates for one variable in terms of another. In this manner it may be possible to reduce many-variable outcomes to single-variable outcomes. In other cases, payoff variables may be mathematically related. Here, too, many-variable outcomes may be reduced to single-variable outcomes. When these fortunate circumstances apply, utility function assessments proceed as in the previous section. However, there may be situations where multiattribute outcomes cannot be adequately described by a single variable. Then, one must either assess multiattribute utility functions for use in decision analysis or proceed without explicit utility functions.

Example 20.2: Multiattribute decision problem A public manager must decide a thorny issue. The attributes of the issue have been reduced to two, namely, net present equivalent benefits NPE in thousands of dollars, and a public acceptability index I, which reflects the distribution of both economic and intangible benefits. The manager must choose between D_1 and D_2. The possible payoffs are shown on the decision tree below. On the basis of net present equivalent benefits, which is an aggregate economic measure, decision D_1 appears best; but on the basis of the public acceptability index, D_2 appears preferable.

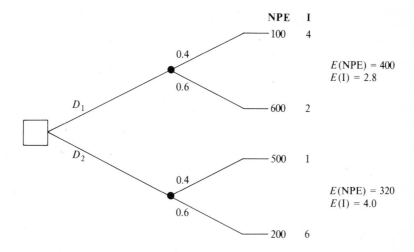

The manager can see that if she values the 80,000 dollar difference in expected net benefits more than the 1.2 unit difference in the acceptability index, she should choose D_1. However, she cannot decide what to do.* If she could only assess a composite utility function for the two-variable outcomes, she could substitute numerical utility values for the two-variable payoffs and make her choice using the maximum expected utility rule. Suppose, for example, she determined that $u(100, 4) = 0.2$, $u(600, 2) = 0.7$, $u(500, 1) = 0.5$, and $u(200, 6) = 0.3$. She would then calculate $Eu(D_1) = 0.5$ and $Eu(D_2) = 0.38$, and conclude that decision D_2 is her best choice.

In the last example the decision maker found it difficult to determine whether the trade-offs implied by selecting one course of action over the other were desirable, without the aid of a utility function. This is typical. Since direct assessments of multiattribute utility functions require precisely these kinds of comparisons, direct assessment is generally impractical. Instead, whenever pos-

---

* Trade-off considerations such as this one are typical. In practice, public officials and staff exercise what is frequently referred to as seasoned judgment and political skill in making such decisions. Their choices reflect their own preferences and perhaps to some degree those of their constituents. Utility theory is the only explicit mechanism for ensuring that such choices are consistent with preferences *and* attitudes toward risks.

Table 20.3 Two-variable utility functions

$u(x, y) = u(x) + u(y)$	separable or additive
$u(x, y) = u(x) + u(y) + Qu(x)u(y)$	quasi-separable
$u(x, y) = u(x)u(y)$	product

sible one composes multiattribute functions from single-attribute functions, which are easier to assess, as illustrated in Table 20.3. The so-called quasi-separable form is the most useful one, because the conditions under which it is applicable are more realistic than those for the others. Furthermore, both the additive and product forms are special cases of the quasi-separable form.

In the remainder of this section the process of assessing two-variable utility functions is explained, and the conditions under which each form is applicable are specified. In decision problems there are best and worst values of each outcome variable. Throughout, the best value is denoted by the subscript B. The worst value is denoted with the subscript W. Furthermore, for convenience all utility functions are scaled to have a value of zero when all the outcome variables are at the worst levels. That is, $u(x_W, y_W) = 0$.

Separable or Additive Utility Functions

A utility function $u(x, y)$ is separable only if the decision options in Figure 20.4 are equally desirable regardless of the values of x and y. That is, the expected utilities for D_1 and D_2 must be equal for every possible combination of x and y values relevant to the decision problem. This is equivalent to the statement that one sees no advantage to a compromise situation paying something for certain—either (x, y_W) or (x_W, y)—over a 50–50 chance at something on the one hand or nothing on the other. To see that these assumptions imply additivity, equate the expected utilities for the two decisions and observe that $u(x_W, y_W) = 0$ by virtue of scaling, and that $u(x, y_W) = u(x)$ while $u(x_W, y) =$

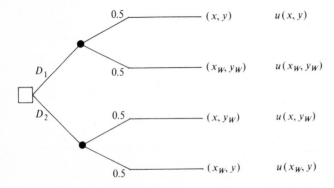

Figure 20.4 Separable utility criterion.

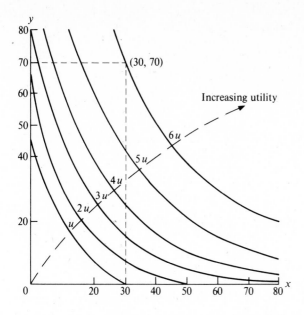

Figure 20.5 Isopreference curves for which $u(x, y)$ is not additive.

$u(y)$, since the values y_W and x_W are fixed and have no influence on the utility level. Thus,

$$0.5[u(x, y) + u(x_W, y_W)] = 0.5[u(x, y_W) + u(x_W, y)]$$

or

$$u(x, y) = u(x) + u(y)$$

Separable utility functions are rare. Here is why. A common pattern for trade-off or isopreference curves is shown in Figure 20.5. Unfortunately these rule out the use of additive utility functions, because $u(30, 70) = 6u$, whereas for an additive utility function, $u(30, 70)$ must equal $u(30, 0) + u(0, 70) = u + 2u = 3u$. Even the linear isopreference curves shown in Figure 20.6 preclude the

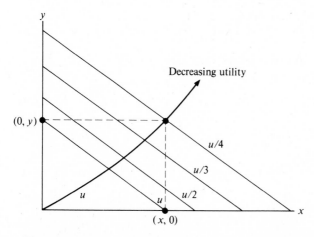

Figure 20.6 Isopreference lines for which $u(x, y)$ is not additive (decreasing utility function).

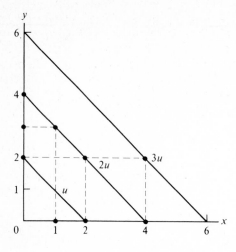

Figure 20.7 Isopreference curves for which $u(x, y)$ is additive.

use of an additive utility function. Here $u(x, y) = u/4$, whereas $u(x, 0) + u(0, y) = 2u$. On the other hand, the uniformly spaced linear isopreference curves shown in Figure 20.7 are consistent with an additive utility function.

In a practical setting one can never be certain that an additive utility function is appropriate. However, the plausibility of that assumption can be established in the process of checking for quasi-additivity.

Quasi-additive Utility Functions

Quasi-additive utility functions are a bit more common. A utility function is quasi-additive if x and y are *mutually utility independent*.

Utility independence is different from stochastic independence. Basically, it

Table 20.4 Mutual utility independence

---

x and y are mutually utility independent if *both* conditions hold:

(a) For any x_1, x_2, one of the following conditions holds for all values of y:

$u(x_1, y) > u(x_2, y)$, or

$u(x_1, y) < u(x_2, y)$, or $\quad\Big\}$ Meaning that utility ranking depends only on x

$u(x_1, y) = u(x_2, y)$,

and

(b) For any y_1, y_2, one of the following conditions holds for all values of x:

$u(x, y_1) > u(x, y_2)$, or

$u(x, y_1) < u(x, y_2)$, or $\quad\Big\}$ Meaning that utility ranking depends only on y

$u(x, y_1) = u(x, y_2)$.

---

means that the direction of change in utility (increase or decrease) as one variable is changed is unaffected by the level at which the other variable is fixed. For example, the variable x is called utility independent of y, if, whenever the payoff (x_1, y_0) is preferable to (x_2, y_0) for some value y_0, it is always true that (x_1, y) is preferable to (x_2, y) regardless of the value of y. Mutual utility independence means that each of the payoff variables is utility independent of the other. These ideas are summarized in Table 20.4.

As a result of utility independence one is free to fix the value of one variable at some convenient level, say, x_0 or y_0, and assess the conditional utility function for the other variable without fear that utility rankings would be different had another level of the first variable been selected. Thus, quasi-additive utility functions, which have the form

$$u(x, y) = u(x) + u(y) + Qu(x)u(y)$$

may be conveniently assessed. The only tricky step is determining the value of the constant Q, which has the form

$$Q = \frac{u(x_1, y_1) - u(x_1, y_0) - u(x_0, y_1)}{u(x_1, y_0)u(x_0, y_1)}$$

But even this step is not difficult. One simply selects convenient outcome values x_1 and y_1 and searches for an equally desirable outcome in the form of (x_2, y_0) or (x_0, y_2). Then since the utility functions $u(x)$ with y held at y_0 and $u(y)$ with x held at x_0 have already been assessed, the unknown utility $u(x_1, y_1)$ is determined from one of these by using the appropriate relation:

$$u(x_1, y_1) = u(x_2, y_0) = u(x_2)$$

or $$u(x_1, y_1) = u(x_0, y_2) = u(y_2)$$

If it should turn out that Q is zero, pick a significantly different pair (x_1, y_1) and recalculate Q. If this second value is also zero, there is a good chance that the two-variable utility function is additive. Check this hypothesis, as in Example 20.1. If it is reinforced, proceed accordingly. If not, calculate a third value of Q and use it, or an average of the three, as the true value in the quasi-separable utility function.

Whenever Q is not zero, one has the option of writing $u(x, y)$ in quasi-additive or product form. Either form will select the same decision option. To write it in product form, rescale the single-variable functions and then take the product as follows:

$$u'(x) = u(x) + \frac{1}{Q} \qquad u'(y) = Qu(y) + 1$$

$$u(x, y) = u'(x)u'(y) + \text{constant}$$

The process of assessing two-variable utility functions is summarized in Figure 20.8 and illustrated in Example 20.2.

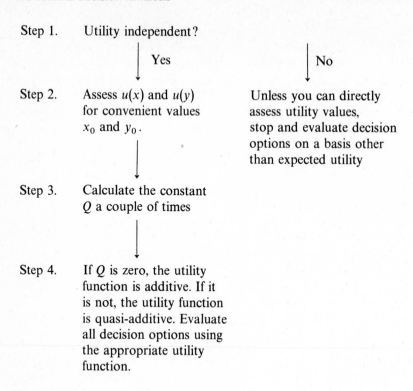

Step 1.   Utility independent?

Yes / No

Step 2.   Assess $u(x)$ and $u(y)$ for convenient values x_0 and y_0.

Unless you can directly assess utility values, stop and evaluate decision options on a basis other than expected utility

Step 3.   Calculate the constant Q a couple of times

Step 4.   If Q is zero, the utility function is additive. If it is not, the utility function is quasi-additive. Evaluate all decision options using the appropriate utility function.

Figure 20.8 Two-variable utility assessment process.

Example 20.3: Establishing the form of $u(x, y)$ Jim has a disorder which can be corrected by an operation. There are three different procedures which can be used. The two factors upon which his decision will turn are cost and comfort. Some possible payoffs, including costs of the operation and values for his comfort index are given below in the form ($, I). The comfort index I varies from zero for least comfortable to unity for most comfortable.

(100, 0.01)	(100, 0.2)	(100, 0.5)	(100, 0.7)	(100, 0.9)
(500, 0.01)	(500, 0.2)	(500, 0.5)	(500, 0.7)	(500, 0.9)
(1000, 0.01)	(1000, 0.2)	(1000, 0.5)	(1000, 0.7)	(1000, 0.9)

The two variables are probably mutually utility independent. Check this on yourself by comparing columns and rows. If your preference for outcomes declines as you look down each column, and increases as you look across each row, costs and comfort are most likely utility independent for you.

Suppose you helped Jim assess his utility functions for money and comfort, with $x_0 = 100$ and $y_0 = 0.01$, and obtained the utility curves shown

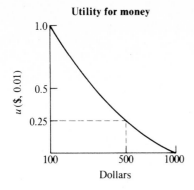

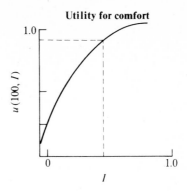

below. As Jim's decision analyst, pick an arbitrary payoff, say, (500, 0.5), and search the first column to see if he values it and any of the three payoffs there equally. If not, search the first row for the same purpose. Suppose Jim decides that he values (500, 0.5) and (100, 0.1) equally. Then $u(500, 0.5) = u(100, 0.1) = 1.0$.* Now calculate,

$$Q = \frac{u(500, 0.5) - u(500, 0.01) - u(100, 0.5)}{u(500, 0.01)u(100, 0.5)}$$

$$= \frac{1.0 - 0.25 - 0.9}{(0.25)(0.9)} = -0.67$$

Since Q is not zero, write

$$u(x, y) = u(x, 0.01) + u(100, y) - 0.67u(x, 0.01)u(100, y)$$

Then $u(100, 0.01) = 0.43$, for example. Using the product form

$$u(x, y) = [u(x, 0.01) - \tfrac{1}{0.67}][-0.67u(100, y) + 1]$$

This form gives $u(100, 0.01) = -0.46$ and $u(100, 0.9) = -0.17$, which leads to the same choice of action as does the quasi-separable form.

One does not like to conclude any discussion on a negative note, but utility theory probably never will be used extensively in public sector decision making. The reason is quite elementary; public managers, policy makers, and the public must rationalize public choices on the basis of who wins and who loses what; and that is best done by considering the possible consequences of alternative courses of actions in their natural units of measurement, or in the form of scenarios. Furthermore, public decisions are made by, or, at least, must be understood by, the public. It is extremely unlikely that people will ever become comfortable using utility functions which assign single numerical values to complex outcomes, thereby obscuring the reality of those outcomes. Also, public decisions

---

* For multiattribute utility functions, one cannot conveniently require that the best case have a utility of 1.0. Thus, less-than-best outcomes may be .given utility values of unity, as in this case.

are group decisions; utility theory applies only to the individual. Consequently, although individuals may evaluate policy options on the basis of their own utility functions, as a basis for debate there is no way to formulate a group utility function from their individual preferences. Finally, operationalization of the theory is just too difficult and the results too time dependent to expect anyone to utilize the theory in fast-moving, complex decision situations.

Perhaps the greatest value of studying utility theory is the conceptual framework it provides for understanding the relative nature of multiattribute consequences, and the need to weigh carefully the sacrifices one must accept in one attribute for gains in another.

20.4 OTHER APPROACHES TO MULTIATTRIBUTE DECISIONS

Prior to and even subsequent to the development of utility theory, practice-minded individuals have proposed a variety of ad hoc procedures for selecting among multiattribute decision options. Among those procedures previously touched upon in this text are: comparisons of projects on benefit-cost measures where benefits may be in terms other than dollars; and explicit, direct comparison and ranking of options on many attributes individually to observe the trade-offs implied by selection (see Chapters 11, 12, 13, and 21). With these and similar approaches the facts and results of analysis form the basis upon which judgmental or subjective comparisons are made by decision makers; to the extent possible, analysts establish the magnitudes and nature of trade-offs implicit in selecting one alternative over another, and decision makers evaluate the relative merits (utility) of those trade-offs.

A number of more formal procedures including simulation modeling and schemes for developing numerical weights for decision outcomes (less sophisticated than utility theory) are discussed by John Van Gigch (see Section 20.6). There are books and articles devoted to a technique called goal programming for multiattribute decision problems, which is something like linear programming with penalty factors built into the objective function to reflect the relative importance of achieving minimum levels of certain attributes.

In all approaches, formal or not, to multiattribute decision analysis the concept of utility provides the rational basis for choices under uncertainty, whether explicitly recognized or not.

20.5 PROBLEMS

20.1 Explain declining marginal utility (MU). Pick the function which is of declining MU.

$$u(x) = ax^2 + bx + c$$

(i) $a < 0, b < 0, c > 0$
(ii) $a > 0, b < 0, c > 0$
(iii) $a > 0, b > 0, c < 0$

20.2 Assess your own utility function by assigning dollar payoffs you would just be willing to accept in lieu of the following lotteries (your certainty equivalent). For convenience set $u(x_w)$ and $u(x_B)$ to suit yourself and calculate the $u(x)$ values for all the dollar payoffs you generate and plot. Are you risk-averse, risk-prone, or risk-neutral?

Payoff, + $10,000 with probability (P)	Payoff, − $1,000 with probability $(1 − P)$	Your certainty equivalent
0.1	0.9	
0.2	0.8	
04	0.6	
0.6	0.4	
0.8	0.2	
1.0	0.0	

20.3 Using the data below indicate the best alternative for the two different utility functions given. Use the expected utility criterion.

Project A Certainty cash flow		Project B Possible cash flows	
		$P = 0.9$	$P = 0.1$
Initial cost	200,000	300,000	300,000
Annual operating revenues	70,000	30,000	300,000
Annual operating costs	10,000	10,000	10,000
Overall tax rate	$t = 50\%$		
Project lifetime	$n = 10$ yrs		
Rate of return	$i = 8\%$		

Utility for money

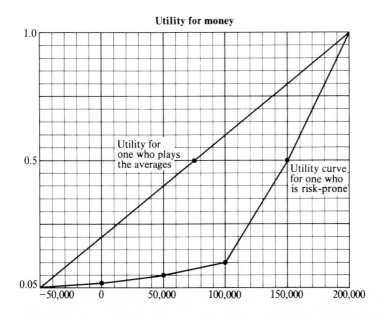

20.4 On graph paper, draw a set of constant utility (isopreference, trade-off, or indifference) curves which are consistent with an additive utility function.

20.5 Suppose you are considering purchasing a used car. You have found two cars which would meet your needs. One is priced at 2000 dollars and the other at 3000 dollars, and you think there is some chance you could get the first for as little as 1800 dollars and the second for as little as 2500 dollars. According to your consumers manual, the frequency of repair rates for the cars, respectively, are about 7 on the average, with a standard deviation of 4, and 5 on the average with a standard deviation of about 2.

   Analyze this problem using decision analysis. You will have to generate your two-variable utility function and develop reasonable probability values for a discrete distribution of payoffs in terms of cost and repair rates.

20.6 You have a personal utility function identical to that pictured. You are currently considering the purchase of an insurance policy to protect yourself from the remote possibility of a loss of 200,000 dollars. The probability that the loss will occur once within any given calendar year is 0.01; the probability of more than one such occurrence in one year is nil.

   (a) What is the maximum annual premium you should be willing to pay? Illustrate your analysis with a decision tree.

   (b) What is the premium an insurance company would have to charge if administrative, selling, income tax, and other costs add 60 percent to the premium required to meet risk only?

   The utility function is

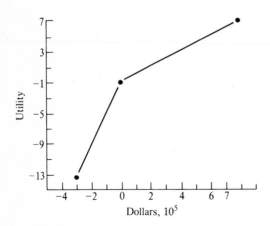

20.7 Discuss the relationship between the criteria of maximizing profit and maximizing utility. Note the conditions under which the two tend to produce (1) identical preferences and (2) different preferences.

20.8 You must decide where to locate the new fire station in your city. The two payoff variables you must consider are cost to the city and response time. The location having the lowest expected response time will probably cost much more. Imagine that first costs of land would range from 10,000 dollars for site A to 30,000 dollars for site B. Response time would range from 20 minutes for site B to 30 minutes for site A. Cost and response time estimates are within 10 percent of actual values.

   (a) Generate your isopreference curves.
   (b) Are the two variables mutually utility independent?
   (c) If they are, what value of Q should be assigned?

20.9 How would you do the last problem, using methods discussed in previous chapters in lieu of a utility function?

20.10 For the following decision problem calculate the minimum expected opportunity loss. What is the *maximum* amount the decision maker should be willing to pay for additional information?

Payoff table

	S_1	S_2
D_1	20	−8
D_2	10	−2
D_3	−10	30
State probabilities	0.7	0.3

20.11 Solve the medical decision analysis problem described by the following data. You must calculate the EVSI and the EVPI. You must identify the best course of action for patient and physician.

A physician makes a preliminary diagnosis of a patient's difficulty. He decides there are only three possible diagnoses: a, b, c. Mentally he concludes that the likelihoods of these being true are as given.

Prior-state probabilities

	State		
Probability	a	b	c
$P(.)$	0.3	0.5	0.2

The physician has four options open to him. He may do nothing D_0. He may proceed to treat the patient for a (Action A), for b (action B), or for c (action C) without further testing D_1. He may postpone treatment until certain test results are returned. In an ideal case, test results would be reported perfectly; call that option D_2. In a real-life situation the tests might be inaccurately reported D_3. The one test available produces positive or negative results only.

The fundamental outcomes to the patient are

Outcomes to patient

	Action			
		D_1		
State	D_0 (do nothing)	A	B	C
a	Death	Cure	Death	Chronic
b	Chronic	Death	Cure	Chronic
c	Cure	Cure	Chronic	Cure

The patient is concerned only about the actual income and has total medical cost coverage. The physician is concerned about making a mistake and being sued for malpractice, his income, and outcome. He may have other considerations as well. Thus, the patient and physician have different utilities for the different actions and outcomes. Their perceived utilities for each strategy are as shown in the table below.

Payoffs

Action-consequence sequence		Utilities	
		Patient	Physician
D_0	a—death	0	0
	b—chronic	80	40
	c—cure	100	95
D_1, A	a—cure	100	100
	b—death	0	0
	c—cure	100	98
D_1, B	a—death	0	0
	b—well	100	100
	c—chronic	80	10
D_1, C	a—chronic	80	10
	b—chronic	80	10
	c—cure	100	95

The laboratory test has been calibrated by determining the likelihoods that it will produce positive (negative) results when a patient is truly in state a, b, or c. These calibration numbers are given in the table below.

Calibration probabilities for test

State	Test	
	Positive result $(+)$	Negative result $(-)$
a	0.9	0.1
b	0.3	0.7
c	0.1	0.9

As mentioned earlier, test results can be incorrectly reported. The chances of this are indicated in the next table.

Conditional probabilities for reported result, given true result

True result	Reported result	
	Positive	Negative
Positive	0.99	0.01
Negative	0.05	0.95

20.6 SUPPLEMENTAL READING

Bell, David E.: "A Utility Function for Time Streams Having Inter-Period Dependencies, *Operations Research*, vol. 25, no. 3, May–June 1977, pp. 448–458.

Carlson, Gerald A.: "A Decision Theoretic Approach to Crop Disease Prediction and Control," unpublished paper.

Ellis, Howard M., and Ralph L. Keeney: "A Rational Approach for Government Decisions Concerning Air Pollution," unpublished paper.

Forst, Brian E.: "Estimating Utility Functions Using Preferences Revealed under Uncertainty," paper presented at the *39th National Meeting of the Operations Research Society of America*, Sheraton-Dallas Hotel, Dallas, Tex., May 1971.

Keeney, Ralph L.: "Multidimensional Utility Functions: Theory, Assessment and Application," Technical Report No. 43, Operations Research Center, M.I.T., Cambridge, October 1969.

————: "Multiplicative Utility Functions," *Operations Research*, vol. 22, no. 1, January–February 1974, p. 22.

———— and Howard Raiffa: *Decisions with Multiple Objectives*, John Wiley & Sons, Inc., New York, 1976.

Kirkwood, Craig W.: "Parametrically Dependent Preferences for Multiattributed Consequences," *Operations Research*, vol. 24, no. 1, January–February, 1976, pp. 92–103.

Petersen, Nancy S.: "An Expected Utility Model for 'Optimal' Selection." (Revised version appears in the *Journal of Educational Statistics*, vol. 1, no. 4, Winter 1976, pp. 333–358.)

Van Gigch, John P.: *Applied General Systems Theory*, 2d ed., Harper and Row, New York, 1978, chap. 11.

Weinstein, Milton C., Donald S. Shepard, and Joseph S. Pliskin: "Decision-Theoretic Approaches to Valuing a Year of Life" (Center for the Analysis of Health Practices, Harvard School of Public Health, Boston), January 1975.

TWENTY-ONE

APPLICATIONS OF DECISION ANALYSIS AND A CASE FOR ANALYSIS

Decision analysis has found numerous applications in medicine, public health and safety, environmental management, public service, and a variety of public and private resource allocation situations. In all cases the fundamental paradigm is the guiding light, but details of technique vary enormously to fit the needs of the specific decision problem. Two recent applications are described in this chapter.

Perhaps the potentially most fruitful area of application is medical diagnosis and treatment. Diagnosis is as much art as science. Diagnostic tests are imperfect, and the efficacy of therapeutic options are often uncertain. In the absence of a systematic procedure for prospectively evaluating the marginal value of each diagnostic step or therapeutic action physicians have had to rely very heavily on their personal experiences when deciding on a course of action. As a result, the quality of health care is quite variable nationally, and even locally. Decision analysis utilizing data generated in scientifically designed controlled studies is likely to induce major improvements in this regard. Diagnostic and treatment protocols for many diseases are likely to be refined to the point where there is general agreement as to the marginal value of each step in the protocol, largely as a consequence of decision-analytic studies.

The breast cancer study described in Section 21.1 is an excellent example of decision analysis in medicine. The main objective of the project is to determine the "best" detection strategies for asymptomatic women according to their (prescreening) epidemiological risk or propensity toward the disease.* The overarching analytical framework of the study only is discussed. This provides a clear picture of the use of the decision analysis paradigm. To adequately cover the details of the substudies for detailed data analysis which flesh out the framework would require a book in itself.

Whereas utility theory is not incorporated into the breast cancer study (at this point), it is at the very heart of the study described in Section 21.2. Dr. Ralph L. Keeney utilized multiattribute utility theory to evaluate 10 alternative pumped storage sites for a large electric utility company. This provided a single measure of desirability which reflected the economic, environmental, and technological perspectives of company decision makers. This application of multiattribute utility theory to complex real-world decision problems is one of very few successful attempts.

21.1 EVALUATING BREAST CANCER DETECTION STRATEGIES

Breast cancer is perhaps the most serious disease now threatening American women. It is the leading cause of death among women between 40 and 60 years of age. It affects between 4 and 7 percent of all women and, of course, the longer a woman lives, the greater are her chances of contracting it. The consequences of the disease are traumatic in both physical and psychological terms. Even a surgical cure is frequently disfiguring and can dramatically alter a woman's perception of herself and her place in life.

Early detection appears to be important. Long-term prognosis is enhanced, at least for women over 50. Disfigurement tends to be less severe when the disease is in an early stage, because the extent of surgery can be reduced. Trauma, too, is minimized, primarily because physical consequences are more promising.

Assuming that early detection is beneficial, the next logical question is what are the best detection strategies? Unfortunately no one can answer this question, yet. In fact, there can be no single answer, because what is best for one woman is not necessarily best for another. The question can only be answered in terms of what is best for particular women in particular circumstances. For example, mammography is the most sensitive screening technique available, especially for older women whose breasts are not so firm as they once were. Very small cancers can be detected by mammography. However, it is a radiological procedure and if it is employed frequently over an extended period of time, it may actually induce new cancers. (A committee of the National Academy of Sciences estimated that a dose of one rad, approximately the dose received

---

* In this chapter "risk" means likelihood of disease. Although this is not consistent with the terminology of decision analysis, where risk refers to expected loss, it is common terminology in medicine.

during a mammographic examination, generates new cancers at the rate of 6 per million women per year after 10 years.) This is an especially important consideration for younger women. On the other hand, both thermography and palpation are basically risk-free techniques in the previous sense. But they are much less sensitive than mammography and often fail to identify an early-stage cancer, thereby diminishing the chance of an effective cure. Clearly there could be circumstances in which mammography should be used in screening younger women, even though it is inherently risky, as when the risk of missing a cancerous lesion outweighs the risk of radiation. Unfortunately, the relative benefits and costs are not well enough understood to permit one to state precisely when to use mammography and when not to, so physicians must rely on personal judgment based on personal experience.

In the early 1960s the Health Insurance Plan of New York (HIP) conducted a randomized, controlled study of the value of screening for breast cancer. The detection techniques used were clinical examination or palpation and (x-ray) mammography. The results suggested that for women over 50 years of age, screening reduced mortality from the disease. To further test this hypothesis and to simultaneously evaluate the individual contributions of three detection modalities, namely, palpation (C), mammography (M), and thermography (T), the National Cancer Institute, in cooperation with the American Cancer Society (NCI/ACS), established 27 detection centers around the nation. In 1973, 280,000 asymptomatic women were enrolled for periodic screening. Each woman was examined by each detection modality at each visit. Each examination was (supposed to be) conducted and interpreted independently so as to avoid confusing the value of one modality with another. The plan was to analyze the data collected after closing out the screening projects six years later.

Circumstances intervened and in 1976 the Schools of Engineering and Medicine at Washington University and the Cancer Research Center (CRC) teamed up to analyze the data for the NCI/ACS detection project at CRC. Major objectives specified for this research were:

- Evaluate the technical quality of each detection modality.
- Develop risk profiles for women, linking medical history and preexamination information to the statistical likelihood that a woman will be found to have the disease.
- Evaluate all possible detection strategies, incorporating various combinations for detection modalities (tests) and logical rules for deciding which alternative modality to employ next, and determine which strategies are best for which risk groups.

Decision analysis is an ideal analytical paradigm for such purposes when embellished by certain statistical procedures such as analysis of cross-classified data to identify significant predictors of disease, regression analysis to group the predictive variables and measure the predictive strength of multivariate groups, and discriminant analysis to calculate epidemiological or pretest as well as posttest likelihoods of disease. The remaining subsections illustrate the use of decision analysis in this context. Important characteristics of the data are discussed, the procedure for enumerating and identifying viable strategies is outlined, and a simplified form of the general evaluation algorithm is presented and illustrated.

Data

The CRC is highly regarded for its data collection procedures. The screening data for its detection project are perhaps the most meticulously collected and recorded data of any of the detection projects. Important qualities which enable us to meet our objectives include:

- Each detection modality is indeed interpreted independently. The only information available to an examiner is that generated by the modality itself.
- One radiologist reads all x-ray mammograms for all women. Another reads all thermogram images. And another physician maintains oversight on all clinical examinations.
- The findings of each examiner are recorded meticulously, including final diagnoses and any recommendations regarding a biopsy.
- Final biopsy recommendations are based on complete reviews of case histories as well as data. All findings, including modified interpretations of individual modalities in the face of additional information, are recorded.
- Pathology reports on all biopsies are included in the data base.
- About 95 percent of the 10,000 women who enrolled in the project are still enrolled after 4 years.

Table 21.1 is suggestive of the kinds of data we have for women. It is no more than suggestive, however, since our data collection forms for recording the findings of each screening visit are more than 10 pages long and include 321 variables.

Table 21.1 Screening and pathology data

Case No.	Pathology	Recent screening (observed features by modality)		
		Thermography	Clinical palpation	Mammography
	Benign mammary dysplasia ⋮ Benign or apparently benign tumours ⋮	Symmetry • Background • Vascular	Mass(s) • Texture • Shape • Mobility	Mass(s) • Shape • Size • Distribution
	Carcinoma ⋮	Temperature variation		
	Sarcoma ⋮	etc.	Skin thickening Nodes Nipple Interpretation	Skin thickening Calcification Interpretation

Detection Strategies

A simplified version of a general model for medical diagnosis and treatment is indicated in Figure 21.1. In the first step of the process, the physician reviews a patient's medical history, physical characteristics, family medical history, and immediate symptoms or indications; and in conjunction with the store of knowledge she has acquired by training and experience, she makes a preliminary diagnosis of the patient's problem. Her diagnosis might consist of one or more alternative causes which have, in her mind, greater or lesser likelihoods of being correct. (One of the alternatives or possibilities could be that there is nothing wrong with the patient.) If the physician is satisfied that she has identified the problem, she might proceed to treat the patient according to some plan, without ordering diagnostic tests. At the other extreme she might delay treatment until she has reports from the laboratory for tests she might order. If she chooses the latter alternative, she must integrate the results of the laboratory tests with her preliminary diagnosis to formulate a revised diagnosis. As with the preliminary diagnosis, the revised diagnosis consists of one or more possible causes with different likelihoods of their being correct. At this time she may or may not revise her original plan on how to proceed in testing or treating the patient. Whatever action she takes must, of course, be predicated on the information she has acquired to that point. Her intermediate alternatives are to begin some form of treatment and request certain diagnostic tests, then revise the treatment appropriately when she has observed its effects and/or when the laboratory test results have been received and interpreted. Of course, whichever alternative she chooses, the physician must monitor the patient's progress and take corrective action as needed; this aspect, too, is included in the feedback loop.

In general this is a very complex decision problem, although in a specific case it may be quite elementary. It contains an element of forecasting outcomes of actions using judgmental probabilities with sample data from laboratory tests and other information sources. It contains a cybernetic element in that decisions are revised based on observations of the patient's progress and laboratory results. And, at each step of the process there are many options for action and many possible interpretations of the data. Although the model in Figure 21.1 includes both diagnostic and treatment phases, the diagnostic phase in breast cancer is sufficiently complex to deserve special attention. Only issues related to diagnosis are covered here.

The diagnostic process can include one or more diagnostic tests. At each step in the process the diagnostician has the option of requesting a biopsy, rescheduling the patient for a reexamination sometime in the future, or ordering an additional diagnostic test. In principle, the diagnostician may take any of these actions regardless of the outcome of the previous test. However, if the *true positive rate* (TP) for a test (or strategy)—the likelihood of a positive examination when the patient has the disease—is greater than the *false positive rate* (FP), the diagnostician does not establish decision rules requiring rescheduling on the basis of a positive test and a biopsy on the basis of a negative test. Furthermore, if the diagnostician makes no use of the information in a test outcome (as when

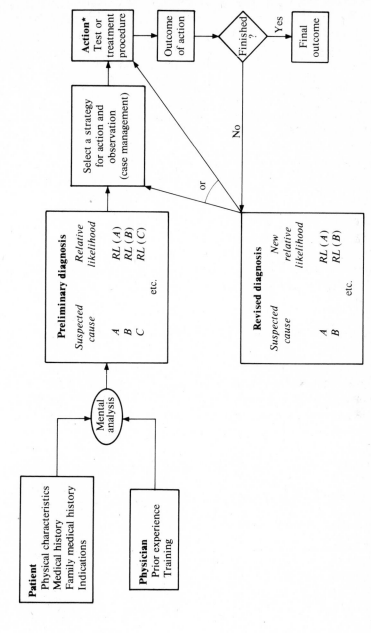

Figure 21.1 General model for diagnosis and treatment. *Note:* *Often treatment and testing proceed simultaneously.

389

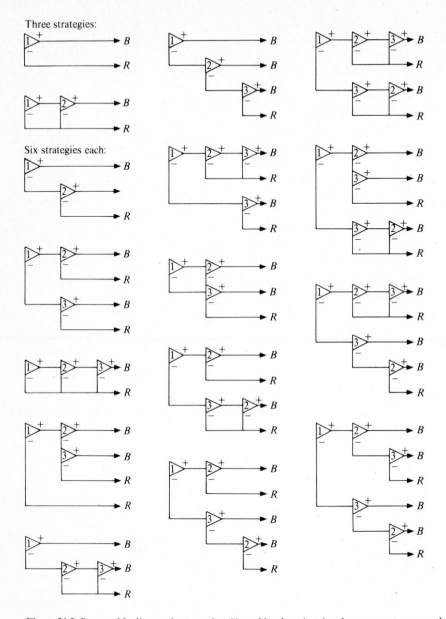

Figure 21.2 Reasonable diagnostic strategies. *Note:* Numbers in triangles represent test numbers.

the decision is made to biopsy regardless of test results), including the test in the diagnostic process merely inflates the cost, with no improvement in diagnostic power. Finally, tests are not repeated. For these reasons, the 16,430 theoretically possible diagnostic strategies incorporating any combination of thermography, mammography, and clinical examination reduces to 92 "reasonable" strategies.

There are three classes of strategies incorporating one, two, and three diagnostic tests. These are illustrated in Figure 21.2. The tests, indicated by a triangle, are labeled 1, 2, 3 to signify only that they are distinct tests. One could substitute M, T, or C for any of these, so that many permutations are possible.

Evaluating Screening Strategies

The true positive (TP) and false positive (FP) rates are statistical approximations to the calibration probabilities for a screening strategy. They depend on the capabilities of clinicians and radiologists and especially on the decision rules established for recommending biopsy. Up-to-date cross tabulations of positive and negative calls for each modality by histological findings provide the basis for calculating TP and FP rates.

Table 21.2 illustrates an efficient summarization of data for calculating calibration probabilities for screening strategies which incorporate any conceivable combination of clinical palpation (C), thermography (T), and mammography (M). Each entry represents an independent interpretation of each modality for a single breast. This is required to avoid confusing the quality of one modality with another. True positive and false positive rates are estimated from Table 21.2 by taking the ratios.

$$TP = \frac{\text{cancerous breasts screened positive}}{\text{total cancerous breasts}}$$

$$FP = \frac{\text{noncancerous breasts screened positive}}{\text{total noncancerous breasts}}$$

Table 21.2 Status by modality indication*
2676 women screened and monitored
126 biopsies on 119 women

	+C (43)				−C (83)				
	+T (28)		−T (15)		+T (67)		−T (16)		Total
	+M	−M	+M	−M	+M	−M	+M	−M	(126)
Positive pathology	8	4	3	2	44	0	8	0	69
Negative pathology	6	10	3	7	23	0	8	0†	57

* +C = Palpable mass; biopsy recommended.
   +T = Abnormal thermogram: localized temperature gradient of at least 2°C, or pronounced asymmetry.
   +M = Suspicious mass or punctate calcifications; biopsy recommended.
   † Assume that all examinations which did not lead to a biopsy fell into this category of all-negative tests. That is, 5352 − 2 × 119, or 5114, all-negative examinations.

For example, the respective rates for the strategy utilizing only mammography are 91 and 0.8 percent.*

Economic costs and the expected level of confirmation of the presence or absence of the disease are important considerations in selecting a strategy. Four related indicators of merit are the confirmatory value or information content of positive and negative screens, expected cost per screen, expected cost per rate of detection, and cost per unit of information expected from a screen. Although these are interrelated measures of merit, each provides insight not attributable to the others. All four measures should be considered when selecting a screening strategy. Additional factors, such as convenience and screening risks which are not explicitly treated, may also be important in certain situations and can be incorporated into the selection process on an ad hoc basis.

Estimating Confirmatory Value

The likelihood $P(C)$ that a woman of a particular age with a specific medical history and epidemiological background will be found to have the disease is estimated using regression techniques and a related procedure called *discriminant analysis*. Stepwise multiple regression permits one to efficiently assess the predictive value of numerous medical and epidemiological features. A variety of features have been implicated in the literature, often without a scientific basis (see Table 21.3). Regression analyses help sort out the wheat from the chaff, so to

---

* Our own data are not included in tables or calculations presented in this chapter unless so indicated. Instead, for illustrative purposes only data published by Agnus Stark and Stanley Way are used (see Section 21.4).

Table 21.3 Risk features implicated in the literature

*Age	(+)	Type of home community	
Race		Relatives with other cancers	(+)
Religion		Previous other breast problems	(+)
Place of birth		Age at first pregnancy	
*Previous x-rays	(+)	Number of pregnancies	(−)
*Relatives with breast cancer		*Age at first birth	(+)
Previous breast cancer	(+)	*Number of live births	(−)
Current breast problems	(+)	*History of breast feeding	(−)
Previous breast biopsies	(+)	*Age at menarche	(−)
Hormone intake	(+)	Other menstrual problems	(+)
Age at menopause	(+)	Prolonged menstrual history	(+)
Previous cancer	(+)	Previous selected other diseases	(+)

Note: The symbol (+) indicates a supposed positive relation between the variable (or its magnitude) and increased risk of contracting the disease. The symbol (−) has the opposite connotation. These presumptions have not been scientifically verified. Testing them is one of the objectives of the Washington University/Cancer Research Center study. An asterisk marks features we have tentatively tested using a sample of 500 cases from our data base.

speak. For example, women whose mothers, sisters, or other female relatives have had breast cancer are generally considered to be at very high risk; the numbers vary from 2 to 47 times the risk of the general female population, depending on the author and his or her data. Yet, regression of the presence of cancer on this feature for 500 women who have had biopsies as a result of screening at the CRC tentatively indicates no significant relation. Similarly, some physicians feel that women who postpone having their first child beyond the age of 30 are at higher risk than women who do not. A simple regression on age at first birth for the 500 women in our sample tentatively suggests just the opposite. Multiple regression on the features preceded by an asterisk in Table 21.3 yields a predictive equation in seven variables for which $R^2 \simeq 22\%$; age and previous x-rays are the dominant variables.

Discriminant analysis permits one to directly calculate $P(C)$ from the list of risk features exhibited by any woman. One can also invoke stepwise algorithms to select the "best" discriminant function for predictive purposes. This would incorporate only the statistically most important risk features. Linear discriminant functions have the form

$$P(C) = P_0 + \alpha_1 X_1 + \cdots + \alpha_n X_n$$

where the X's symbolize risk features.

Screening provides an opportunity to refine the estimate. The degree of refinement offered depends only on the prevalence rate and the TP and FP rates for the strategy employed.

The refinement process is illustrated in Figure 21.3. The symbols $P(C/+)$ and $P(\bar{C}/-)$ represent the likelihoods of cancer and no cancer given positive and negative screens, respectively. The refinement algorithms, based on Bayes' rule, are:

$$P(C/+) = \left[1 + \left(\frac{\text{FP}}{\text{TP}} \frac{P(\bar{C})}{P(C)} \right) \right]^{-1}$$

$$P(C_0/-) = \left[1 + \left(\frac{1 - \text{TP}}{1 - \text{FP}} \frac{P(C)}{P(\bar{C})} \right) \right]^{-1}$$

They have equivalent graphic representations, as illustrated in Figure 21.4. The confirmatory values of positive and negative screens are approximated by the differences:

$$[P(C/+) - P(C)] \sim \text{Confirmatory value of positive screen}$$

$$[P(\bar{C}/-) - P(\bar{C})] \sim \text{Confirmatory value of negative screen}$$

The process of calculating confirmatory values is simple. With statistical estimates for TP and FP rates and an estimate of $P(C)$, one can simply read the refined likelihood estimates $P(C/+)$ and $P(\bar{C}/-)$ from a nomogram (see Figure 21.4) and mentally calculate the confirmatory values of positive and negative screens. Or, one can utilize a calculator programmed with the equations above.

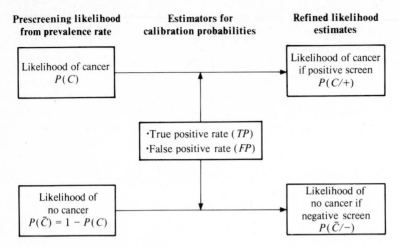

Figure 21.3 Likelihood refinement by screening.

Assuming that current TP and FP rates are in the calculator memory, the confirmatory values for numerous strategies can be displayed instantly upon entry of the prevalence rate.

These results can be organized to indicate which strategies imply the greatest confirmatory value for women within risk groups. Medical advisors can commit the results to memory. For example, strategies incorporating only clinical exam-

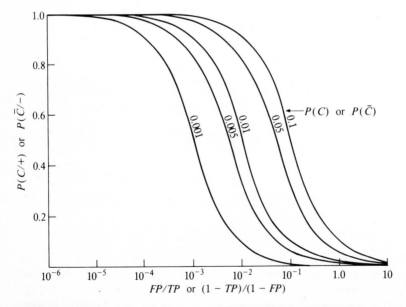

Figure 21.4 Nomogram for revising estimates of the likelihoods that the disease is present or absent.

ination and thermography might provide as much information as strategies incorporating mammography, when screening young, low-risk women, where $P(C)$ is less than 0.5 percent. Mammography, on the other hand, might be essential for screening older, higher-risk women, where $P(C)$ is between 2 and 3 percent. Women in the first group might be counseled only on the relative merits of the more effective strategies utilizing clinical examination and thermography; mammography might be offered only if requested. Women in the higher-risk group might be advised of the merits of strategies incorporating mammography.

Estimating Economic Measures of Merit

For any strategy, the expected screening charge is simply the sum of products of unit charges per biopsy or screening modality times the statistical proportion of women who utilize each procedure. The proportions come from tabulations such as Table 21.2. The calculational expression is

$$E(\text{cost/screen}) = P_B \times C_B + \sum_{\text{all } i} P_i \times C_i$$

where P_B is the biopsy rate and P_i is the screening rate for modality i within the strategy.

Although expected screening charge alone is a crude economic indicator of quality, when taken in combination with measures of technical quality it can provide a relative indication of the cost to benefit ratio for strategies. For example, the ratio of expected screening charge to the detection rate reflects the expected cost per cancer detected by risk category. Also, the ratios of expected charge to positive and negative confirmatory values reflect the unit costs of information expected from a screen. These measures should play important roles in the strategy selection process, regardless of who pays the bill.

Other Measures of Quality

Utility theory has a potential role in evaluating screening strategies. Ideally, the best strategy is one which offers a person maximum utility. However, the technical problems one would encounter in this particular application may well outweigh the potential value of the theory. At least for the short term, more subjective and direct balancing of the multiple effectiveness measures described above is necessary.

Strategy Selection Illustrated

By way of illustration the 14 plausible strategies diagramed in Figure 21.5 are evaluated. Neither the calibration probabilities nor the economic data utilized can be taken as representative of any specific clinical setting. The comparison is only illustrative.

Elementary decision rules for proceeding to biopsy

Single modality   Multiple modality

s_1 – C positive s_4 – C or T positive s_7 – C and T positive

s_2 – T positive s_5 – C or M positive s_8 – C and M positive

s_3 – M positive s_6 – T or M positive s_9 – T and M positive

s_{10} – At least two of C, T, and M positive

More complex decision rules for proceeding to biopsy

Multiple modality

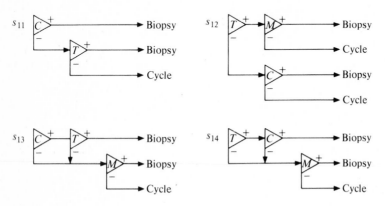

Figure 21.5 Some plausible strategies utilizing clinical palpation (C), thermography (T), and mammography (M).

Screening data for 5171 examinations were summarized in Table 21.2. The decision rules appropriate to these data are:

$+M$ ~ biopsy recommended; suspicious mass or punctate calcification

$+C$ ~ biopsy recommended; palpable mass

$+T$ ~ thermogram indicated localized temperature gradient of 2°C or more, asymmetric pattern

The results of the analysis for all 14 strategies are summarized in Table 21.4. The confirmatory values CV of positive and negative screens are based on a prevalence rate of 2.3 percent $[P(C) = 0.023]$, which reflects the risk profiles of women screened. The economic indicators are based on charges per procedure of 40 dollars for clinical examination, 50 dollars for thermogram, 48 dollars for mammogram, and 400 dollars for biopsy. For strategy 3, which incorporates only mammography, the basic calculations using the data from Figure 21.2 are:

(a) TP $= (8 + 3 + 44 + 8)/69 = 0.91$ FP $= 40/5171 = 0.008$

(b) Positive CV $= 0.73 - 0.023$ Negative CV $= 0.998 - 0.977$

(c) E(cost/screen) $= 400 \times (103/2676) + 48 \times (2676/2676) \simeq 63$

(d) Expected cost per detection rate $= 63 \times 69/62 = 70*$

---

* 62 cancers detected by mammography, 69 cancers total, and 63 dollars per screen.

Table 21.4 Relative merits of screening strategies
Prevalence rate $P(C) = 0.023$

Strategy no.	Calibration, %		Confirmatory value (CV), %		Expected costs			
	TP	FP	Positive	Negative	$/Screen	$/Detection rate	$/CV positive	$/CV negative
1	25	0.7	43.7	0.5	46	168	1.05	92.00
2	81	0.8	67.7	1.8	64	71	0.96	35.56
3	91	0.8	70.7	2.1	63	70	0.89	30.00
4	88	1.0	64.7	2.0	106	107	1.64	53.00
5*	100	1.1	65.7	2.3	106	95	1.61	46.09
6	97	1.0	67.7	2.3	116	106	1.71	50.43
7	17	1.3	21.7	0.3	94	483	4.33	313.33
8	16	0.2	62.7	0.3	91	509	1.45	303.33
9	75	0.6	72.7	1.7	110	121	1.51	64.70
10	86	0.8	69.7	2.0	154	161	2.21	77.00
11	88	1.0	64.7	2.0	105	107	1.62	52.50
12	84	0.6	74.7	1.9	104	113	1.39	54.74
13†	97	1.0	67.7	2.2	106	97	1.57	48.18
14†	97	1.0	67.7	2.2	114	104	1.68	51.82

* A TP rate of 100 percent for this strategy is a statistical accident. The true rate is most likely high, but not 100 percent.
† These strategies are equivalent technically but not economically. Which is best depends on the relative costs of the thermogram and the clinical examination.

Clearly, strategies 1 and 7 are greatly inferior; requiring positive clinical findings as a condition for biopsy is not advisable for women in this high-risk group. Strategies 3, 9, and 12 are superior in terms of information provided, although 2, 6, 10, 12, 13, and 14 are close runners-up. Strategies 1, 2, 3, 5, and 13 are the least expensive per screen. When cost per detection and costs per unit of information are factored into the comparison, the best six strategies appear to be numbers 2, 3, 8, 9, 12, and 13. Strategy 12 looks particularly good. Its cost per screen is relatively low; the confirmatory value of a positive screen is larger than for any other strategy and mammography is used as a backup for thermography, which reduces overall exposure to radiation.

21.2 PUMPED STORAGE SITE SELECTION

In 1977 Dr. Ralph L. Keeney of Woodward-Clyde Consultants directed an evaluation of alternative sites for pumped storage hydroelectric generation for a large electric utility company. The company had decided that it could meet growing peak demand levels for electric power by pumping water into storage reservoirs at elevated locations during nonpeak hours and releasing it to generate additional power during peak demand periods. This being a less expensive alternative than building additional power plants, the company undertook a search for desirable reservoir sites. Dr. Keeney's responsibility was to identify and rank potential sites. The methodology he employed was utility theory.

The first step was to identify specific considerations which might bear on site selection. Areas of consideration were public health and safety, environmental effects, socioeconomic effects, system economics, quality of service, and public acceptance of sites. Many of the considerations were relatively unchanged from site to site, such as archeological significance. These, therefore, were discarded as useless ranking factors. Other factors of a sociopolitical nature were eliminated for other reasons. The factors found to be most important were cost, transmission line efficiency, and environmental impacts. The four attributes selected for the analysis following extensive discussions with the client were

x_1, first-year cost
x_2, transmission line distance
x_3, pinyon-juniper forest
x_4, riparian community

Teams of engineers, biologists, and others representing the electric utility company (in some cases working for Woodward-Clyde) estimated values for each of the four attributes for each site. To reflect the undesirability of running power lines through some areas, mileages through those areas were multiplied by a relative impact factor which had the effect of increasing equivalent mileage in negatively impacted areas. The factors ranged from 1.0 for rangeland to 10.0

for national parks and wildlife preserves. This was the basic data set for ranking sites.

Working with electric company personnel, Dr. Keeney determined that:

- Trade-offs between first-year costs x_1 and transmission line distance x_2 were unaffected by assumed levels of x_3 and x_4. This implied that (x_1, x_2) were preferentially independent of forest type and riparian life (x_3, x_4). The same was true for other pairs of the attributes. Thus preferential independence was established.
- First-year costs of 68.75 million dollars for certain was equivalent to a 50–50 chance of 50 million dollars or 75 million dollars, regardless of the values of the mileage x_2, forest x_3, and riparian x_4 factors. Thus, x_1 was utility independent of the other attributes.
- Experts were asked to select between two options: A, having a 50–50 chance at minimum cost and low mileage versus maximum cost and a higher mileage; B, having a 50–50 chance at minimum cost and the higher mileage versus maximum cost and the lower mileage. They invariably selected B so as to avoid any chance of facing the worst case on both attributes. This precluded the use of an additive utility function (see Chapter 20).

Although an additive utility function was not appropriate, these determinations were consistent with a quasi-additive structure which in turn has a multiplicative equivalent, as was shown in Chapter 20. In this case the utility function was of the form*

$$u(x_1, x_2, x_3, x_4) = \frac{1}{k} \prod_{i=1}^{4} [1 + kk_i u_i(x_i)] - \frac{1}{k}$$

Thus the assessment process reduced to assessing individual utility functions $u(x)$ for each of the attributes and determining the constants K, k_1, k_2, k_3, and k_4.

Each utility function was scaled between zero and one. Each was checked for its general shape and risk aversion versus risk proneness, and a single midvalue was assessed. The three utility values and the general shape for each were incorporated into exponential functions of the form

$$u(x) = b[1 - e^{r(x - x_{max})}]$$

where b and r were determined by substituting the minimum and then the midrange values of x and solving the two equations simultaneously.

The scaling factors k and k_i were calculated via a complex sequence of comparing hypothetical situations and searching for points of indifference. This produced five equations which were solved simultaneously to obtain:

$$k_1 = 0.716 \qquad k_3 = 0.014 \qquad k = -0.534$$
$$k_2 = 0.382 \qquad k_4 = 0.077$$

---

* The symbol $\prod_{i=1}^{4}$ means multiply the four terms bracketed.

Using the four-dimensional multiplicative utility function above, 10 sites were ranked. The sensitivity of rankings to variations in attitudes toward uniqueness of environmental situations, uncertainties in estimates, and changes in attitudes toward risk was checked. Rankings were very sensitive to environmental attitudes, relatively stable in the face of uncertainties in the other factors, and practically impervious to variations in attitudes toward economic risk.

21.3 EPSDT/MEDICAID IMPLEMENTATION IN STATE X: A CASE FOR ANALYSIS

In 1967 Title XIX of the Social Security Act was amended to require that an Early and Periodic Screening, Diagnosis and Treatment (EPSDT) program be implemented by states. The target population of the program was all individuals under the age of 21 years who were eligible for Medicaid. The purpose of the program was to reduce the cyclical dependency of people on the welfare system by improving their health status. The legislation specified that a state agency must provide:

Medical screening services to assess the physical and mental health status of eligible individuals. The screening procedures used could be quick and simple, but should be determined through consultation with health experts. The screening was to record at least: health and developmental history; physical growth factors; defects of ears, nose, throat, mouth (including teeth and gums); cardiac abnormalities; anemia; sickle-cell trait; lead poisoning; diabetes; infections and other urinary tract conditions; nutritional and immunization status.

Diagnostic services to determine the nature or cause of any abnormality found in screening. This could be done at the time of screening.

Treatment within the amount, duration, and scope of the state's Medicaid plan. Treatment for visual and hearing defects, including eyeglasses and hearing aids, must be provided as must services to maintain or restore dental health.

However, it was 1972 before the federal government issued regulations and guidelines for implementation of the EPSDT program. All states were required to offer EPSDT services. They were also required to establish active programs to inform the eligible populations of the availability of these services and encourage them to make use of them, and programs to enlist public and private clinics and practitioners throughout the state as service providers. The states were further required to establish procedures for ensuring that eligibles having abnormalities identified during the screening (and diagnosis) received the proper treatment. Finally, states were required to submit monthly reports to the Department of Health, Education, and Welfare (HEW), Social and Rehabilitation Services (SRS) indicating monthly totals for the number of children:

1. Screened
2. Referred for diagnosis and/or treatment for:
   a. Eye problems
   b. Hearing deficiencies
   c. Dental problems
   d. Sickle-cell anemia
   e. Lead poisoning
   f. All other conditions

These data were to be provided for children under 6 years old and for those between 6 and 21 years old separately. The records from which these data were derived, it was suggested, should be such that detailed benefit-cost analyses could be based on them. Each state's program was to be fully implemented by February 7, 1974 unless a state had an HEW-approved plan for phasing it in, as follows:

Phase 1 Full service to children under 6 years old plus an approved plan for full implementation by February 7, 1974.
Phase 2 Full service to all eligibles below 21 years old by June 1975.

Thus it took the federal structure about five years to transform an ill-defined legislative objective into an action program with reasonably well-defined objectives and program elements. The EPSDT program as it is now defined is summarized in Figure 21.6.

The State's Response

State X had a new Department of Social Services, directed by a widely respected leader recruited from another state, when the HEW regulations were issued. The Department was an umbrella agency for a wide range of human services and was charged with striking at the roots of dependency, disability, and abuse of society's rules. Among the more than 100 programs administered by the Department were Aid to Dependent Children (ADC) and EPSDT/Medicaid. The Department also had the responsibility for comprehensive health planning for the state: planning, coordinating, and funding support for the Office of Economic Opportunity (OEO) programs and the delivery of certain medical services in the area of maternal and child care.

The Department of Social Services had five administrative or staff units and five operating units, as shown in Figure 21.7. Of particular interest for the purposes of this case are two staff units: Finance, and Planning and Research; and three operating units: Family Services, and Health and Special Services. The EPSDT program overlaps into each of these units.

The Division of Finance was responsible for accounting for all state and federal funds, procurement, budgeting, and auditing. The Division of Planning and Research had the responsibility of developing a statewide social services

Legislative objective: Through improved health status help to break the chain of succeeding generations' dependency on welfare

Federal response: EPSDT program

Eligible population: Welfare dependents and Medicaid eligibles up to 21 years old

Debilitating abnormalities: Sight, hearing, dental, lead poisoning, sickle-cell, other

Other abnormalities: Not defined

Mandated program elements:

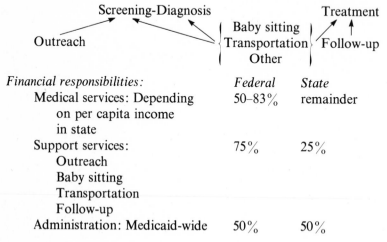

Financial responsibilities:	*Federal*	*State*
Medical services: Depending on per capita income in state	50–83%	remainder
Support services: Outreach Baby sitting Transportation Follow-up	75%	25%
Administration: Medicaid-wide	50%	50%

Management responsibilities: Planning, organizing, staffing, directing, monitoring, and controlling are shared by HEW and the states:

HEW/SRS:

1. Overall program definition and redefinition.
2. Review and approve states' implementation plans and proposals for change.
3. HEW officials will inspect each state's program at the appropriate times to see if it is in full compliance. If a state is not in compliance, it will be penalized financially. Specifically, if a state chooses to implement the program for all age groups immediately and it does not accomplish this on time, it will lose 1 percent of *all* its Medicaid funds for each month after the 2-year period until it is in compliance. If it chooses to phase the program in, it will lose 10 percent of all its Medicaid funds for each month until compliance, after the $2\frac{1}{2}$-year implementation period. In the latter case, noncompliance for more than 6 months beyond the implementation period will result in a federal court suit against state officials. In the former case, noncompliance for 1 year and 6 months would result in a federal suit.
4. Set annual program objectives, monitor their achievement, and take corrective action as necessary (including funding adjustments ranging from penalties

Figure 21.6 The EPSDT program.

to goal-directed incentives, such as extra funds for general purpose computers).

5. Establish reporting requirements and funding procedures.

State:

1. Develop plan for implementation of the program:
   a. Appropriate organization structure
   b. Professional staffing (including vendors)
   c. Case tracking system, data gathering system and report system
   d. Timetable for implementation
2. Day-to-day program management (objectives setting, action planning, organizing, staffing, directing, monitoring, and controlling)

Figure 21.6 *Continued.*

plan for the Department, implementing Management-by-Objectives (MBO), coordinating research required by federal reporting and grant management, and coordinating program performance measurements for planning and budgeting functions. The Division of Special Services was responsible for implementing the Federal Partnership for Health Act, for comprehensive health planning in the state, for providing support and coordination for comprehensive health planning and community action agencies in the state, and for implementing the Federal Economic Opportunity Act (which included *some* medical screening of children). The Division of Health was responsible for certain maternal and child care services such as dental health and communicable disease control and maintained some district health units.

The Division of Family Services was to be formally charged with operating responsibility for the EPSDT program. This division was divided into three sections: Income Maintenance (IM), Medical Services (MS), and Social Services (SS), and each was responsible for some aspect of the EPSDT program. Income Maintenance was responsible for investigating and certifying the eligibility of applicants for a wide range of programs, including those of ADC and Medicaid. It did this through its county and local offices. Medical Services was the Medicaid (Title XIX) agency. That is, it maintained files on eligible Medicaid recipients and service vendors and verified the legitimacy of vendors' bills for services. It did not pay the bills but took care of all details prior to that; Finance wrote the check at the behest of Medical Services. The director's dilemma was to determine which section should be assigned primary responsibility for managing the program. This is now your dilemma.

Since EPSDT was a Medicaid program, it might have made sense to put it in MS, which had sole responsibility for all other facets of Medicaid. The only difficulty was that EPSDT required the provision of screening, diagnosis, and

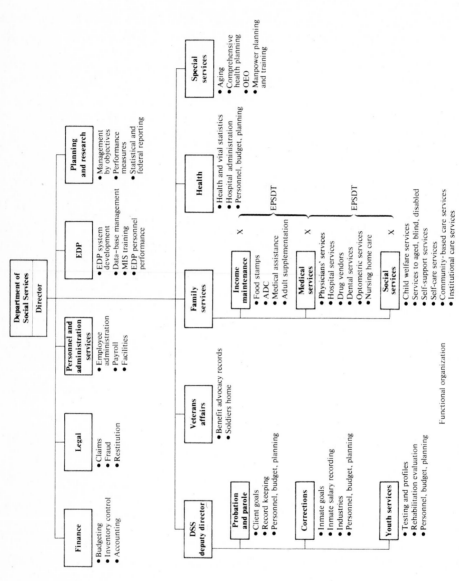

Figure 21.7 Departmental organization.

treatment services, and the necessary supporting social services to get children examined and treated. All the support services were within the purview of SS, which was the professional social worker unit, and IM, which had all welfare programs within its administrative bag and did much of the eligibility determination for clients in conjunction with the professional social workers. The role of MS had always been to process and authorize payment of Medicare and Medicaid bills and to maintain a file of eligible Medicaid physicians.

A brief description of the processes of providing eligible clients with screening, diagnosis, and treatment services and of paying the providers will indicate how these organizational units are interlaced by the EPSDT program (see Figure 21.8). When at the time of application or case review at a county office of the IM section of the Family Services Division, a family is determined eligible for welfare, the interviewer inquires whether or not children in the family have had a physical examination in the last year. If a child has not, he or she is eligible for screening, diagnosis, and treatment under the EPSDT program. Parents who decide to take advantage of the program are given a MS-40 form to present along with their identification card to a participating screening unit when they take their child for an examination. This form is filled out by personnel of the screening and diagnosis unit and forwarded to the MS section at the state level for payment. There it is checked for compliance with established screening and diagnosis reporting rules and to verify that the screening and diagnosis (S & D) units and the patient are participants in the EPSDT program. If discrepancies are found, appropriate correspondence is attached to the MS-40 and returned to the S & D unit(s) for correction. Processing begins again when the MS-40 is corrected and returned. When the MS section has determined that everything is in order and payment should be made for S & D, the Finance Division is given the necessary paperwork and makes payment.

After diagnosis, an eligible patient is referred to a participating treatment (T) unit (this may be the same unit as the S & D unit in some cases). All treatment is billed separately through the MS which notifies SS case workers in the county of such cases, and the latter are expected to follow up to ensure that proper treatment is received promptly. Medical Services in effect decides for case workers which cases need attention.

In addition to the responsibility to follow up a diagnosis to ensure treatment, case workers have the responsibility of ensuring that transportation, baby sitting or older-person sitting, and other services are provided when necessary to ensure that eligible children actually receive the screening and diagnosis as well as treatment prescribed by the EPSDT régulations.

The Decision Problem

The chief of MS would really like to manage the EPSDT program. The chiefs of IM and SS are not so enthusiastic, since their people are already overworked and their budgets are frozen. On the other hand, most of the essential activities, upon which the success of the program will depend, will have to be done by IM

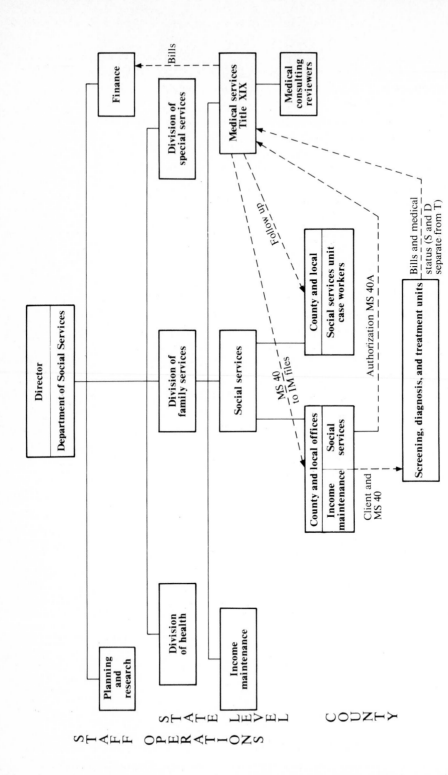

Figure 21.8 Case processing.

and SS people, regardless of which division manages the program. The director is aware of these facts and also knows that the chief of MS is a much more effective manager than either of the other two.

A fourth alternative being considered by the director is to create an EPSDT program management team of about five people, some taken from each of the sections in the Division of Family Services and some from other divisions. The program manager (PM) for the team would report directly to the director and would have a separate budget agreed to by all involved divisions. The team would be responsible for implementing the program, in cooperation with the divisions, and managing it for at least 3 years. The team members would be essentially 90 percent of the program. The director has decided that if she goes with the team approach, the only reasonable thing would be to begin full implementation of the program immediately. If she goes with any of the others, she might phase in the EPSDT program or try for immediate full implementation; she has not decided yet.

The director has analyzed many factors and concluded that:

1. If she established a team, there would be a 0.5 probability that it would meet the HEW deadline, a 0.25 probability of slipping the deadline by 2 months, a 0.2 probability of slipping it by 4 months, and a 0.05 probability of missing it by as much as 6 months.
2. If she implemented the program within MS, IM, or SS, the odds would be about as shown in Table 21.5.

The director believes that the chances for success of the EPSDT and all other programs will be significantly enhanced if she can get a management information system installed. She has petitioned the legislature for the money to install the system. The legislature will decide in 3 months. If the system is authorized, it will be fully installed in the subsequent 3 months. The odds of the legislature's approving the request are estimated by the director as seven in favor to three against. If the system is approved, the director believes that Table 21.6 will more accurately represent the likelihoods of completion dates.

Table 21.5 Likelihoods for completion dates

	Income maintenance		Social services		Medical services	
	Phase in	Full	Phase in	Full	Phase in	Full
Success	0.7	0.4	0.8	0.3	0.5	0.7
Slip by 3 months	0.2	NA	0.1	NA	0.3	NA
Slip by 6 months	0.1	0.3	0.1	0.3	0.2	0.1
Slip by 12 months	NA	0.2	NA	0.2	NA	0.1
Slip by 18 months	NA	0.1	NA	0.2	NA	0.1

Table 21.6 Posterior likelihoods for completion dates

	Income maintenance		Social services		Medical services	
	Phase in	Full	Phase in	Full	Phase in	Full
Success	0.8	0.5	0.8	0.3	0.7	0.9
Slip by 3 months	0.1	NA	0.1	NA	0.2	NA
Slip by 6 months	0.1	0.2	0.1	0.3	0.1	0.05
Slip by 12 months	NA	0.2	NA	0.2	NA	0.03
Slip by 18 months	NA	0.1	NA	0.2	NA	0.02

The director has explained to her staff that she would be most happy if the program were implemented in 1 year and least happy if it took $2\frac{1}{2}$ years, only 25 percent less happy if it took 2 years, and 50 percent less happy if it took 2 years and 3 months.

You are a member of the director's staff and it is your job to help her decide on a course of action. Draw a decision tree and find the director's optimal decision. Discuss the validity of the director's comments for approximating her utility function. That is, what questions should be asked of her to try to get a more accurate estimate and speculate on how this estimated utility function might differ from the more inaccurate version? What change in probability estimates would make the director opt for the "runner-up" alternative? What change in the utility function could cause the director to opt for the "runner-up"? What is the probability that the state will be sued by the federal government for noncompliance if the management information system is approved versus if it is not approved? What is the total probability that the state will be sued by the federal government? What is the likelihood that the state will lose more than 1 million dollars of Medicaid funds over the first 24 months of the program, if it ordinarily receives 10 million dollars per month?

21.4 SUPPLEMENTAL READING

Keeney, R. L.: *Evaluation of Proposed Pumped Storage Sites*, Project Report, Woodward-Clyde Consultants, Three Embarcadero Center, Suite 700, San Francisco, 94111, November 1977.
—— and H. Raiffa: *Decisions with Multiple Objectives*, John Wiley & Sons, Inc., New York, 1976.
Gohagan, John Kenneth (PI): " *B/C (Benefit/Cost) Evaluation of Breast Cancer Detection Strategies*," No. HSCA-03256-01, National Center for Health Services Research/National Institutes of Health, USDHEW, February 1978.
Stark, A. M., and S. Way: "The Screening of Well Women for Early Detection of Breast Cancer Using Clinical Examination with Thermography and Mammography," *Cancer*, vol. 33, June 1974, pp. 1671–1678.

APPENDIXES

THE FEDERAL STATISTICAL SYSTEM

The federal government of the United States is probably the world's largest generator, collector, and user of statistical data. Almost every federal agency is substantially involved in at least one of these activities. Figure A1.1 was developed by the Bureau of the Budget, the predecessor of the Office of Management and Budget (OMB), to illustrate the extent of the federal statistical system as of 1963. Although no updated version of this diagram exists, the odds are that the system is even more comprehensive today. Coordination planning and policy development for the federal system is increasingly the responsibility of the Statistical Policy Division of OMB.

The federal system extends to the state level via cooperative data collection and distribution programs. Cooperative programs are supported by the departments of: Labor; Agriculture; Health, Education, and Welfare; and Commerce. The usual practice is for the federal agency to provide salary support directly or through program grants. Statistical standards and practices are established by the federal agencies.

Numerous large data banks have been developed by federal agencies over the years. With the exceptions of classified defense and national security data, the responsible agencies strive to make their data banks accessible to the public. The Bureau of the Census, U.S. Department of Commerce, and many other agencies publish tables of data accompanied by explanations and even analyses in some instances. In addition they publish guides to their programs of data collection and their publications. Access to their computerized data bases is generally available to legitimate users such as universities and corporations with the capacity to handle large volumes of data stored on magnetic tape.

To learn in detail what is available from federal agencies one needs only to visit the government documents section of a good city or university library. For a brief overview of the system write to the Statistical Policy Division, Office of Management and Budget, Executive Office of the President, 726 Jackson Place, N.W. Washington, D.C., 20503, and request a copy of *Framework for Planning U.S. Federal Statistics, 1978–1989.*

The Federal Statistical System

1. GENERAL COORDINATION

The statistical organization of the Federal Government has developed in a decentralized pattern, with different agencies having responsibilities for the collection, compilation or analysis of statistical data in specified areas.

A central statistical office is therefore required to achieve an integrated and accurate system of governmental statistics and to prevent duplication.

OFFICE OF STATISTICAL STANDARDS
BUREAU OF THE BUDGET

Develops a single coordinated statistical system in the Federal Government. Prevents duplication of statistics and reduces reporting burdens. Develops and enforces standards for the quality and comparability of data.

---

2. GENERAL-PURPOSE STATISTICAL AGENCIES

The primary function of these agencies is collection, compilation and publication of statistics for general use.

Each of these agencies is responsible for the regular collection, analysis and publication of data in specified fields.

As a group, they account for a large proportion of Federal statistical activities.

STATISTICAL REPORTING SERVICE
(Dept. of Agriculture)

Statistics on crop and livestock production; prices received and paid by farmers; farm employment and wage rates. Special surveys.

BUREAU OF LABOR STATISTICS
(Dept. of Labor)

Statistics on labor force, employment and unemployment; earnings, manhours and wage rates; productivity; industrial injuries; industrial relations; wholesale and retail prices; urban consumer price indexes; foreign labor conditions.

BUREAU OF THE CENSUS
(Dept. of Commerce)

Censuses of population, housing, agriculture, manufactures, mineral industries, business, governments. Current statistics in these areas, and in foreign trade, shipping construction.

NATIONAL CENTER FOR HEALTH STATISTICS
(Dept. of Health, Education and Welfare)

Statistics on morbidity, health care, and demographic, social and economic factors related to health; and on births, deaths, marriages and divorces.

---

3. ANALYTIC AND RESEARCH AGENCIES

These agencies are major users of data collected by the public-purpose agencies, the administrative and regulatory agencies, and private sources. They are particularly important in the compilation, analysis and interpretation of statistics, though some direct collection of statistics for general use is involved in their activities.

COUNCIL OF ECONOMIC ADVISERS
(Executive Office of the President)

Analyses of economic data on the general economic situation, and appraisal of related Federal activities. Assists the President in preparation of his economic reports to the Congress.

ECONOMIC RESEARCH SERVICE
(Dept. of Agriculture)

Analyses of economic situation and outlook for farm products. Research and statistical studies on farm population, prices and income; food consumption and marketing margins; agricultural finance, farm costs and returns, and agricultural productivity.

OFFICE OF BUSINESS ECONOMICS
(Dept. of Commerce)

Estimates of national income, gross national product, inter-industry sales and purchases, and related data on business developments; analyses of business trends; estimates of the balance of international payments and of foreign investments.

BOARD OF GOVERNORS
OF THE FEDERAL RESERVE SYSTEM

Analyzes economic and credit conditions. Prepares and publishes statistics on money, banking, international finance, industrial production, consumer credit, and department store activities for general public use.

HOUSEHOLD ECONOMICS RESEARCH DIVISION
(ARS, Dept. of Agriculture)

Statistics of household food consumption, dietary adequacy and rural family living.

BUSINESS AND DEFENSE SERVICES ADMIN.
(Dept. of Commerce)

Analyzes and disseminates information on the condition and levels of business activity in specific industries and trades.

BUREAU OF INTERNATIONAL COMMERCE
(Dept. of Commerce)

Analyses of data compiled by the Census Bureau on U.S. foreign trade, and of data on trade of other countries.

BUREAU OF MINES
(Dept. of the Interior)

Current statistics on production, consumption and stocks of metals, mineral fuels and other nonmetallic minerals; employment and injuries in mineral industries. Studies and analyses in these areas.

4. ADMINISTRATIVE, REGULATORY AND DEFENSE AGENCIES

These agencies provide many important and valuable statistical series used by the public and other agencies of the Government. They collect statistical information primarily as part of their administrative and operating responsibilities, such as the collection of taxes or regulation of public utilities, though a few of them have specialized statistical responsibilities.

OTHER AGENCIES AND UNITS LOCATED IN EXECUTIVE DEPARTMENTS

AGRICULTURE
a. AMS (Marketing Services Division)
b. Agricultural Research Service (various branches)
c. Commodity Exchange Authority
d. Farmers Home Administration
e. Forest Service
f. Rural Electrification Administration

COMMERCE
a. Area Redevelopment Administration
b. Bureau of Public Roads
c. Bureau of Standards
d. Maritime Administration
e. Weather Bureau

TREASURY
a. Bureau of Customs
b. Bureau of the Mint
c. Comptroller of the Currency
d. Fiscal Service
e. Internal Revenue Service
f. Office of International Finance

LABOR
a. Bureau of Apprenticeship and Training
b. Bureau of Employees' Compensation
c. Bureau of Employment Security
d. Bureau of Labor Standards
e. Wage, Hour and Public Contracts Division
f. Women's Bureau

HEALTH, EDUCATION AND WELFARE
a. Office of Education
b. Office of Vocational Rehabilitation
c. Public Health Service
d. Social Security Administration
e. Welfare Administration

JUSTICE
a. Antitrust Division
b. Bureau of Prisons
c. Federal Bur. of Investigation
d. Immigration and Naturalization Service

DEFENSE
a. Army, Navy and Air Force produce economic statistics on expenditures, contracts, and purchases
b. Board of Engineers for Rivers and Harbors
c. Office of Civil Defense

INTERIOR
a. Bureau of Indian Affairs
b. Bureau of Land Management
c. Bureau of Reclamation
d. Fish and Wildlife Service
e. Nat'l Park Service

STATE
a. Agency for International Development
b. Foreign Service Reporting
c. Passport Office

INDEPENDENT OFFICES

CIVIL AERONAUTICS BOARD	FEDERAL POWER COMMISSION	RAILROAD RETIREMENT BOARD
FARM CREDIT ADMINISTRATION	FEDERAL TRADE COMMISSION	SECURITIES AND EXCHANGE COMMISSION
FEDERAL AVIATION AGENCY	HOUSING AND HOME FINANCE AGENCY	SELECTIVE SERVICE SYSTEM
FEDERAL COMMUNICATIONS COMMISSION	INTERSTATE COMMERCE COMMISSION	U.S. CIVIL SERVICE COMMISSION
FEDERAL DEPOSIT INSURANCE CORPORATION	NATIONAL SCIENCE FOUNDATION	U.S. TARIFF COMMISSION
FEDERAL HOME LOAN BANK BOARD	OFFICE OF EMERGENCY PLANNING	VETERANS ADMINISTRATION

EXECUTIVE OFFICE OF THE PRESIDENT
Bureau of the Budget
June 1963
692-031 O - 63 (In pocket)

EXPLANATORY NOTE

The various agencies and units are arranged on this chart to indicate the essential structure of the statistical activities of the Federal Government. Only those agencies and units whose statistical functions are readily identified are included. Some of the agencies might well be listed under several of the major categories, but to prevent excessive detail, they are classified under a single category. While all agencies have some performance or operational data, this chart considers such statistics only when they have also been adapted for general analytic use. A more detailed description of the system is presented in "Statistical Services of the United States Government".

Figure A1-1 The federal statistical system.

413

TWO

PROBABILITY DISTRIBUTIONS

List of tables:

- Individual terms of the binomial distribution
- The normal distribution
- Critical values of the χ^2 distribution
- Critical values of the Student's-t distribution
- Critical values of the F distribution
- Fractiles of the beta distribution

Table A2.1 Binomial probability mass functions*

$$p(x) = \binom{n}{x} p^x (1 - p)^{n-x}$$

								p						
n	x	0.01	0.05	0.10	0.20	0.30	0.40	0.50	0.60	0.70	0.80	0.90	0.95	0.99
2	0	9801	9025	8100	6400	4900	3600	2500	1600	900	400	100	25	1
	1	198	950	1800	3200	4200	4800	5000	4800	4200	3200	1800	950	198
	2	1	25	100	400	900	1600	2500	3600	4900	6400	8100	9025	9801
3	0	9703	8574	7290	5120	3430	2160	1250	640	270	80	10	1	0
	1	294	1354	2430	3840	4410	4320	3750	2880	1890	960	270	71	3
	2	3	71	270	960	1890	2880	3750	4320	4410	3840	2430	1354	294
	3	0	1	10	80	270	640	1250	2160	3430	5120	7290	8574	9703
4	0	9606	8145	6561	4096	2401	1296	625	256	81	16	1	0	0
	1	388	1715	2916	4096	4116	3456	2500	1536	756	256	36	5	0
	2	6	135	486	1536	2646	3456	3750	3456	2646	1536	486	135	6
	3	0	5	36	256	756	1536	2500	3456	4116	4096	2916	1715	388
	4	0	0	1	16	81	256	625	1296	2401	4096	6561	8145	9606
5	0	9510	7738	5905	3277	1681	778	313	102	24	3	0	0	0
	1	480	2036	3280	4096	3601	2592	1563	768	283	64	4	0	0
	2	10	214	729	2048	3087	3456	3125	2304	1323	512	81	11	0
	3	0	11	81	512	1323	2304	3125	3456	3087	2048	729	214	10
	4	0	0	4	64	283	768	1563	2592	3601	4096	3280	2036	480
	5	0	0	0	3	24	102	313	778	1681	3277	5905	7738	9510
6	0	9415	7351	5314	2621	1176	467	156	41	7	1	0	0	0
	1	571	2321	3543	3932	3025	1866	938	369	102	15	1	0	0
	2	14	305	984	2458	3241	3110	2344	1382	595	154	12	1	0
	3	0	21	146	819	1852	2765	3125	2765	1852	819	146	21	0
	4	0	1	12	154	595	1382	2344	3110	3241	2458	984	305	14
	5	0	0	1	15	102	369	938	1866	3025	3932	3543	2321	571
	6	0	0	0	1	7	41	156	467	1176	2621	5314	7351	9415
7	0	9321	6983	4783	2097	824	280	78	16	2	0	0	0	0
	1	659	2573	3720	3670	2471	1306	547	172	36	4	0	0	0
	2	20	406	1240	2753	2613	1641	774	250	43	2	0	0	0
	3	0	36	230	1147	2269	2903	2734	1935	972	287	26	2	0
	4	0	2	26	287	972	1935	2734	2903	2269	1147	230	36	0
	5	0	0	2	43	250	774	1641	2613	3177	2753	1240	406	20
	6	0	0	0	4	36	172	547	1306	2471	3670	3720	2573	659
	7	0	0	0	0	2	16	78	280	824	2097	4783	6983	9321
8	0	9227	6634	4305	1678	576	168	39	7	1	0	0	0	0
	1	746	2793	3826	3355	1977	896	313	79	12	1	0	0	0
	2	26	515	1488	2936	2965	2090	1094	413	100	11	0	0	0
	3	1	54	331	1468	2541	2787	2188	1239	467	92	4	0	0
	4	0	4	46	459	1361	2322	2734	2322	1361	459	46	4	0
	5	0	0	4	92	467	1239	2188	2787	2541	1468	331	54	1
	6	0	0	0	11	100	413	1094	2090	2965	2936	1488	515	26
	7	0	0	0	1	12	79	313	896	1977	3355	3826	2793	746
	8	0	0	0	0	1	7	39	168	576	1678	4305	6634	9227

Table A2.1—*Continued*

n	x	0.01	0.05	0.10	0.20	0.30	0.40	0.50	0.60	0.70	0.80	0.90	0.95	0.99
9	0	9135	6302	3874	1342	404	101	20	3	0	0	0	0	0
	1	830	2985	3874	3020	1556	605	176	35	4	0	0	0	0
	2	34	629	1722	3020	2668	1612	703	212	39	3	0	0	0
	3	1	77	446	1762	2668	2508	1641	743	210	28	1	0	0
	4	0	6	74	661	1715	2508	2461	1672	735	165	8	0	0
	5	0	0	8	165	735	1672	2461	2508	1715	661	74	6	0
	6	0	0	1	28	210	743	1641	2508	2668	1762	446	77	1
	7	0	0	0	3	39	212	703	1612	2668	3020	1722	629	34
	8	0	0	0	0	4	35	176	605	1556	3020	3874	2985	830
	9	0	0	0	0	0	3	20	101	404	1342	3874	6302	9135
10	0	9044	5987	3487	1074	282	60	10	1	0	0	0	0	0
	1	914	3151	3874	2684	1211	403	98	16	1	0	0	0	0
	2	42	746	1937	3020	2335	1209	439	106	14	1	0	0	0
	3	1	105	574	2013	2668	2150	1172	425	90	8	0	0	0
	4	0	10	112	881	2001	2508	2051	1115	368	55	1	0	0
	5	0	1	15	264	1029	2007	2461	2007	1029	264	15	1	0
	6	0	0	1	55	368	1115	2051	2508	2001	881	112	10	0
	7	0	0	0	8	90	425	1172	2150	2668	2013	574	105	1
	8	0	0	0	1	14	106	439	1209	2335	3020	1937	746	42
	9	0	0	0	0	1	16	98	403	1211	2684	3874	3151	914
	10	0	0	0	0	0	1	10	60	282	1074	3487	5987	9044
11	0	8953	5688	3138	859	198	36	5	0	0	0	0	0	0
	1	995	3293	3835	2362	932	266	54	7	0	0	0	0	0
	2	50	867	2131	2953	1998	887	269	52	5	0	0	0	0
	3	2	137	710	2215	2568	1774	806	234	37	2	0	0	0
	4	0	14	158	1107	2201	2365	1611	701	173	17	0	0	0
	5	0	1	25	388	1321	2207	2256	1471	566	97	3	0	0
	6	0	0	3	97	566	1471	2256	2207	1321	388	25	1	0
	7	0	0	0	17	173	701	1611	2365	2201	1107	158	14	0
	8	0	0	0	2	37	234	806	1774	2568	2215	710	137	2
	9	0	0	0	0	5	52	269	887	1998	2953	2131	867	50
	10	0	0	0	0	0	7	54	266	932	2362	3835	3293	995
	11	0	0	0	0	0	0	5	36	198	859	3138	5688	8953
12	0	8864	5404	2824	687	138	22	2	0	0	0	0	0	0
	1	1074	3413	3766	2062	712	174	29	3	0	0	0	0	0
	2	60	988	2301	2835	1678	639	161	25	2	0	0	0	0
	3	2	173	852	2362	2397	1419	537	125	15	1	0	0	0
	4	0	21	213	1329	2311	2128	1208	420	78	5	0	0	0
	5	0	2	38	532	1585	2270	1934	1009	291	33	0	0	0
	6	0	0	5	155	792	1766	2256	1766	792	155	5	0	0
	7	0	0	0	33	291	1009	1934	2270	1585	532	38	2	0
	8	0	0	0	5	78	420	1208	2128	2311	1329	213	21	0
	9	0	0	0	1	15	125	537	1419	2397	2362	852	173	2
	10	0	0	0	0	2	25	161	639	1678	2835	2301	988	60
	11	0	0	0	0	0	3	29	174	712	2062	3766	3413	1074
	12	0	0	0	0	0	0	2	22	138	687	2824	5404	8864

Table A2.1—*Continued*

n	x	0.01	0.05	0.10	0.20	0.30	0.40	0.50	0.60	0.70	0.80	0.90	0.95	0.99
								p						
13	0	8775	5133	2542	550	97	13	1	0	0	0	0	0	0
	1	1152	3512	3672	1787	540	113	16	1	0	0	0	0	0
	2	70	1109	2448	2680	1388	453	95	12	1	0	0	0	0
	3	3	214	997	2457	2181	1107	349	65	6	0	0	0	0
	4	0	28	277	1535	2337	1845	873	243	34	1	0	0	0
	5	0	3	55	691	1803	2214	1571	656	142	11	0	0	0
	6	0	0	8	230	1030	1968	2095	1312	442	58	1	0	0
	7	0	0	1	58	442	1312	2095	1968	1030	230	8	0	0
	8	0	0	0	11	142	656	1571	2214	1803	691	55	3	0
	9	0	0	0	1	34	243	873	1845	2337	1535	277	28	0
	10	0	0	0	0	6	65	349	1107	2181	2457	997	214	3
	11	0	0	0	0	1	12	95	453	1388	2680	2448	1109	70
	12	0	0	0	0	0	1	16	113	540	1787	3672	3512	1152
	13	0	0	0	0	0	0	1	13	97	550	2542	5133	8775
14	0	8687	4877	2288	440	68	8	1	0	0	0	0	0	0
	1	1229	3593	3559	1539	407	73	9	1	0	0	0	0	0
	2	81	1229	2570	2501	1134	317	56	5	0	0	0	0	0
	3	3	259	1142	2501	1943	845	222	33	2	0	0	0	0
	4	0	37	349	1720	2290	1549	611	136	14	0	0	0	0
	5	0	4	78	860	1963	2066	1222	408	66	3	0	0	0
	6	0	0	13	322	1262	2066	1833	918	232	20	0	0	0
	7	0	0	2	92	618	1574	2095	1574	618	92	2	0	0
	8	0	0	0	20	232	918	1833	2066	1262	322	13	0	0
	9	0	0	0	3	66	408	1222	2066	1963	860	78	4	0
	10	0	0	0	0	14	136	611	1549	2290	1720	349	37	0
	11	0	0	0	0	2	33	222	845	1943	2501	1142	259	3
	12	0	0	0	0	0	5	56	317	1134	2501	2570	1229	81
	13	0	0	0	0	0	1	9	73	407	1539	3559	3593	1229
	14	0	0	0	0	0	0	1	8	68	440	2288	4877	8687
15	0	8601	4633	2059	352	47	5	0	0	0	0	0	0	0
	1	1303	3658	3432	1319	305	47	5	0	0	0	0	0	0
	2	92	1348	2669	2309	916	219	32	3	0	0	0	0	0
	3	4	307	1285	2501	1700	634	139	16	1	0	0	0	0
	4	0	49	428	1876	2186	1268	417	74	6	0	0	0	0
	5	0	6	105	1032	2061	1859	916	245	30	1	0	0	0
	6	0	0	19	430	1472	2066	1527	612	116	7	0	0	0
	7	0	0	3	138	811	1771	1964	1181	348	35	0	0	0
	8	0	0	0	35	348	1181	1964	1771	811	138	3	0	0
	9	0	0	0	7	116	612	1527	2066	1472	430	19	0	0
	10	0	0	0	1	30	245	916	1859	2061	1032	105	6	0
	11	0	0	0	0	6	74	417	1268	2186	1876	428	49	0
	12	0	0	0	0	1	16	139	634	1700	2501	1285	307	4
	13	0	0	0	0	0	3	32	219	916	2309	2669	1348	92
	14	0	0	0	0	0	0	5	47	305	1319	3432	3658	1303
	15	0	0	0	0	0	0	0	5	47	352	2059	4633	8601

Table A2.1—*Continued*

								p						
n	x	0.01	0.05	0.10	0.20	0.30	0.40	0.50	0.60	0.70	0.80	0.90	0.95	0.99
16	0	8515	4401	1853	281	33	3	0	0	0	0	0	0	0
	1	1376	3706	3294	1126	228	30	2	0	0	0	0	0	0
	2	104	1463	2745	2111	732	150	18	1	0	0	0	0	0
	3	5	359	1423	2463	1465	468	85	8	0	0	0	0	0
	4	0	61	514	2001	2040	1014	278	40	2	0	0	0	0
	5	0	8	137	1201	2099	1623	667	142	13	0	0	0	0
	6	0	1	28	550	1649	1983	1222	392	56	2	0	0	0
	7	0	0	4	197	1010	1889	1746	840	185	12	0	0	0
	8	0	0	1	55	487	1417	1964	1417	487	55	1	0	0
	9	0	0	0	12	185	840	1746	1889	1010	197	4	0	0
	10	0	0	0	2	56	392	1222	1983	1649	550	28	1	0
	11	0	0	0	0	13	142	667	1623	2099	1201	137	8	0
	12	0	0	0	0	2	40	278	1014	2040	2001	514	61	0
	13	0	0	0	0	0	8	85	468	1465	2463	1423	359	5
	14	0	0	0	0	0	1	18	150	732	2111	2745	1463	104
	15	0	0	0	0	0	0	2	30	228	1126	3294	3706	1376
	16	0	0	0	0	0	0	0	3	33	281	1853	4401	8515
17	0	8429	4181	1668	225	23	2	0	0	0	0	0	0	0
	1	1447	3741	3150	957	169	19	.1	0	0	0	0	0	0
	2	117	1575	2800	1914	581	102	10	1	0	0	0	0	0
	3	6	415	1556	2393	1245	341	52	4	0	0	0	0	0
	4	0	76	605	2093	1868	796	182	21	1	0	0	0	0
	5	0	10	175	1361	2081	1379	472	81	6	0	0	0	0
	6	0	1	39	680	1784	1839	944	242	26	1	0	0	0
	7	0	0	7	267	1201	1927	1484	571	95	4	0	0	0
	8	0	0	1	84	644	1606	1855	1070	276	21	0	0	0
	9	0	0	0	21	276	1070	1855	1606	644	84	1	0	0
	10	0	0	0	4	95	571	1484	1927	1201	267	7	0	0
	11	0	0	0	1	26	242	944	1839	1784	680	39	1	0
	12	0	0	0	0	6	81	472	1379	2081	1361	175	10	0
	13	0	0	0	0	1	21	182	796	1868	2093	605	76	0
	14	0	0	0	0	0	4	52	341	1245	2393	1556	415	6
	15	0	0	0	0	0	1	10	102	581	1914	2800	1575	117
	16	0	0	0	0	0	0	1	19	169	957	3150	3741	1447
	17	0	0	0	0	0	0	0	2	23	225	1668	4181	8429
18	0	8345	3972	1501	180	16	1	0	0	0	0	0	0	0
	1	1517	3763	3002	811	126	12	1	0	0	0	0	0	0
	2	130	1683	2835	1723	458	69	6	0	0	0	0	0	0
	3	7	473	1680	2297	1046	246	31	2	0	0	0	0	0
	4	0	93	700	2153	1681	614	117	11	0	0	0	0	0
	5	0	14	218	1507	2017	1146	327	45	2	0	0	0	0
	6	0	2	52	816	1873	1655	708	145	12	0	0	0	0
	7	0	0	10	350	1376	1892	1214	374	46	1	0	0	0
	8	0	0	2	120	811	1734	1669	771	149	8	0	0	0
	9	0	0	0	33	386	1284	1855	1284	386	33	0	0	0
	10	0	0	0	8	149	771	1669	1734	811	120	2	0	0
	11	0	0	0	1	46	374	1214	1892	1376	350	10	0	0
	12	0	0	0	0	12	145	708	1655	1873	816	52	2	0
	13	0	0	0	0	2	45	327	1146	2017	1507	218	14	0
	14	0	0	0	0	0	11	117	614	1681	2153	700	93	0

Table A2.1—*Continued*

n	x	0.01	0.05	0.10	0.20	0.30	0.40	0.50	0.60	0.70	0.80	0.90	0.95	0.99
	15	0	0	0	0	0	2	31	246	1046	2297	1680	473	7
	16	0	0	0	0	0	0	6	69	458	1723	2835	1683	130
	17	0	0	0	0	0	0	1	12	126	811	3002	3763	1517
	18	0	0	0	0	0	0	0	1	16	180	1501	3972	8345
19	0	8262	3774	1351	144	11	1	0	0	0	0	0	0	0
	1	1586	3774	2852	685	93	8	0	0	0	0	0	0	0
	2	144	1787	2852	1540	358	46	3	0	0	0	0	0	0
	3	8	533	1796	2182	869	175	18	1	0	0	0	0	0
	4	0	112	798	2182	1491	467	74	5	0	0	0	0	0
	5	0	18	266	1636	1916	933	222	24	1	0	0	0	0
	6	0	2	69	955	1916	1451	518	85	5	0	0	0	0
	7	0	0	14	443	1525	1797	961	237	22	0	0	0	0
	8	0	0	2	166	981	1797	1442	532	77	3	0	0	0
	9	0	0	0	51	514	1464	1762	976	220	13	0	0	0
	10	0	0	0	13	220	976	1762	1464	514	51	0	0	0
	11	0	0	0	3	77	532	1442	1797	981	166	2	0	0
	12	0	0	0	0	22	237	961	1797	1525	443	14	0	0
	13	0	0	0	0	5	85	518	1451	1916	955	69	2	0
	14	0	0	0	0	1	24	222	933	1916	1636	266	18	0
	15	0	0	0	0	0	5	74	467	1491	2182	798	112	0
	16	0	0	0	0	0	1	18	175	869	2182	1796	533	8
	17	0	0	0	0	0	0	3	46	358	1540	2852	1787	144
	18	0	0	0	0	0	0	0	8	93	685	2852	3774	1586
	19	0	0	0	0	0	0	0	1	11	144	1351	3774	8262
20	0	8179	3585	1216	115	8	0	0	0	0	0	0	0	0
	1	1652	3774	2702	576	68	5	0	0	0	0	0	0	0
	2	159	1887	2852	1369	278	31	2	0	0	0	0	0	0
	3	10	596	1901	2054	716	123	11	0	0	0	0	0	0
	4	0	133	898	2182	1304	350	46	3	0	0	0	0	0
	5	0	22	319	1746	1789	746	148	13	0	0	0	0	0
	6	0	3	89	1091	1916	1244	370	49	2	0	0	0	0
	7	0	0	20	545	1643	1659	739	146	10	0	0	0	0
	8	0	0	4	222	1144	1797	1201	355	39	1	0	0	0
	9	0	0	1	74	654	1597	1602	710	120	5	0	0	0
	10	0	0	0	20	308	1171	1762	1171	308	20	0	0	0
	11	0	0	0	5	120	710	1602	1597	654	74	1	0	0
	12	0	0	0	1	39	355	1201	1797	1144	222	4	0	0
	13	0	0	0	0	10	146	739	1659	1643	545	20	0	0
	14	0	0	0	0	2	49	370	1244	1916	1091	89	3	0
	15	0	0	0	0	0	13	148	746	1789	1746	319	22	0
	16	0	0	0	0	0	3	46	350	1304	2182	898	133	0
	17	0	0	0	0	0	0	11	123	716	2054	1901	596	10
	18	0	0	0	0	0	0	2	31	278	1369	2852	1887	159
	19	0	0	0	0	0	0	0	5	68	576	2702	3774	1652
	20	0	0	0	0	0	0	0	0	8	115	1216	3585	8179

* Table computed by Dr. Edward L. Spitznagel. Reproduced with his permission.

Table A2.2 Standard normal areas*

$P(Z \le z)$ = area shaded

	0.00	0.01	0.02	0.03	0.04	0.05	0.06	0.07	0.08	0.09
0.0	0.5000	0.5040	0.5080	0.5120	0.5160	0.5199	0.5239	0.5279	0.5319	0.5359
0.1	0.5398	0.5438	0.5478	0.5517	0.5557	0.5596	0.5636	0.5675	0.5714	0.5753
0.2	0.5793	0.5832	0.5871	0.5910	0.5948	0.5987	0.6026	0.6064	0.6103	0.6141
0.3	0.6179	0.6217	0.6255	0.6293	0.6331	0.6368	0.6406	0.6443	0.6480	0.6517
0.4	0.6554	0.6591	0.6628	0.6664	0.6700	0.6736	0.6772	0.6808	0.6844	0.6879
0.5	0.6915	0.6950	0.6985	0.7019	0.7054	0.7088	0.7123	0.7157	0.7190	0.7224
0.6	0.7257	0.7291	0.7324	0.7357	0.7389	0.7422	0.7454	0.7486	0.7517	0.7549
0.7	0.7580	0.7611	0.7642	0.7673	0.7704	0.7734	0.7764	0.7794	0.7823	0.7852
0.8	0.7881	0.7910	0.7939	0.7967	0.7995	0.8023	0.8051	0.8078	0.8106	0.8133
0.9	0.8159	0.8186	0.8212	0.8238	0.8264	0.8289	0.8315	0.8340	0.8365	0.8389
1.0	0.8413	0.8438	0.8461	0.8485	0.8508	0.8531	0.8554	0.8577	0.8599	0.8621
1.1	0.8643	0.8665	0.8686	0.8708	0.8729	0.8749	0.8770	0.8790	0.8810	0.8830
1.2	0.8849	0.8869	0.8888	0.8907	0.8925	0.8944	0.8962	0.8980	0.8997	0.9015
1.3	0.9032	0.9049	0.9066	0.9082	0.9099	0.9115	0.9131	0.9147	0.9162	0.9177
1.4	0.9192	0.9207	0.9222	0.9236	0.9251	0.9265	0.9279	0.9292	0.9306	0.9319
1.5	0.9332	0.9345	0.9357	0.9370	0.9382	0.9394	0.9406	0.9418	0.9429	0.9441
1.6	0.9452	0.9463	0.9474	0.9484	0.9495	0.9505	0.9515	0.9525	0.9535	0.9545
1.7	0.9554	0.9564	0.9573	0.9582	0.9591	0.9599	0.9608	0.9616	0.9625	0.9633
1.8	0.9641	0.9649	0.9656	0.9664	0.9671	0.9678	0.9686	0.9693	0.9699	0.9706
1.9	0.9713	0.9719	0.9726	0.9732	0.9738	0.9744	0.9750	0.9756	0.9761	0.9767
2.0	0.9772	0.9778	0.9783	0.9788	0.9793	0.9798	0.9803	0.9808	0.9812	0.9817
2.1	0.9821	0.9826	0.9830	0.9834	0.9838	0.9842	0.9846	0.9850	0.9854	0.9857
2.2	0.9861	0.9864	0.9868	0.9871	0.9875	0.9878	0.9881	0.9884	0.9887	0.9890
2.3	0.9893	0.9896	0.9898	0.9901	0.9904	0.9906	0.9909	0.9911	0.9913	0.9916
2.4	0.9918	0.9920	0.9922	0.9925	0.9927	0.9929	0.9931	0.9932	0.9934	0.9936
2.5	0.9938	0.9940	0.9941	0.9943	0.9945	0.9946	0.9948	0.9949	0.9951	0.9952
2.6	0.9953	0.9955	0.9956	0.9957	0.9959	0.9960	0.9961	0.9962	0.9963	0.9964
2.7	0.9965	0.9966	0.9967	0.9968	0.9969	0.9970	0.9971	0.9972	0.9973	0.9974
2.8	0.9974	0.9975	0.9976	0.9977	0.9977	0.9978	0.9979	0.9979	0.9980	0.9981
2.9	0.9981	0.9982	0.9982	0.9983	0.9984	0.9984	0.9985	0.9985	0.9986	0.9986
3.0	0.9987	0.9987	0.9987	0.9988	0.9988	0.9989	0.9989	0.9989	0.9990	0.9990
3.1	0.9990	0.9991	0.9991	0.9991	0.9992	0.9992	0.9992	0.9992	0.9993	0.9993
3.2	0.9993	0.9993	0.9994	0.9994	0.9994	0.9994	0.9994	0.9995	0.9995	0.9995
3.3	0.9995	0.9995	0.9995	0.9996	0.9996	0.9996	0.9996	0.9996	0.9996	0.9997
3.4	0.9997	0.9997	0.9997	0.9997	0.9997	0.9997	0.9997	0.9997	0.9997	0.9998

* By permission from A. M. Mood, *Introduction to the Theory of Statistics*, McGraw-Hill Book Company, Inc., New York, 1950.

Table A2.3 Critical values of the χ^2 distribution*

v is the parameter (degrees of freedom); P is the probability that χ^2 exceeds
the critical value k

v \ P	0.05	0.01	0.001	0.975	0.025
1	3.841	6.635	10.828	0.0009821	5.024
2	5.991	9.210	13.816	0.05064	7.378
3	7.815	11.345	16.266	0.2158	9.348
4	9.488	13.277	18.467	0.4844	11.143
5	11.070	15.086	20.515	0.8312	12.833
6	12.592	16.812	22.458	1.237	14.449
7	14.067	18.475	24.322	1.690	16.013
8	15.507	20.090	26.124	2.180	17.535
9	16.919	21.666	27.877	2.700	19.023
10	18.307	23.209	29.588	3.247	20.483
11	19.675	24.725	31.264	3.816	21.920
12	21.026	26.217	32.909	4.404	23.337
13	22.362	27.688	34.528	5.009	24.736
14	23.685	29.141	36.123	5.629	26.119
15	24.996	30.578	37.697	6.262	27.488
16	26.296	32.000	39.252	6.908	28.845
17	27.587	33.409	40.790	7.564	30.191
18	28.869	34.805	42.312	8.231	31.526
19	30.144	36.191	43.820	8.907	32.852
20	31.410	37.566	45.315	9.591	34.170
21	32.671	38.932	46.797	10.283	35.479
22	33.924	40.289	48.268	10.982	36.781
23	35.172	41.638	49.728	11.689	38.076
24	36.415	42.980	51.179	12.401	39.364
25	37.652	44.314	52.620	13.120	40.646
26	38.885	45.642	54.052	13.844	41.923
27	40.113	46.963	55.476	14.573	43.195
28	41.337	48.278	56.892	15.308	44.461
29	42.557	49.580	58.301	16.047	45.722
30	43.773	50.892	59.703	16.791	46.979
31	44.985	52.191	61.098	17.539	48.232
32	46.194	53.486	62.487	18.291	49.480
33	47.400	54.776	63.870	19.047	50.725
34	48.602	56.061	65.247	19.806	51.966
35	49.802	57.342	66.619	20.569	53.203
36	50.998	58.619	67.985	21.336	54.437
37	52.192	59.893	69.346	22.106	55.668
38	53.384	61.162	70.703	22.878	56.896
39	54.572	62.428	72.055	23.654	58.120
40	55.758	63.691	73.402	24.433	59.342
41	56.942	64.950	74.745	25.215	60.561
42	58.124	66.206	76.084	25.999	61.777

Table A2.3—*Continued*

v \ P	0.05	0.01	0.001	0.975	0.025
43	59.304	67.459	77.419	26.785	62.990
44	60.481	68.710	78.750	27.575	64.201
45	61.656	69.957	80.077	28.366	65.410
46	62.830	71.201	81.400	29.160	66.617
47	64.001	72.443	82.720	29.956	67.821
48	65.171	73.683	84.037	30.755	69.023
49	66.339	74.919	85.351	31.555	70.222
50	67.505	76.154	86.661	32.357	71.420
60	79.082	88.379	99.607	40.482	83.298
70	90.531	100.425	112.317	48.758	95.023
80	101.879	112.329	124.839	57.153	106.629
90	113.145	124.116	137.208	65.647	118.136
100	124.342	135.807	149.449	74.222	129.561

Note: For $v > 100$, $\sqrt{2\chi^2} - \sqrt{2v-1}$ is approximately a standard normal variable.

* Table computed by Dr. Edward L. Spitznagel. Reproduced with his permission.

Table A2.4 Critical values of Student's-t distribution*

v is the parameter (degrees of freedom); P is the probability that t exceeds the critical value k

P v	0.05	0.01	0.001	0.1	0.02	0.002	0.5
1	12.706	63.657	636.619	6.314	31.821	318.309	1.0000
2	4.303	9.925	31.599	2.920	6.965	22.327	0.8165
3	3.182	5.841	12.924	2.353	4.541	10.215	0.7649
4	2.777	4.604	8.610	2.132	3.747	7.173	0.7407
5	2.571	4.032	6.869	2.015	3.365	5.893	0.7267
6	2.447	3.708	5.959	1.943	3.143	5.208	0.7175
7	2.365	3.500	5.408	1.895	2.998	4.786	0.7111
8	2.306	3.355	5.041	1.860	2.896	4.501	0.7064
9	2.262	3.250	4.781	1.833	2.821	4.297	0.7027
10	2.228	3.169	4.587	1.812	2.764	4.144	0.6998
11	2.201	3.106	4.437	1.796	2.718	4.025	0.6974
12	2.179	3.055	4.318	1.782	2.681	3.930	0.6955
13	2.160	3.012	4.221	1.771	2.650	3.852	0.6938
14	2.145	2.977	4.140	1.761	2.624	3.787	0.6924
15	2.131	2.947	4.073	1.753	2.602	3.733	0.6912
16	2.120	2.921	4.015	1.746	2.583	3.686	0.6901
17	2.110	2.898	3.965	1.740	2.567	3.646	0.6892
18	2.101	2.878	3.922	1.734	2.552	3.610	0.6884
19	2.093	2.861	3.883	1.729	2.539	3.579	0.6876
20	2.086	2.845	3.850	1.725	2.528	3.552	0.6870
21	2.080	2.831	3.819	1.721	2.510	3.527	0.6864
22	2.074	2.819	3.792	1.717	2.508	3.505	0.6858
23	2.069	2.807	3.768	1.714	2.500	3.485	0.6853
24	2.064	2.797	3.745	1.711	2.492	3.467	0.6848
25	2.060	2.787	3.725	1.708	2.485	3.450	0.6844
26	2.056	2.779	3.707	1.706	2.479	3.435	0.6840
27	2.052	2.771	3.690	1.703	2.473	3.421	0.6837
28	2.048	2.763	3.674	1.701	2.467	3.408	0.6834
29	2.045	2.756	3.659	1.699	2.462	3.396	0.6830
30	2.042	2.750	3.646	1.697	2.457	3.385	0.6820
31	2.040	2.744	3.633	1.696	2.453	3.375	0.6825
32	2.037	2.738	3.622	1.694	2.449	3.365	0.6822
33	2.035	2.733	3.611	1.692	2.445	3.356	0.6820
34	2.032	2.728	3.601	1.691	2.441	3.348	0.6818
35	2.030	2.724	3.591	1.690	2.438	3.340	0.8616
36	2.028	2.719	3.582	1.688	2.434	3.333	0.6814
37	2.026	2.715	3.574	1.687	2.431	3.326	0.6812
38	2.024	2.712	3.566	1.686	2.429	3.319	0.6810
39	2.023	2.708	3.558	1.685	2.426	3.313	0.6808
40	2.021	2.704	3.551	1.684	2.423	3.307	0.6807
41	2.020	2.701	3.544	1.683	2.421	3.301	0.6805

Table A2.4—*Continued*

v \ P	0.05	0.01	0.001	0.1	0.02	0.002	0.5
42	2.018	2.698	3.538	1.682	2.418	3.296	0.6804
43	2.017	2.695	3.532	1.681	2.416	3.291	0.6802
44	2.015	2.692	3.526	1.680	2.414	3.286	0.6801
45	2.014	2.690	3.520	1.679	2.412	3.281	0.6800
46	2.013	2.687	3.515	1.679	2.410	3.277	0.6799
47	2.012	2.685	3.510	1.678	2.408	3.273	0.6797
48	2.011	2.682	3.505	1.677	2.407	3.269	0.6796
49	2.010	2.680	3.500	1.677	2.405	3.265	0.6795
50	2.009	2.678	3.496	1.676	2.403	3.261	0.6794
51	2.008	2.676	3.492	1.675	2.402	3.258	0.6793
52	2.007	2.674	3.488	1.675	2.400	3.255	0.6792
53	2.006	2.672	3.484	1.674	2.399	3.251	0.6791
54	2.005	2.670	3.480	1.674	2.397	3.248	0.6791
55	2.004	2.668	3.476	1.673	2.396	3.245	0.6790
56	2.003	2.667	3.473	1.673	2.395	3.242	0.6789
57	2.002	2.665	3.470	1.672	2.394	3.239	0.6788
58	2.002	2.663	3.466	1.672	2.392	3.237	0.6787
59	2.001	2.662	3.463	1.671	2.391	3.234	0.6787
60	2.000	2.660	3.460	1.671	2.390	3.232	0.6786
61	2.000	2.659	3.457	1.670	2.389	3.229	0.6785
62	1.999	2.657	3.454	1.670	2.388	3.227	0.6785
63	1.998	2.656	3.452	1.669	2.387	3.225	0.6784
64	1.998	2.655	3.449	1.669	2.386	3.223	0.6783
65	1.997	2.654	3.447	1.669	2.385	3.220	0.6783
66	1.997	2.652	3.444	1.668	2.384	3.218	0.6782
67	1.996	2.651	3.442	1.668	2.383	3.216	0.6782
68	1.995	2.650	3.439	1.668	2.382	3.214	0.6781
69	1.995	2.649	3.437	1.667	2.382	3.213	0.6781
70	1.994	2.648	3.435	1.667	2.381	3.211	0.6780
71	1.994	2.647	3.433	1.667	2.380	3.209	0.6780
72	1.993	2.646	3.431	1.666	2.379	3.207	0.6779
73	1.993	2.645	3.429	1.666	2.379	3.206	0.6779
74	1.993	2.644	3.427	1.666	2.378	3.204	0.6778
75	1.992	2.643	3.425	1.665	2.377	3.202	0.6778
76	1.992	2.642	3.423	1.665	2.376	3.201	0.6777
77	1.991	2.641	3.421	1.665	2.376	3.199	0.6777
78	1.991	2.640	3.420	1.665	2.375	3.198	0.6776
79	1.990	2.640	3.418	1.664	2.374	3.197	0.6776
80	1.990	2.639	3.416	1.664	2.374	3.195	0.6776
81	1.990	2.638	3.615	1.664	2.373	3.194	0.6775
82	1.989	2.637	3.413	1.664	2.373	3.193	0.6775
83	1.989	2.636	3.412	1.663	2.372	3.191	0.6775
84	1.989	2.636	3.410	1.663	2.372	3.190	0.6774
85	1.988	2.635	3.409	1.663	2.371	3.189	0.6774

Table A2.4—*Continued*

P v	0.05	0.01	0.001	0.1	0.02	0.002	0.5
86	1.988	2.634	3.407	1.663	2.370	3.188	0.6774
87	1.988	2.634	3.406	1.663	2.370	3.187	0.6773
88	1.987	2.633	3.405	1.662	2.369	3.185	0.6773
89	1.987	2.632	3.403	1.662	2.369	3.184	0.6773
90	1.987	2.632	3.402	1.662	2.368	3.183	0.6772
91	1.986	2.631	3.401	1.662	2.368	3.182	0.6772
92	1.986	2.630	3.399	1.662	2.368	3.181	0.6772
93	1.986	2.630	3.398	1.661	2.367	3.180	0.6771
94	1.986	2.629	3.397	1.661	2.367	3.179	0.6771
95	1.985	2.629	3.396	1.661	2.366	3.178	0.6771
96	1.985	2.628	3.395	1.661	2.366	3.177	0.6771
97	1.985	2.627	3.394	1.661	2.365	3.176	0.6770
98	1.984	2.627	3.393	1.661	2.365	3.175	0.6770
99	1.984	2.626	3.392	1.660	2.365	3.175	0.6770
100	1.984	2.626	3.390	1.660	2.364	3.174	0.6770
110	1.982	2.621	3.381	1.659	2.361	3.166	0.6767
120	1.980	2.617	3.373	1.658	2.358	3.160	0.6765
130	1.978	2.614	3.367	1.657	2.355	3.154	0.6764
140	1.977	2.611	3.361	1.656	2.353	3.149	0.6762
150	1.976	2.609	3.357	1.655	2.351	3.145	0.6761
160	1.975	2.607	3.352	1.654	2.350	3.142	0.6760
170	1.974	2.605	3.349	1.654	2.348	3.139	0.6759
180	1.973	2.603	3.345	1.653	2.347	3.136	0.6759
190	1.973	2.602	3.342	1.653	2.346	3.134	0.6758
200	1.972	2.601	3.340	1.653	2.345	3.131	0.6757
210	1.971	2.599	3.337	1.652	2.344	3.129	0.6757
220	1.971	2.598	3.335	1.652	2.343	3.128	0.6756
230	1.970	2.597	3.333	1.652	2.343	3.126	0.6756
240	1.970	2.596	3.332	1.651	2.342	3.125	0.6755
250	1.969	2.596	3.330	1.651	2.341	3.123	0.6755
260	1.969	2.595	3.328	1.651	2.341	3.122	0.6754
270	1.969	2.594	3.327	1.651	2.340	3.121	0.6754
280	1.968	2.594	3.326	1.650	2.340	3.120	0.6754
290	1.968	2.593	3.324	1.650	2.339	3.119	0.6753
300	1.968	2.592	3.323	1.650	2.339	3.118	0.6753
310	1.968	2.592	3.322	1.650	2.338	3.117	0.6753
320	1.967	2.591	3.321	1.650	2.338	3.116	0.6753
330	1.967	2.591	3.320	1.649	2.338	3.115	0.6752
340	1.967	2.590	3.319	1.649	2.337	3.114	0.6752
350	1.967	2.590	3.319	1.649	2.337	3.114	0.6752
360	1.967	2.590	3.318	1.649	2.337	3.113	0.6752
370	1.966	2.589	3.317	1.649	2.336	3.112	0.6752
380	1.966	2.589	3.316	1.649	2.336	3.112	0.6751
390	1.966	2.588	3.316	1.649	2.336	3.111	0.6751
400	1.966	2.588	3.315	1.649	2.336	3.111	0.6751

Table A2.4—*Continued*

ν \ P	0.05	0.01	0.001	0.1	0.02	0.002	0.5
410	1.966	2.588	3.314	1.649	2.335	3.110	0.6751
420	1.966	2.588	3.314	1.648	2.335	3.110	0.6751
430	1.965	2.587	3.313	1.648	2.335	3.109	0.6751
440	1.965	2.587	3.313	1.648	2.335	3.109	0.6750
450	1.965	2.587	3.312	1.648	2.335	3.108	0.6750
460	1.965	2.587	3.312	1.648	2.334	3.108	0.6750
470	1.965	2.586	3.311	1.648	2.334	3.108	0.6750
480	1.965	2.586	3.311	1.648	2.334	3.107	0.6750
490	1.965	2.586	3.310	1.648	2.334	3.107	0.6750
500	1.965	2.586	3.310	1.648	2.334	3.107	0.6750
∞	1.960	2.576	3.291	1.645	2.326	3.090	0.6745

* Table computed by Dr. Edward L. Spitznagel. Reproduced with his permission.

Table A2.5 Critical values of F with v_1 and v_2 degrees of freedom.* One-sided critical region

v_1 is the degrees of freedom for numerator; v_2 is the degrees of freedom for denominator

5% Level

v_2 \ v_1	1	2	3	4	5	6	7	8	9	10	15	25	50	100	∞
1	161	200	216	225	230	234	237	239	241	242	246	249	252	253	254
2	18.51	19.00	19.16	19.25	19.30	19.33	19.36	19.37	19.38	19.39	19.43	19.45	19.47	19.49	19.50
3	10.13	9.55	9.28	9.12	9.01	8.94	8.88	8.84	8.81	8.78	8.70	8.64	8.58	8.56	8.53
4	7.71	6.94	6.59	6.39	6.26	6.16	6.09	6.04	6.00	5.96	5.85	5.77	5.70	5.66	5.63
5	6.61	5.79	5.41	5.19	5.05	4.95	4.88	4.82	4.78	4.74	4.62	4.53	4.44	4.40	4.36
6	5.99	5.14	4.76	4.53	4.39	4.28	4.21	4.15	4.10	4.06	3.94	3.84	3.75	3.71	3.67
7	5.59	4.74	4.35	4.12	3.97	3.87	3.79	3.73	3.68	3.63	3.50	3.41	3.32	3.28	3.23
8	5.32	4.46	4.07	3.84	3.69	3.58	3.50	3.44	3.39	3.34	3.21	3.11	3.03	2.98	2.93
9	5.12	4.26	3.86	3.63	3.48	3.37	3.29	3.23	3.18	3.13	3.00	2.89	2.80	2.76	2.71
10	4.96	4.10	3.71	3.48	3.33	3.22	3.14	3.07	3.02	2.97	2.84	2.73	2.64	2.59	2.54
15	4.54	3.68	3.29	3.06	2.90	2.79	2.70	2.64	2.59	2.55	2.41	2.28	2.18	2.12	2.07
25	4.24	3.38	2.99	2.76	2.60	2.49	2.41	2.34	2.28	2.24	2.08	1.95	1.84	1.77	1.71
50	4.03	3.18	2.79	2.56	2.40	2.29	2.20	2.13	2.07	2.02	1.87	1.73	1.60	1.52	1.44
100	3.94	3.09	2.70	2.46	2.30	2.19	2.10	2.03	1.97	1.92	1.77	1.62	1.48	1.39	1.28
500	3.86	3.02	2.62	2.39	2.23	2.12	2.03	1.96	1.90	1.85	1.69	1.53	1.38	1.28	1.12
1,000	3.85	3.00	2.61	2.38	2.22	2.10	2.02	1.95	1.89	1.84	1.67	1.52	1.36	1.26	1.08
∞	3.84	2.99	2.60	2.37	2.21	2.09	2.01	1.94	1.88	1.83	1.66	1.51	1.35	1.24	1.00

1% Level

v_2 \ v_1	1	2	3	4	5	6	7	8	9	10	15	25	50	100	∞
1	4,052	4,999	5,403	5,625	5,764	5,859	5,928	5,981	6,022	6,056	6,158	6,238	6,302	6,334	6,366
2	98.49	99.01	99.17	99.25	99.30	99.33	99.34	99.36	99.38	99.40	99.43	99.46	99.48	99.49	99.50
3	34.12	30.81	29.46	28.71	28.24	27.91	27.67	27.49	27.34	27.23	26.87	26.58	26.30	26.23	26.12
4	21.20	18.00	16.69	15.98	15.52	15.21	14.98	14.80	14.66	14.54	14.19	13.91	13.69	13.57	13.46
5	16.26	13.27	12.06	11.39	10.97	10.67	10.45	10.27	10.15	10.05	9.72	9.45	9.24	9.13	9.02
6	13.74	10.92	9.78	9.15	8.75	8.47	8.26	8.10	7.98	7.87	7.56	7.30	7.09	6.99	6.88
7	12.25	9.55	8.45	7.85	7.46	7.19	7.00	6.84	6.71	6.62	6.31	6.05	5.85	5.75	5.65
8	11.26	8.65	7.59	7.01	6.63	6.37	6.19	6.03	5.91	5.82	5.52	5.27	5.06	4.96	4.86
9	10.56	8.02	6.99	6.42	6.06	5.80	5.62	5.47	5.35	5.26	4.96	4.72	4.51	4.41	4.31
10	10.04	7.56	6.55	5.99	5.64	5.39	5.21	5.06	4.95	4.85	4.56	4.32	4.12	4.01	3.91
15	8.68	6.36	5.42	4.89	4.56	4.32	4.14	4.00	3.89	3.80	3.52	3.27	3.07	2.97	2.87
25	7.77	5.57	4.68	4.18	3.86	3.63	3.46	3.32	3.21	3.13	2.85	2.61	2.40	2.29	2.17
50	7.17	5.06	4.20	3.72	3.41	3.18	3.02	2.88	2.78	2.70	2.42	2.17	1.94	1.82	1.68
100	6.90	4.82	3.98	3.51	3.20	2.99	2.82	2.69	2.59	2.51	2.22	1.96	1.73	1.59	1.43
500	6.69	4.65	3.82	3.36	3.06	2.84	2.68	2.55	2.45	2.36	2.07	1.81	1.56	1.41	1.18
1,000	6.66	4.62	3.80	3.34	3.04	2.82	2.66	2.53	2.43	2.34	2.05	1.79	1.54	1.38	1.11
∞	6.64	4.60	3.78	3.32	3.02	2.80	2.64	2.51	2.41	2.32	2.03	1.77	1.52	1.36	1.00

* By permission from G. R. Wadsworth and J. G. Bryan, *Introduction to Probability and Random Variables*, McGraw–Hill Book Company, Inc., New York, 1960.

Table A2.6 Fractiles of the beta distribution*

Column headings = $P(X \leq x)$. Table entries are x. The left two columns give the parameters of the distribution

t	r	0.01	0.05	0.10	0.15	0.20	0.25	0.30	0.35	0.40	0.45	0.50	0.55	0.60	0.65	0.70	0.75	0.80	0.85	0.90	0.95	0.99
2	1	0100	0500	1000	1500	2000	2500	3000	3500	4000	4500	5000	5500	6000	6500	7000	7500	8000	8500	9000	9500	9900
3	1	0050	0253	0513	0780	1056	1340	1633	1938	2254	2584	2929	3292	3675	4084	4523	5000	5528	6127	6838	7764	9000
4	1	0033	0170	0345	0527	0717	0914	1121	1338	1566	1807	2063	2337	2632	2953	3306	3700	4152	4687	5358	6316	7846
4	2	0589	1354	1958	2444	2871	3264	3633	3986	4329	4666	5000	5334	5671	6014	6367	6736	7129	7556	8042	8646	9411
5	1	0025	0127	0260	0398	0543	0694	0853	1021	1199	1388	1591	1810	2047	2308	2599	2929	3313	3777	4377	5271	6838
5	2	0420	0976	1426	1794	2123	2430	2724	3010	3292	3573	3857	4147	4445	4756	5084	5437	5825	6265	6795	7514	8591
6	1	0020	0102	0209	0320	0436	0559	0688	0825	0971	1127	1294	1476	1674	1894	2140	2421	2752	3157	3690	4507	6019
6	2	0327	0764	1122	1419	1686	1938	2180	2418	2656	2895	3138	3389	3650	3925	4220	4542	4902	5321	5839	6574	7779
6	3	1056	1893	2466	2899	3266	3594	3898	4186	4463	4733	5000	5267	5537	5814	6102	6406	6734	7101	7534	8107	8944
7	1	0017	0085	0174	0267	0365	0468	0577	0693	0816	0948	1091	1246	1416	1605	1818	2063	2353	2711	3187	3930	5358
7	2	0268	0628	0926	1174	1399	1612	1818	2022	2226	2433	2644	2864	3094	3339	3604	3895	4224	4613	5103	5818	7057
7	3	0847	1532	2009	2374	2686	2969	3233	3486	3731	3973	4214	4458	4708	4967	5239	5532	5854	6222	6668	7287	8269
8	1	0014	0073	0149	0229	0314	0403	0497	0597	0704	0819	0943	1078	1227	1393	1580	1797	2054	2374	2803	3482	4821
8	2	0227	0534	0788	1001	1195	1380	1559	1737	1916	2098	2285	2480	2685	2905	3143	3409	3709	4067	4526	5207	6434
8	3	0708	1288	1696	2011	2283	2531	2763	2987	3206	3423	3641	3863	4092	4331	4586	4867	5168	5523	5962	6587	7637
8	4	1423	2253	2786	3176	3501	3788	4052	4301	4539	4771	5000	5229	5461	5699	5948	6212	6499	6824	7214	7747	8577
9	1	0013	0064	0131	0201	0275	0353	0436	0524	0619	0720	0830	0950	1082	1230	1397	1591	1822	2111	2501	3123	4377
9	2	0197	0464	0686	0873	1044	1206	1365	1523	1682	1844	2011	2186	2371	2570	2786	3027	3304	3635	4062	4707	5899
9	3	0608	1111	1469	1746	1986	2206	2413	2614	2811	3007	3205	3408	3618	3839	4075	4332	4621	4959	5382	5997	7068
9	4	1210	1929	2397	2742	3032	3291	3530	3756	3975	4189	4402	4616	4835	5061	5299	5555	5837	6159	6554	7108	8018
10	1	0011	0057	0116	0179	0245	0315	0389	0467	0552	0645	0741	0849	0968	1101	1252	1428	1637	1901	2257	2831	4005
10	2	0174	0410	0608	0774	0926	1072	1214	1355	1499	1645	1796	1955	2123	2304	2501	2723	2978	3285	3684	4291	5440
10	3	0533	0977	1295	1542	1758	1955	2142	2324	2502	2681	2862	3048	3242	3446	3665	3905	4177	4496	4901	5496	6563
10	4	1053	1688	2104	2414	2675	2910	3127	3335	3535	3733	3931	4131	4336	4550	4776	5020	5291	5605	5994	6551	7500
10	5	1710	2514	3010	3367	3661	3920	4156	4378	4590	4796	5000	5204	5410	5622	5844	6080	6339	6633	6994	7486	8290
11	1	0010	0051	0104	0161	0220	0284	0350	0422	0498	0580	0670	0767	0876	0997	1134	1295	1487	1728	2057	2589	3690
11	2	0155	0368	0545	0695	0833	0964	1093	1221	1351	1485	1623	1767	1921	2087	2269	2474	2710	2996	3368	3942	5044
11	3	0475	0873	1158	1381	1576	1756	1926	2092	2255	2419	2586	2757	2936	3126	3330	3554	3809	4111	4496	5069	6117
11	4	0932	1500	1876	2156	2394	2609	2808	2999	3184	3367	3551	3738	3930	4131	4345	4577	4837	5139	5517	6066	7029
11	5	1504	2224	2673	2998	3268	3507	3726	3932	4131	4325	4517	4710	4907	5111	5325	5555	5809	6100	6458	6965	7817

12	1	0009	0047	0095	0147	0201	0258	0319	0384	0454	0529	0611	0700	0799	0910	1037	1184	1361	1584	1889	2384	3421
12	2	0141	0333	0495	0631	0756	0876	0994	1111	1230	1353	1480	1613	1755	1908	2077	2266	2486	2753	3102	3644	4698
12	3	0428	0788	1048	1251	1429	1593	1750	1902	2052	2204	2358	2517	2683	2860	3050	3261	3501	3786	4152	4701	5723
12	4	0837	1351	1692	1949	2167	2364	2548	2724	2896	3067	3238	3413	3593	3782	3984	4205	4452	4742	5108	5644	6604
12	5	1344	1996	2405	2704	2953	3173	3377	3570	3755	3938	4119	4302	4489	4684	4889	5111	5357	5642	5995	6502	7378
12	6	1940	2713	3177	3508	3779	4016	4232	4434	4627	4815	5000	5185	5373	5566	5768	5984	6221	6492	6823	7287	8060
13	1	0008	0043	0087	0135	0184	0237	0293	0353	0417	0486	0561	0644	0735	0838	0955	1091	1255	1462	1746	2209	3187
13	2	0128	0305	0452	0577	0693	0803	0911	1020	1130	1242	1360	1483	1615	1757	1914	2091	2296	2546	2875	3387	4395
13	3	0390	0719	0957	1143	1307	1459	1603	1744	1883	2024	2167	2315	2470	2635	2814	3012	3238	3508	3855	4381	5373
13	4	0759	1229	1542	1778	1979	2162	2332	2496	2656	2815	2976	3140	3309	3488	3679	3888	4124	4401	4753	5273	6222
13	5	1215	1810	2187	2463	2693	2898	3088	3268	3443	3614	3785	3958	4136	4321	4518	4731	4968	5245	5590	6091	6976
13	6	1746	2453	2882	3189	3441	3663	3866	4057	4240	4418	4595	4772	4953	5140	5336	5547	5779	6047	6377	6848	7651
14	1	0008	0039	0081	0124	0170	0219	0271	0326	0385	0449	0519	0596	0681	0776	0885	1011	1164	1358	1623	2058	2983
14	2	0118	0281	0417	0532	0639	0741	0841	0942	1044	1149	1258	1373	1496	1629	1775	1941	2133	2368	2678	3163	4128
14	3	0358	0660	0880	1053	1204	1345	1479	1610	1740	1871	2004	2143	2288	2443	2611	2798	3011	3267	3598	4101	5062
14	4	0695	1127	1416	1635	1822	1991	2150	2303	2453	2602	2753	2907	3067	3235	3416	3615	3839	4105	4443	4946	5878
14	5	1108	1657	2005	2261	2476	2668	2845	3014	3178	3340	3502	3666	3835	4011	4199	4403	4631	4899	5234	5726	6609
14	6	1588	2240	2637	2923	3160	3368	3559	3740	3913	4082	4251	4420	4593	4773	4963	5167	5394	5657	5982	6452	7271
14	7	2129	2870	3309	3618	3870	4090	4290	4477	4656	4829	5000	5171	5344	5523	5710	5910	6130	6382	6691	7130	7871
15	1	0007	0037	0075	0115	0158	0203	0252	0303	0358	0418	0483	0554	0634	0722	0824	0943	1086	1267	1517	1926	2803
15	2	0110	0260	0387	0494	0593	0688	0781	0875	0970	1068	1170	1278	1393	1517	1655	1810	1992	2213	2507	2967	3891
15	3	0331	0611	0815	0975	1117	1248	1373	1495	1617	1739	1865	1995	2132	2277	2436	2612	2814	3057	3372	3854	4783
15	4	0640	1040	1309	1513	1688	1846	1995	2138	2279	2419	2561	2706	2857	3017	3189	3377	3592	3845	4170	4657	5567
15	5	1019	1527	1851	2090	2291	2471	2638	2797	2952	3104	3258	3413	3574	3742	3921	4117	4336	4594	4920	5400	6274
15	6	1457	2061	2432	2699	2921	3117	3298	3468	3633	3794	3954	4116	4282	4455	4637	4835	5055	5311	5631	6096	6920
15	7	1947	2636	3046	3336	3574	3782	3972	4151	4321	4487	4651	4816	4984	5157	5339	5535	5751	6001	6309	6750	7512
16	1	0007	0034	0070	0108	0148	0190	0235	0283	0335	0391	0452	0518	0593	0676	0771	0883	1017	1188	1423	1810	2644
16	2	0102	0242	0360	0461	0553	0642	0729	0817	0906	0998	1094	1195	1303	1420	1550	1697	1868	2077	2356	2794	3679
16	3	0307	0568	0759	0909	1041	1163	1281	1395	1510	1625	1743	1866	1995	2133	2282	2450	2641	2872	3173	3634	4532
16	4	0594	0967	1218	1408	1572	1720	1860	1995	2128	2260	2394	2531	2675	2826	2989	3169	3373	3616	3928	4398	5285
16	5	0944	1417	1720	1944	2132	2302	2459	2609	2755	2900	3045	3193	3346	3506	3678	3865	4076	4325	4640	5108	5969
16	6	1346	1909	2256	2507	2716	2902	3072	3234	3390	3544	3697	3851	4010	4176	4352	4543	4756	5005	5317	5774	6597
16	7	1795	2437	2822	3096	3321	3518	3699	3869	4032	4191	4348	4507	4669	4836	5013	5204	5415	5660	5965	6404	7177
16	8	2287	3000	3415	3707	3944	4150	4337	4512	4679	4841	5000	5159	5321	5488	5663	5850	6056	6293	6585	7000	7713
17	1	0006	0032	0066	0101	0138	0178	0220	0266	0314	0367	0424	0487	0557	0635	0725	0830	0957	1118	1340	1708	2501
17	2	0095	0227	0337	0431	0518	0602	0684	0766	0850	0937	1027	1122	1224	1335	1457	1596	1758	1957	2222	2640	3488
17	3	0287	0531	0710	0850	0975	1090	1200	1308	1416	1525	1637	1752	1874	2005	2147	2306	2488	2709	2996	3438	4305
17	4	0554	0903	1138	1317	1471	1611	1743	1870	1995	2121	2247	2378	2514	2658	2813	2985	3180	3413	3712	4166	5029
17	5	0878	1321	1606	1816	1994	2154	2303	2445	2583	2721	2859	3000	3145	3299	3463	3642	3845	4085	4389	4844	5690
17	6	1251	1778	2104	2341	2539	2714	2876	3030	3178	3324	3471	3619	3771	3930	4099	4283	4489	4731	5035	5483	6299
17	7	1665	2267	2629	2888	3101	3289	3461	3623	3779	3931	4082	4235	4391	4553	4724	4909	5116	5355	5654	6090	6866
17	8	2117	2786	3178	3455	3680	3877	4056	4224	4384	4540	4694	4849	5006	5168	5339	5522	5726	5960	6250	6666	7393

Table A2.6—Continued

t	r	0.01	0.05	0.10	0.15	0.20	0.25	0.30	0.35	0.40	0.45	0.50	0.55	0.60	0.65	0.70	0.75	0.80	0.85	0.90	0.95	0.99
18	1	0006	0030	0062	0095	0130	0168	0208	0250	0296	0346	0400	0459	0525	0599	0684	0783	0903	1056	1267	1616	2373
18	2	0090	0213	0317	0406	0488	0566	0644	0722	0801	0883	0968	1058	1154	1259	1375	1507	1661	1850	2102	2501	3316
18	3	0254	0490	0667	0799	0916	1025	1129	1231	1333	1436	1542	1652	1764	1892	2027	2178	2352	2562	2837	3262	4099
18	4	0519	0846	1068	1237	1382	1514	1639	1760	1879	1997	2118	2242	2371	2509	2657	2821	3008	3231	3519	3956	4796
18	5	0822	1238	1506	1705	1873	2024	2165	2300	2432	2562	2694	2828	2968	3114	3271	3451	3639	3869	4164	4605	5434
18	6	1168	1664	1972	2196	2383	2549	2703	2849	2991	3131	3270	3412	3558	3711	3874	4051	4251	4485	4781	5219	6025
18	7	1552	2119	2461	2707	2909	3088	3252	3406	3555	3702	3847	3994	4144	4300	4466	4646	4846	5080	5374	5803	6577
18	8	1971	2601	2973	3235	3450	3638	3809	3970	4124	4275	4423	4573	4725	4883	5049	5229	5426	5658	5945	6360	7094
18	9	2422	3108	3504	3780	4004	4199	4376	4541	4698	4850	5000	5150	5302	5459	5624	5801	5996	6220	6496	6892	7578
19	1	0005	0028	0058	0090	0123	0159	0196	0236	0280	0327	0378	0434	0496	0567	0647	0741	0855	1000	1201	1533	2257
19	2	0085	0201	0299	0383	0460	0535	0608	0682	0757	0834	0915	1000	1092	1192	1302	1427	1574	1754	1995	2377	3160
19	3	0254	0470	0629	0754	0865	0968	1066	1163	1260	1358	1458	1562	1673	1791	1920	2064	2230	2431	2694	3103	3912
19	4	0488	0797	1006	1166	1304	1429	1547	1661	1775	1888	2002	2121	2249	2375	2510	2674	2853	3067	3344	3767	4583
19	5	0772	1164	1418	1606	1765	1909	2043	2170	2297	2422	2547	2676	2809	2949	3100	3265	3453	3676	3960	4389	5199
19	6	1096	1563	1855	2067	2245	2404	2550	2688	2825	2958	3092	3228	3368	3515	3672	3843	4036	4263	4550	4978	5772
19	7	1454	1990	2314	2547	2740	2910	3067	3212	3357	3497	3637	3779	3923	4074	4235	4409	4604	4832	5118	5540	6309
19	8	1844	2440	2792	3042	3247	3427	3591	3742	3894	4039	4182	4327	4474	4628	4789	4964	5159	5385	5667	6078	6814
19	9	2263	2912	3288	3552	3767	3954	4124	4279	4434	4582	4727	4873	5022	5175	5336	5510	5702	5923	6198	6594	7290
20	1	0005	0027	0055	0085	0117	0150	0186	0224	0265	0310	0358	0412	0471	0538	0614	0704	0812	0950	1141	1459	2152
20	2	0080	0190	0283	0363	0436	0507	0576	0646	0717	0791	0868	0949	1036	1131	1236	1355	1495	1668	1898	2264	3018
20	3	0240	0445	0595	0714	0817	0916	1010	1096	1194	1289	1383	1482	1587	1700	1823	1961	2120	2312	2565	2958	3741
20	4	0461	0753	0951	1103	1233	1353	1465	1574	1682	1789	1899	2012	2130	2255	2391	2541	2713	2919	3186	3594	4387
20	5	0728	1099	1339	1518	1670	1806	1934	2055	2176	2295	2415	2538	2666	2801	2943	3105	3285	3500	3775	4191	4983
20	6	1032	1475	1751	1953	2123	2274	2414	2545	2676	2804	2932	3061	3197	3339	3490	3655	3841	4061	4338	4758	5538
20	7	1368	1875	2183	2405	2589	2752	2902	3041	3180	3315	3449	3585	3725	3871	4026	4195	4384	4606	4886	5300	6060
20	8	1734	2297	2633	2871	3067	3239	3397	3543	3688	3827	3966	4106	4249	4397	4554	4725	4915	5136	5413	5819	6553
20	9	2124	2739	3101	3351	3556	3736	3899	4049	4199	4342	4483	4625	4769	4919	5076	5246	5435	5653	5925	6319	7020
20	10	2540	3201	3579	3843	4056	4241	4408	4561	4713	4858	5000	5142	5287	5435	5592	5759	5944	6157	6421	6799	7460
21	1	0005	0026	0053	0081	0111	0143	0177	0213	0252	0294	0341	0391	0448	0511	0584	0670	0773	0905	1088	1391	2057
21	2	0076	0181	0269	0344	0414	0481	0547	0614	0682	0752	0825	0903	0986	1076	1176	1291	1424	1589	1810	2161	2888
21	3	0227	0422	0564	0677	0777	0870	0959	1047	1135	1223	1315	1410	1510	1618	1736	1868	2026	2205	2448	2826	3583
21	4	0436	0714	0902	1046	1170	1284	1391	1495	1598	1701	1805	1913	2026	2147	2277	2421	2586	2785	3042	3437	4207
21	5	0688	1041	1269	1440	1584	1714	1836	1953	2068	2182	2297	2415	2537	2666	2806	2959	3133	3340	3607	4010	4783
21	6	0975	1396	1659	1851	2013	2157	2291	2418	2542	2665	2788	2911	3043	3183	3325	3484	3665	3878	4173	4556	5321
21	7	1292	1773	2067	2278	2453	2610	2753	2889	3021	3150	3279	3406	3545	3686	3836	4000	4184	4400	4673	5078	5829
21	8	1634	2171	2491	2719	2906	3071	3223	3365	3503	3637	3771	3906	4041	4189	4341	4507	4692	4908	5180	5580	6309
21	9	2001	2586	2929	3171	3368	3541	3698	3846	3988	4126	4263	4400	4541	4686	4840	5008	5191	5405	5673	6064	6766
21	10	2390	3020	3382	3635	3840	4018	4180	4331	4476	4616	4754	4893	5034	5180	5333	5498	5680	5891	6152	6531	7199

Group 22 (columns 1–11):

```
1969 2768 3439 4041 4598 5120 5613 6082 6528 6953 7358
1329 2067 2706 3292 3844 4370 4871 5359 5828 6281 6719
1038 1729 2340 2910 3452 3973 4466 4966 5449 5905 6356
0864 1518 2107 2662 3194 3710 4211 4700 5178 5646 6104
0738 1360 1929 2471 2994 3503 4001 4489 4967 5438 5900
0639 1232 1783 2312 2826 3329 3823 4308 4787 5258 5723
0557 1122 1656 2173 2678 3175 3664 4147 4624 5097 5564
0488 1026 1543 2044 2544 3034 3519 3999 4475 4946 5415
0427 0940 1440 1933 2420 2903 3382 3859 4333 4804 5273
0373 0860 1344 1824 2302 2778 3252 3725 4196 4667 5135
0325 0786 1253 1721 2189 2657 3126 3594 4063 4531 5000
0281 0717 1166 1621 2079 2539 3001 3465 3930 4397 4865
0240 0650 1081 1522 1969 2421 2877 3335 3797 4261 4727
0203 0585 0997 1424 1860 2302 2750 3203 3660 4121 4585
0168 0521 0914 1325 1748 2180 2620 3066 3517 3974 4436
0136 0458 0828 1222 1631 2052 2482 2920 3366 3818 4277
0106 0394 0740 1114 1507 1914 2333 2762 3199 3646 4100
0077 0328 0644 0995 1369 1759 2154 2582 3010 3448 3896
0050 0256 0537 0858 1206 1575 1962 2363 2778 3205 3644
0024 0172 0401 0688 0988 1324 1682 2057 2450 2858 3281
0005 0072 0216 0414 0653 0925 1223 1546 1891 2257 2642
```

Group 23 (columns 1–11):

```
1889 2658 3159 3887 4426 4933 5412 5864 6304 6720 7119
1273 1981 2595 3159 3698 4198 4685 5155 5609 6048 6475
0994 1656 2242 2789 3310 3801 4297 4768 5229 5675 6112
0826 1453 2017 2550 3061 3537 4037 4508 4966 5419 5862
0705 1301 1846 2365 2867 3355 3833 4302 4762 5215 5660
0611 1178 1705 2212 2705 3187 3660 4126 4585 5038 5486
0533 1073 1584 2078 2562 3038 3506 3969 4427 4880 5328
0466 0981 1458 1958 2433 2902 3366 3825 4281 4733 5182
0408 0898 1376 1817 2313 2775 3234 3690 4143 4594 5043
0356 0822 1284 1743 2200 2655 3108 3561 4011 4460 4909
0310 0751 1197 1641 2091 2538 2986 3433 3881 4329 4776
0268 0684 1098 1518 1985 2425 2866 3309 3753 4198 4644
0230 0620 1032 1453 1880 2311 2746 3183 3623 4066 4510
0194 0558 0961 1359 1775 2197 2624 3056 3491 3930 4372
0161 0498 0872 1264 1667 2079 2498 2923 3353 3788 4228
0130 0437 0790 1166 1556 1956 2366 2783 3207 3637 4073
0101 0376 0706 1062 1436 1824 2223 2631 3047 3471 3902
0074 0313 0615 0949 1304 1676 2061 2458 2864 3280 3705
0048 0242 0517 0817 1149 1500 1867 2248 2642 3046 3462
0023 0164 0382 0646 0941 1260 1599 1956 2327 2713 3113
0005 0069 0206 0394 0621 0879 1162 1468 1795 2138 2501
```

Group 24 (columns 1–11):

```
1815 2557 3181 3745 4267 4758 5224 5669 6094 6502 6892
1221 1902 2492 3036 3549 4039 4510 4964 5405 5832 6249
0953 1588 2156 2678 3176 3663 4131 4586 5029 5462 5885
0792 1393 1935 2446 2938 3414 3878 4331 4775 5210 5637
0676 1247 1759 2268 2750 3220 3679 4130 4573 5009 5438
0585 1128 1634 2120 2593 3054 3511 3959 4406 4836 5267
0510 1028 1517 1992 2456 2912 3362 3806 4246 4681 5112
0446 0939 1404 1876 2331 2780 3225 3666 4103 4538 4969
0391 0860 1318 1769 2215 2658 3098 3535 3969 4402 4832
0341 0787 1229 1669 2106 2542 2976 3409 3841 4271 4701
0297 0719 1147 1573 2001 2430 2858 3286 3713 4145 4572
0257 0655 1066 1481 1899 2320 2742 3165 3590 4016 4443
0220 0594 0988 1390 1799 2213 2626 3044 3465 3887 4312
0186 0531 0873 1246 1646 2013 2509 2921 3337 3756 4178
0154 0476 0834 1209 1594 1980 2388 2793 3204 3619 4038
0124 0418 0756 1114 1487 1870 2260 2658 3062 3472 3887
0097 0360 0705 1015 1372 1745 2123 2511 2903 3312 3722
0070 0299 0587 0906 1246 1601 1968 2352 2732 3128 3532
0046 0234 0489 0781 1097 1432 1782 2152 2518 2903 3297
0022 0157 0365 0617 0898 1193 1525 1863 2216 2582 2961
0004 0066 0196 0376 0593 0838 1107 1397 1705 2031 2374
```

Group 25 (columns 1–12):

```
1746 2462 3066 3612 4118 4595 5048 5481 5897 6295 6678 7047
1173 1829 2398 2936 3418 3891 4347 4787 5214 5629 6032 6424
0915 1526 2069 2575 3056 3525 3976 4416 4845 5264 5674 6076
0760 1337 1859 2351 2824 3283 3730 4167 4596 5016 5429 5835
0649 1197 1700 2179 2642 3094 3542 3971 4398 4819 5233 5642
0561 1083 1569 2036 2491 2936 3373 3804 4229 4649 5064 5475
0489 0986 1456 1912 2357 2796 3228 3656 4079 4498 4913 5324
0428 0901 1356 1797 2237 2669 3096 3520 3940 4357 4772 5184
0375 0825 1264 1697 2126 2550 2972 3392 3809 4223 4639 5050
0327 0755 1179 1601 2020 2438 2855 3270 3684 4096 4510 4921
0285 0690 1099 1509 1919 2331 2743 3151 3562 3971 4384 4795
0246 0628 1022 1421 1821 2228 2634 3041 3449 3858 4258 4668
0211 0569 0947 1333 1724 2119 2517 2917 3320 3724 4131 4539
0178 0512 0873 1246 1626 2013 2403 2798 3196 3597 4007 4407
0148 0456 0799 1158 1527 1904 2287 2675 3067 3464 3864 4268
0119 0401 0724 1068 1424 1790 2164 2544 2931 3322 3718 4120
0093 0345 0646 0972 1668 2031 2431 2781 3167 3558 3955
0067 0287 0563 0868 1192 1532 1882 2243 2612 2989 3374 3766
0046 0224 0468 0750 1050 1369 1703 2049 2406 2772 3148 3532
0022 0150 0350 0590 0859 1149 1457 1780 2116 2460 2824 3194
0004 0063 0188 0360 0566 0800 1056 1332 1625 1935 2260 2599
```

Table A2.6—Continued

t	r	0.01	0.05	0.10	0.15	0.20	0.25	0.30	0.35	0.40	0.45	0.50	0.55	0.60	0.65	0.70	0.75	0.80	0.85	0.90	0.95	0.99
26	1	0004	0020	0042	0065	0089	0114	0142	0171	0202	0236	0273	0314	0360	0411	0470	0539	0624	0731	0880	1129	1682
26	2	0060	0144	0215	0275	0331	0385	0438	0492	0546	0603	0662	0725	0793	0866	0948	1041	1151	1287	1469	1761	2375
26	3	0180	0335	0449	0540	0620	0695	0767	0838	0909	0981	1055	1133	1216	1303	1400	1509	1635	1788	1991	2310	2959
26	4	0345	0566	0717	0832	0933	1024	1111	1196	1279	1364	1449	1538	1631	1730	1838	1958	2096	2263	2480	2817	3488
26	5	0542	0823	1006	1143	1260	1366	1465	1561	1655	1749	1844	1941	2043	2150	2267	2396	2543	2722	2947	3296	3979
26	6	0765	1101	1312	1468	1599	1717	1827	1932	2034	2136	2238	2343	2451	2566	2689	2824	2978	3161	3397	3754	4443
26	7	1010	1395	1632	1804	1947	2075	2194	2306	2416	2524	2632	2743	2857	2977	3105	3246	3405	3593	3833	4195	4884
26	8	1273	1704	1962	2149	2303	2440	2566	2685	2800	2914	3027	3142	3260	3384	3517	3661	3823	4015	4258	4622	5311
26	9	1553	2024	2303	2502	2666	2810	2942	3066	3186	3304	3422	3540	3662	3789	3924	4071	4236	4429	4673	5036	5711
26	10	1848	2356	2653	2863	3034	3184	3322	3451	3575	3696	3816	3937	4062	4191	4328	4476	4642	4836	5080	5439	6100
26	11	2156	2699	3011	3230	3408	3564	3705	3838	3965	4088	4211	4334	4460	4590	4728	4877	5042	5235	5477	5832	6476
26	12	2479	3051	3377	3604	3788	3947	4092	4227	4356	4482	4605	4730	4856	4987	5124	5273	5438	5629	5867	6214	6837
26	13	2814	3414	3751	3985	4173	4335	4482	4619	4749	4876	5000	5124	5251	5381	5518	5665	5827	6015	6249	6586	7186
27	1	0004	0020	0040	0062	0085	0110	0136	0164	0195	0227	0263	0302	0346	0396	0453	0519	0600	0704	0848	1088	1623
27	2	0058	0138	0206	0264	0318	0370	0421	0473	0526	0580	0637	0698	0763	0834	0913	1002	1108	1239	1415	1698	2293
27	3	0173	0322	0432	0519	0596	0668	0737	0806	0874	0944	1015	1090	1181	1254	1348	1453	1575	1723	1920	2229	2859
27	4	0331	0543	0688	0800	0896	0985	1068	1151	1230	1311	1394	1480	1570	1666	1770	1886	2020	2181	2392	2719	3372
27	5	0520	0790	0966	1098	1211	1313	1409	1501	1591	1682	1773	1868	1966	2070	2183	2308	2450	2621	2843	3182	3849
27	6	0734	1056	1260	1412	1536	1650	1756	1857	1956	2054	2153	2256	2359	2470	2590	2721	2870	3048	3277	3626	4300
27	7	0968	1338	1566	1732	1870	1994	2108	2217	2323	2427	2532	2639	2750	2866	2991	3128	3282	3465	3700	4054	4729
27	8	1220	1633	1883	2062	2212	2344	2465	2581	2692	2802	2912	3024	3139	3259	3388	3528	3687	3873	4111	4470	5140
27	9	1487	1940	2209	2401	2559	2698	2826	2947	3063	3178	3292	3407	3525	3649	3781	3924	4087	4274	4513	4870	5535
27	10	1768	2257	2544	2746	2911	3058	3191	3316	3436	3554	3671	3789	3910	4036	4170	4315	4478	4668	4907	5262	5916
27	11	2062	2584	2886	3098	3271	3421	3559	3688	3811	3931	4051	4171	4294	4421	4556	4702	4865	5055	5293	5643	6284
27	12	2369	2921	3236	3456	3634	3789	3930	4061	4187	4310	4431	4552	4676	4804	4939	5085	5245	5436	5671	6016	6639
27	13	2688	3266	3593	3820	4002	4161	4304	4437	4565	4688	4810	4932	5056	5184	5319	5464	5625	5811	6043	6379	6982
28	1	0004	0019	0039	0060	0082	0106	0131	0158	0187	0219	0253	0291	0334	0381	0436	0500	0579	0679	0817	1050	1568
28	2	0056	0133	0199	0255	0306	0356	0406	0455	0506	0559	0614	0672	0735	0803	0880	0966	1069	1195	1366	1640	2217
28	3	0166	0310	0415	0499	0574	0643	0710	0776	0842	0909	0978	1050	1123	1209	1299	1401	1519	1663	1853	2153	2766
28	4	0318	0522	0662	0770	0862	0948	1029	1107	1185	1263	1343	1426	1517	1606	1709	1819	1948	2105	2309	2627	3264
28	5	0500	0759	0929	1057	1165	1264	1356	1445	1533	1620	1709	1800	1895	1996	2105	2226	2364	2530	2745	3076	3727
28	6	0705	1015	1211	1356	1478	1588	1690	1788	1883	1978	2074	2172	2274	2382	2497	2625	2770	2943	3166	3506	4166
28	7	0929	1285	1505	1665	1799	1918	2029	2134	2237	2338	2440	2544	2651	2764	2885	3018	3168	3347	3575	3921	4584
28	8	1170	1568	1809	1983	2127	2255	2372	2484	2592	2699	2805	2914	3026	3143	3268	3405	3559	3742	3974	4323	4984
28	9	1426	1862	2122	2308	2461	2596	2719	2837	2950	3060	3171	3283	3399	3519	3647	3787	3944	4129	4364	4714	5370
28	10	1695	2166	2443	2639	2800	2941	3070	3192	3308	3423	3537	3652	3770	3893	4023	4165	4324	4511	4746	5095	5742
28	11	1976	2479	2771	2976	3144	3290	3424	3549	3669	3786	3903	4020	4140	4265	4397	4540	4699	4886	5120	5466	6102
28	12	2268	2801	3106	3319	3492	3643	3780	3908	4031	4150	4268	4387	4508	4634	4767	4910	5069	5255	5488	5829	6450
28	13	2572	3131	3448	3668	3845	4000	4139	4270	4394	4515	4634	4754	4876	5001	5134	5277	5436	5620	5849	6184	6787
28	14	2887	3470	3796	4021	4203	4360	4501	4633	4759	4880	5000	5120	5241	5367	5499	5640	5797	5979	6204	6530	7113

29	1	0004	0018	0038	0058	0079	0102	0127	0153	0181	0211	0245	0281	0322	0368	0421	0483	0559	0655	0789	1015	1517
29	2	0054	0128	0192	0245	0295	0344	0391	0439	0488	0539	0592	0649	0709	0775	0849	0933	1032	1155	1319	1585	2146
29	3	0160	0298	0400	0481	0553	0620	0685	0748	0812	0877	0944	1013	1087	1167	1254	1353	1467	1606	1791	2082	2679
29	4	0306	0503	0638	0742	0831	0918	0992	1068	1143	1219	1296	1376	1460	1550	1647	1756	1882	2034	2232	2542	3162
29	5	0481	0731	0895	1018	1124	1218	1307	1393	1478	1563	1648	1737	1829	1927	2033	2150	2284	2445	2655	2977	3613
29	6	0678	0977	1166	1306	1424	1530	1629	1724	1816	1908	2001	2096	2195	2299	2412	2536	2677	2845	3062	3394	4039
29	7	0893	1237	1449	1604	1733	1848	1956	2058	2157	2255	2354	2454	2559	2668	2786	2915	3062	3236	3458	3797	4447
29	8	1125	1509	1741	1909	2049	2172	2286	2395	2499	2603	2706	2812	2920	3034	3156	3290	3440	3618	3845	4187	4837
29	9	1370	1791	2042	2221	2370	2500	2621	2734	2844	2952	3059	3168	3281	3398	3523	3660	3813	3994	4224	4567	5214
29	10	1627	2082	2350	2540	2696	2831	2958	3076	3190	3301	3412	3524	3639	3759	3887	4025	4181	4364	4598	4938	5578
29	11	1896	2383	2665	2864	3026	3169	3298	3420	3537	3651	3765	3879	3997	4118	4248	4388	4544	4728	4958	5300	5930
29	12	2176	2691	2987	3193	3361	3508	3641	3766	3886	4002	4118	4234	4353	4476	4606	4746	4903	5086	5316	5654	6271
29	13	2467	3007	3314	3528	3701	3851	3987	4114	4236	4354	4471	4588	4707	4831	4961	5102	5258	5440	5667	6000	6602
29	14	2767	3331	3648	3867	4044	4197	4335	4464	4587	4706	4824	4941	5061	5184	5314	5454	5609	5789	6013	6338	6922
30	1	0003	0018	0036	0056	0077	0099	0122	0147	0175	0204	0236	0272	0311	0356	0407	0467	0540	0633	0763	0981	1468
30	2	0052	0124	0185	0237	0285	0332	0378	0424	0472	0521	0572	0627	0685	0749	0820	0902	0998	1116	1276	1534	2079
30	3	0155	0288	0386	0464	0534	0599	0661	0721	0784	0847	0911	0979	1051	1128	1212	1308	1418	1554	1733	2016	2596
30	4	0295	0485	0615	0715	0802	0882	0957	1031	1104	1177	1252	1329	1411	1498	1592	1698	1820	1967	2150	2461	3066
30	5	0463	0705	0863	0982	1083	1175	1262	1345	1427	1509	1592	1678	1767	1862	1965	2079	2209	2366	2570	2884	3505
30	6	0653	0942	1120	1260	1374	1476	1572	1664	1754	1843	1933	2025	2121	2222	2332	2452	2589	2753	2965	3289	3920
30	7	0860	1192	1397	1547	1672	1783	1887	1986	2082	2178	2273	2372	2473	2579	2694	2819	2962	3130	3349	3680	4317
30	8	1083	1453	1678	1841	1976	2096	2206	2311	2413	2514	2614	2717	2822	2933	3052	3182	3329	3503	3725	4060	4699
30	9	1318	1725	1968	2142	2285	2412	2529	2639	2745	2850	2955	3061	3171	3285	3407	3540	3690	3867	4092	4429	5066
30	10	1565	2005	2264	2448	2599	2732	2854	2969	3079	3188	3296	3405	3517	3634	3758	3895	4046	4226	4452	4790	5422
30	11	1823	2293	2567	2760	2918	3056	3182	3301	3414	3526	3637	3748	3863	3982	4108	4246	4399	4579	4806	5143	5767
30	12	2091	2589	2876	3077	3240	3383	3513	3634	3751	3865	3977	4091	4207	4328	4455	4593	4749	4928	5154	5488	6101
30	13	2369	2893	3191	3398	3566	3713	3845	3970	4088	4205	4318	4433	4550	4672	4800	4938	5092	5271	5496	5825	6426
30	14	2657	3203	3511	3724	3896	4046	4181	4307	4427	4544	4659	4775	4892	5014	5142	5280	5433	5611	5832	6156	6741
30	15	2953	3520	3837	4055	4230	4382	4518	4646	4767	4884	5000	5116	5233	5354	5482	5618	5770	5945	6163	6480	7047
31	1	0003	0017	0035	0054	0074	0095	0118	0143	0169	0197	0228	0263	0301	0344	0393	0452	0522	0613	0739	0950	1423
31	2	0050	0120	0177	0229	0276	0321	0365	0410	0456	0513	0553	0606	0663	0725	0794	0873	0965	1081	1236	1486	2016
31	3	0149	0278	0373	0449	0516	0579	0639	0699	0758	0819	0881	0947	1016	1091	1173	1265	1373	1504	1678	1953	2519
31	4	0285	0469	0594	0691	0775	0852	0925	0996	1067	1138	1210	1286	1365	1449	1541	1644	1762	1905	2093	2386	2976
31	5	0447	0681	0833	0948	1047	1136	1219	1300	1380	1459	1540	1623	1709	1802	1901	2012	2139	2292	2490	2796	3403
31	6	0630	0909	1086	1217	1327	1426	1519	1608	1695	1782	1869	1959	2052	2150	2257	2374	2507	2667	2874	3190	3808
31	7	0830	1150	1348	1493	1615	1723	1824	1919	2013	2105	2199	2294	2392	2496	2607	2730	2869	3034	3247	3570	4195
31	8	1044	1402	1620	1777	1909	2024	2132	2234	2332	2430	2528	2628	2731	2839	2954	3081	3224	3394	3611	3939	4567
31	9	1270	1663	1899	2067	2206	2329	2443	2550	2654	2756	2858	2961	3068	3179	3298	3428	3575	3748	3968	4299	4927
31	10	1508	1933	2184	2363	2509	2638	2757	2869	2976	3082	3187	3294	3403	3518	3639	3772	3921	4096	4319	4651	5274
31	11	1755	2211	2476	2663	2816	2951	3074	3189	3300	3409	3517	3626	3738	3854	3978	4112	4263	4440	4663	4994	5612
31	12	2013	2495	2773	2968	3127	3266	3392	3511	3625	3736	3846	3957	4071	4189	4314	4450	4601	4784	5001	5331	5939
31	13	2280	2787	3076	3278	3442	3584	3714	3835	3951	4064	4176	4288	4403	4522	4648	4784	4936	5113	5334	5661	6258
31	14	2555	3085	3384	3592	3759	3905	4037	4160	4278	4393	4506	4619	4734	4854	4980	5116	5267	5443	5662	5984	6567
31	15	2839	3389	3697	3909	4081	4229	4363	4487	4606	4722	4835	4949	5065	5184	5310	5445	5595	5769	5985	6301	6868

* Table computed by Dr. Edward L. Spitznagel. Reproduced with his permission.

THREE

INTEREST TABLES*

Definitions:

$$(a/f) = i/[(1 + i)^n - 1]$$

$$(a/p) = i(1 + i)^n/[(1 + i)^n - 1]$$

$$(a/g) = [1 - n(a/f)]/i$$

$$(p/f) = 1/(1 + i)^n$$

$$(p/a) = 1/(a/p)$$

$$(p/g) = (p/a)(a/g)$$

$$(f/p) = 1/(p/f)$$

$$(f/a) = 1/(a/f)$$

$$(f/g) = (f/a)(a/g)$$

---

* Tables A3.1 through A3.24 were prepared for the author by Victor R. Conocchioli.

Table A3.1 $i = 1\%$

n	(a/f)	(a/p)	(a/g)	(p/f)	(p/a)	(p/g)	(f/p)	(f/a)	(f/g)
1	1.000	1.010	0.0	0.990	0.99	0.0	1.01	1.00	0.0
2	0.498	0.508	0.49	0.980	1.97	0.97	1.02	2.01	0.99
3	0.330	0.340	0.99	0.971	2.94	2.90	1.03	3.03	2.99
4	0.246	0.256	1.48	0.961	3.90	5.77	1.04	4.06	6.01
5	0.196	0.206	1.97	0.951	4.85	9.57	1.05	5.10	10.06
6	0.163	0.173	2.46	0.942	5.80	14.28	1.06	6.15	15.16
7	0.139	0.149	2.95	0.933	6.73	19.87	1.07	7.21	21.30
8	0.121	0.131	3.44	0.923	7.65	26.32	1.08	8.29	28.50
9	0.107	0.117	3.93	0.914	8.57	33.63	1.09	9.37	36.78
10	0.096	0.106	4.41	0.905	9.47	41.77	1.10	10.46	46.14
11	0.086	0.096	4.89	0.896	10.37	50.73	1.12	11.57	56.60
12	0.079	0.089	5.37	0.887	11.25	60.48	1.13	12.68	68.15
13	0.072	0.082	5.85	0.879	12.13	71.02	1.14	13.81	80.83
14	0.067	0.077	6.33	0.870	13.00	82.32	1.15	14.95	94.63
15	0.062	0.072	6.81	0.861	13.86	94.37	1.16	16.10	109.56
16	0.058	0.068	7.28	0.853	14.72	107.16	1.17	17.26	125.65
17	0.054	0.064	7.75	0.844	15.56	120.66	1.18	18.43	142.90
18	0.051	0.061	8.22	0.836	16.40	134.87	1.20	19.61	161.32
19	0.048	0.058	8.69	0.828	17.22	149.76	1.21	20.81	180.93
20	0.045	0.055	9.16	0.820	18.04	165.33	1.22	22.02	201.73
21	0.043	0.053	9.63	0.811	18.86	181.55	1.23	23.24	223.73
22	0.041	0.051	10.09	0.803	19.66	198.41	1.24	24.47	246.96
23	0.039	0.049	10.56	0.795	20.45	215.90	1.26	25.71	271.42
24	0.037	0.047	11.02	0.788	21.24	234.01	1.27	26.97	297.12
25	0.035	0.045	11.48	0.780	22.02	252.72	1.28	28.24	324.09
26	0.034	0.044	11.93	0.772	22.79	272.01	1.30	29.52	352.32
27	0.032	0.042	12.39	0.764	23.56	291.88	1.31	30.82	381.84
28	0.031	0.041	12.84	0.757	24.31	312.31	1.32	32.13	412.64
29	0.030	0.040	13.30	0.749	25.06	333.28	1.33	33.45	444.76
30	0.029	0.039	13.75	0.742	25.81	354.79	1.35	34.78	478.19
35	0.024	0.034	15.98	0.706	29.41	469.92	1.42	41.66	665.67
40	0.020	0.030	18.17	0.672	32.83	596.59	1.49	48.88	888.21
45	0.018	0.028	20.32	0.639	36.09	733.41	1.56	56.48	1147.60
50	0.016	0.026	22.43	0.608	39.19	879.09	1.64	64.46	1445.72
500	0.000	0.010	96.52	0.007	99.31	9585.41	144.72	14372.05	*********

Table A3.2 $i = 2\%$

n	(a/f)	(a/p)	(a/g)	(p/f)	(p/a)	(p/g)	(f/p)	(f/a)	(f/g)
1	1.000	1.020	0.00	0.980	0.98	0.00	1.02	1.00	0.00
2	0.495	0.515	0.49	0.961	1.94	0.96	1.04	2.02	1.00
3	0.327	0.347	0.99	0.942	2.88	2.84	1.06	3.06	3.02
4	0.243	0.263	1.47	0.924	3.81	5.61	1.08	4.12	6.07
5	0.192	0.212	1.96	0.906	4.71	9.23	1.10	5.20	10.20
6	0.159	0.179	2.44	0.888	5.60	13.67	1.13	6.31	15.40
7	0.135	0.155	2.92	0.871	6.47	18.89	1.15	7.43	21.70
8	0.117	0.137	3.39	0.853	7.33	24.87	1.17	8.58	29.14
9	0.103	0.123	3.87	0.837	8.16	31.56	1.20	9.75	37.72
10	0.091	0.111	4.34	0.820	8.98	38.94	1.22	10.95	47.47
11	0.082	0.102	4.80	0.804	9.79	46.98	1.24	12.17	58.42
12	0.075	0.095	5.26	0.788	10.58	55.66	1.27	13.41	70.59
13	0.068	0.088	5.72	0.773	11.35	64.93	1.29	14.68	84.00
14	0.063	0.083	6.18	0.758	12.11	74.78	1.32	15.97	98.67
15	0.058	0.078	6.63	0.743	12.85	85.18	1.35	17.29	114.65
16	0.054	0.074	7.08	0.728	13.58	96.11	1.37	18.64	131.94
17	0.050	0.070	7.52	0.714	14.29	107.54	1.40	20.01	150.57
18	0.047	0.067	7.97	0.700	14.99	119.44	1.43	21.41	170.58
19	0.044	0.064	8.41	0.686	15.68	131.79	1.46	22.84	191.99
20	0.041	0.061	8.84	0.673	16.35	144.58	1.49	24.30	214.83
21	0.039	0.059	9.27	0.660	17.01	157.77	1.52	25.78	239.13
22	0.037	0.057	9.70	0.647	17.66	171.35	1.55	27.30	264.91
23	0.035	0.055	10.13	0.634	18.29	185.30	1.58	28.84	292.20
24	0.033	0.053	10.55	0.622	18.91	199.60	1.61	30.42	321.05
25	0.031	0.051	10.97	0.610	19.52	214.23	1.64	32.03	351.46
26	0.030	0.050	11.39	0.598	20.12	229.17	1.67	33.67	383.49
27	0.028	0.048	11.80	0.586	20.71	244.40	1.71	35.34	417.16
28	0.027	0.047	12.21	0.574	21.28	259.91	1.74	37.05	452.50
29	0.026	0.046	12.62	0.563	21.84	275.68	1.78	38.79	489.55
30	0.025	0.045	13.02	0.552	22.40	291.68	1.81	40.57	528.34
35	0.020	0.040	14.99	0.500	25.00	374.85	2.00	49.99	749.64
40	0.017	0.037	16.89	0.453	27.36	461.95	2.21	60.40	1019.99
45	0.014	0.034	18.70	0.410	29.49	551.52	2.44	71.89	1344.50
50	0.012	0.032	20.44	0.372	31.42	642.32	2.69	84.58	1728.81
500	0.000	0.020	49.97	0.000	50.00	2498.62	19951.72	********	*********

Table A3.3 $i = 3\%$

n	(a/f)	(a/p)	(a/g)	(p/f)	(p/a)	(p/g)	(f/p)	(f/a)	(f/g)
1	1.000	1.030	0.00	0.971	0.97	0.00	1.03	1.00	0.00
2	0.493	0.523	0.49	0.943	1.91	0.94	1.06	2.03	1.00
3	0.324	0.354	0.98	0.915	2.83	2.77	1.09	3.09	3.03
4	0.239	0.269	1.46	0.888	3.72	5.44	1.13	4.18	6.12
5	0.188	0.218	1.94	0.863	4.58	8.89	1.16	5.31	10.30
6	0.155	0.185	2.41	0.837	5.42	13.07	1.19	6.47	15.61
7	0.131	0.161	2.88	0.813	6.23	17.95	1.23	7.66	22.08
8	0.112	0.142	3.34	0.789	7.02	23.48	1.27	8.89	29.74
9	0.098	0.128	3.80	0.766	7.79	29.61	1.30	10.16	38.63
10	0.087	0.117	4.26	0.744	8.53	36.31	1.34	11.46	48.79
11	0.078	0.108	4.70	0.722	9.25	43.53	1.38	12.81	60.26
12	0.070	0.100	5.15	0.701	9.95	51.24	1.43	14.19	73.06
13	0.064	0.094	5.59	0.681	10.63	59.42	1.47	15.62	87.25
14	0.059	0.089	6.02	0.661	11.30	68.01	1.51	17.09	102.87
15	0.054	0.084	6.45	0.642	11.94	77.00	1.56	18.60	119.96
16	0.050	0.080	6.87	0.623	12.56	86.34	1.60	20.16	138.56
17	0.046	0.076	7.29	0.605	13.17	96.02	1.65	21.76	158.71
18	0.043	0.073	7.71	0.587	13.75	106.01	1.70	23.41	180.47
19	0.040	0.070	8.12	0.570	14.32	116.27	1.75	25.12	203.89
20	0.037	0.067	8.52	0.554	14.88	126.79	1.81	26.87	229.00
21	0.035	0.065	8.92	0.538	15.41	137.54	1.86	28.68	255.87
22	0.033	0.063	9.32	0.522	15.94	148.50	1.92	30.54	284.55
23	0.031	0.061	9.71	0.507	16.44	159.65	1.97	32.45	315.08
24	0.029	0.059	10.10	0.492	16.94	170.97	2.03	34.43	347.53
25	0.027	0.057	10.48	0.478	17.41	182.43	2.09	36.46	381.96
26	0.026	0.056	10.85	0.464	17.88	194.02	2.16	38.55	418.42
27	0.025	0.055	11.23	0.450	18.33	205.72	2.22	40.71	456.97
28	0.023	0.053	11.59	0.437	18.76	217.53	2.29	42.93	497.68
29	0.022	0.052	11.96	0.424	19.19	229.41	2.36	45.22	540.61
30	0.021	0.051	12.31	0.412	19.60	241.35	2.43	47.57	585.83
35	0.017	0.047	14.04	0.355	21.49	301.62	2.81	60.46	848.71
40	0.013	0.043	15.65	0.307	23.11	361.74	3.26	75.40	1180.00
45	0.011	0.041	17.16	0.264	24.52	420.62	3.78	92.72	1590.61
50	0.009	0.039	18.56	0.228	25.73	477.47	4.38	112.79	2093.16
500	0.000	0.030	33.33	0.000	33.33	1111.10	********	********	*********

Table A3.4 $i = 4\%$

n	(a/f)	(a/p)	(a/g)	(p/f)	(p/a)	(p/g)	(f/p)	(f/a)	(f/g)
1	1.000	1.040	0.00	0.962	0.96	0.00	1.04	1.00	0.00
2	0.490	0.530	0.49	0.925	1.89	0.92	1.08	2.04	1.00
3	0.320	0.360	0.97	0.889	2.78	2.70	1.12	3.12	3.04
4	0.235	0.275	1.45	0.855	3.63	5.27	1.17	4.25	6.16
5	0.185	0.225	1.92	0.822	4.45	8.55	1.22	5.42	10.41
6	0.151	0.191	2.39	0.790	5.24	12.51	1.27	6.63	15.82
7	0.127	0.167	2.84	0.760	6.00	17.07	1.32	7.90	22.46
8	0.109	0.149	3.29	0.731	6.73	22.18	1.37	9.21	30.36
9	0.094	0.134	3.74	0.703	7.44	27.80	1.42	10.58	39.57
10	0.083	0.123	4.18	0.676	8.11	33.88	1.48	12.01	50.15
11	0.074	0.114	4.61	0.650	8.76	40.38	1.54	13.49	62.16
12	0.067	0.107	5.03	0.625	9.39	47.25	1.60	15.03	75.64
13	0.060	0.100	5.45	0.601	9.99	54.45	1.67	16.63	90.67
14	0.055	0.095	5.87	0.577	10.56	61.96	1.73	18.29	107.30
15	0.050	0.090	6.27	0.555	11.12	69.74	1.80	20.02	125.59
16	0.046	0.086	6.67	0.534	11.65	77.74	1.87	21.82	145.61
17	0.042	0.082	7.07	0.513	12.17	85.96	1.95	23.70	167.44
18	0.039	0.079	7.45	0.494	12.66	94.35	2.03	25.65	191.13
19	0.036	0.076	7.83	0.475	13.13	102.89	2.11	27.67	216.78
20	0.034	0.074	8.21	0.456	13.59	111.56	2.19	29.78	244.45
21	0.031	0.071	8.58	0.439	14.03	120.34	2.28	31.97	274.23
22	0.029	0.069	8.94	0.422	14.45	129.20	2.37	34.25	306.20
23	0.027	0.067	9.30	0.406	14.86	138.13	2.46	36.62	340.45
24	0.026	0.066	9.65	0.390	15.25	147.10	2.56	39.08	377.06
25	0.024	0.064	9.99	0.375	15.62	156.10	2.67	41.65	416.15
26	0.023	0.063	10.33	0.361	15.98	165.12	2.77	44.31	457.79
27	0.021	0.061	10.66	0.347	16.33	174.14	2.88	47.08	502.10
28	0.020	0.060	10.99	0.333	16.66	183.14	3.00	49.97	549.19
29	0.019	0.059	11.31	0.321	16.98	192.12	3.12	52.97	599.15
30	0.018	0.058	11.63	0.308	17.29	201.06	3.24	56.08	652.12
35	0.014	0.054	13.12	0.253	18.66	244.88	3.95	73.65	966.30
40	0.011	0.051	14.48	0.208	19.79	286.53	4.80	95.03	1375.63
45	0.008	0.048	15.70	0.171	20.72	325.40	5.84	121.03	1900.73
50	0.007	0.047	16.81	0.141	21.48	361.16	7.11	152.67	2566.67
500	0.000	0.040	25.00	0.000	25.00	625.00	********	********	*********

Table A3.5 $i = 5\%$

n	(a/f)	(a/p)	(a/g)	(p/f)	(p/a)	(p/g)	(f/p)	(f/a)	(f/g)
1	1.000	1.050	0.00	0.952	0.95	0.00	1.05	1.00	0.00
2	0.488	0.538	0.49	0.907	1.86	0.91	1.10	2.05	1.00
3	0.317	0.367	0.97	0.864	2.72	2.63	1.16	3.15	3.05
4	0.232	0.282	1.44	0.823	3.55	5.10	1.22	4.31	6.20
5	0.181	0.231	1.90	0.784	4.33	8.24	1.28	5.53	10.51
6	0.147	0.197	2.36	0.746	5.08	11.97	1.34	6.80	16.04
7	0.123	0.173	2.80	0.711	5.79	16.23	1.41	8.14	22.84
8	0.105	0.155	3.24	0.677	6.46	20.97	1.48	9.55	30.98
9	0.091	0.141	3.68	0.645	7.11	26.12	1.55	11.03	40.53
10	0.080	0.130	4.10	0.614	7.72	31.65	1.63	12.58	51.55
11	0.070	0.120	4.51	0.585	8.31	37.50	1.71	14.21	64.13
12	0.063	0.113	4.92	0.557	8.86	43.62	1.80	15.92	78.34
13	0.056	0.106	5.32	0.530	9.39	49.98	1.89	17.71	94.25
14	0.051	0.101	5.71	0.505	9.90	56.55	1.98	19.60	111.96
15	0.046	0.096	6.10	0.481	10.38	63.28	2.08	21.58	131.56
16	0.042	0.092	6.47	0.458	10.84	70.16	2.18	23.66	153.14
17	0.039	0.089	6.84	0.436	11.27	77.14	2.29	25.84	176.80
18	0.036	0.086	7.20	0.416	11.69	84.20	2.41	28.13	202.63
19	0.033	0.083	7.56	0.396	12.09	91.32	2.53	30.54	230.77
20	0.030	0.080	7.90	0.377	12.46	98.48	2.65	33.07	261.30
21	0.028	0.078	8.24	0.359	12.82	105.66	2.79	35.72	294.37
22	0.026	0.076	8.57	0.342	13.16	112.84	2.93	38.50	330.08
23	0.024	0.074	8.90	0.326	13.49	120.00	3.07	41.43	368.59
24	0.022	0.072	9.21	0.310	13.80	127.14	3.23	44.50	410.02
25	0.021	0.071	9.52	0.295	14.09	134.22	3.39	47.73	454.52
26	0.020	0.070	9.83	0.281	14.38	141.25	3.56	51.11	502.24
27	0.018	0.068	10.12	0.268	14.64	148.22	3.73	54.67	553.35
28	0.017	0.067	10.41	0.255	14.90	155.11	3.92	58.40	608.02
29	0.016	0.066	10.69	0.243	15.14	161.91	4.12	62.32	666.42
30	0.015	0.065	10.97	0.231	15.37	168.62	4.32	66.44	728.74
35	0.011	0.061	12.25	0.181	16.37	200.58	5.52	90.32	1106.35
40	0.008	0.058	13.38	0.142	17.16	229.54	7.04	120.80	1615.91
45	0.006	0.056	14.36	0.111	17.77	255.31	8.98	159.69	2293.88
50	0.005	0.055	15.22	0.087	18.26	277.91	11.47	209.34	3186.79
500	0.000	0.050	20.00	0.000	20.00	400.00	********	********	*********

Table A3.6 $i = 6\%$

n	(a/f)	(a/p)	(a/g)	(p/f)	(p/a)	(p/g)	(f/p)	(f/a)	(f/g)
1	1.000	1.060	0.00	0.943	0.94	0.00	1.06	1.00	0.00
2	0.485	0.545	0.49	0.890	1.83	0.89	1.12	2.06	1.00
3	0.314	0.374	0.96	0.840	2.67	2.57	1.19	3.18	3.06
4	0.229	0.289	1.43	0.792	3.47	4.94	1.26	4.37	6.24
5	0.177	0.237	1.88	0.747	4.21	7.93	1.34	5.64	10.62
6	0.143	0.203	2.33	0.705	4.92	11.46	1.42	6.98	16.25
7	0.119	0.179	2.77	0.665	5.58	15.45	1.50	8.39	23.23
8	0.101	0.161	3.20	0.627	6.21	19.84	1.59	9.90	31.62
9	0.087	0.147	3.61	0.592	6.80	24.58	1.69	11.49	41.52
10	0.076	0.136	4.02	0.558	7.36	29.60	1.79	13.18	53.01
11	0.067	0.127	4.42	0.527	7.89	34.87	1.90	14.97	66.19
12	0.059	0.119	4.81	0.497	8.38	40.34	2.01	16.87	81.16
13	0.053	0.113	5.19	0.469	8.85	45.96	2.13	18.88	98.03
14	0.048	0.108	5.56	0.442	9.29	51.71	2.26	21.01	116.91
15	0.043	0.103	5.93	0.417	9.71	57.55	2.40	23.28	137.93
16	0.039	0.099	6.28	0.394	10.11	63.46	2.54	25.67	161.20
17	0.035	0.095	6.62	0.371	10.48	69.40	2.69	28.21	186.87
18	0.032	0.092	6.96	0.350	10.83	75.36	2.85	30.91	215.09
19	0.030	0.090	7.29	0.331	11.16	81.30	3.03	33.76	245.99
20	0.027	0.087	7.60	0.312	11.47	87.23	3.21	36.79	279.75
21	0.025	0.085	7.91	0.294	11.76	93.11	3.40	39.99	316.53
22	0.023	0.083	8.22	0.278	12.04	98.94	3.60	43.39	356.53
23	0.021	0.081	8.51	0.262	12.30	104.70	3.82	47.00	399.92
24	0.020	0.080	8.79	0.247	12.55	110.38	4.05	50.81	446.91
25	0.018	0.078	9.07	0.233	12.78	115.97	4.29	54.86	497.73
26	0.017	0.077	9.34	0.220	13.00	121.47	4.55	59.16	552.59
27	0.016	0.076	9.60	0.207	13.21	126.86	4.82	63.70	611.74
28	0.015	0.075	9.86	0.196	13.41	132.14	5.11	68.53	675.45
29	0.014	0.074	10.10	0.185	13.59	137.31	5.42	73.64	743.97
30	0.013	0.073	10.34	0.174	13.76	142.36	5.74	79.06	817.61
35	0.009	0.069	11.43	0.130	14.50	165.74	7.69	111.43	1273.87
40	0.006	0.066	12.36	0.097	15.05	185.95	10.29	154.76	1912.64
45	0.005	0.065	13.14	0.073	15.46	203.11	13.76	212.74	2795.64
50	0.003	0.063	13.80	0.054	15.76	217.46	18.42	290.33	4005.46
500	0.000	0.060	16.67	0.000	16.67	277.78	********	********	*********

Table A3.7 $i = 7\%$

n	(a/f)	(a/p)	(a/g)	(p/f)	(p/a)	(p/g)	(f/p)	(f/a)	(f/g)
1	1.000	1.070	0.00	0.935	0.93	0.00	1.07	1.00	0.00
2	0.483	0.553	0.48	0.873	1.81	0.87	1.14	2.07	1.00
3	0.311	0.381	0.95	0.816	2.62	2.51	1.23	3.21	3.07
4	0.225	0.295	1.42	0.763	3.39	4.79	1.31	4.44	6.28
5	0.174	0.244	1.86	0.713	4.10	7.65	1.40	5.75	10.72
6	0.140	0.210	2.30	0.666	4.77	10.98	1.50	7.15	16.48
7	0.116	0.186	2.73	0.623	5.39	14.71	1.61	8.65	23.63
8	0.097	0.167	3.15	0.582	5.97	18.79	1.72	10.26	32.28
9	0.083	0.153	3.55	0.544	6.52	23.14	1.84	11.98	42.54
10	0.072	0.142	3.95	0.508	7.02	27.72	1.97	13.82	54.52
11	0.063	0.133	4.33	0.475	7.50	32.47	2.10	15.78	68.34
12	0.056	0.126	4.70	0.444	7.94	37.35	2.25	17.89	84.12
13	0.050	0.120	5.06	0.415	8.36	42.33	2.41	20.14	102.01
14	0.044	0.114	5.42	0.388	8.75	47.37	2.58	22.55	122.15
15	0.040	0.110	5.76	0.362	9.11	52.45	2.76	25.13	144.70
16	0.036	0.106	6.09	0.339	9.45	57.53	2.95	27.89	169.83
17	0.032	0.102	6.41	0.317	9.76	62.59	3.16	30.84	197.71
18	0.029	0.099	6.72	0.296	10.06	67.62	3.38	34.00	228.55
19	0.027	0.097	7.02	0.277	10.34	72.60	3.62	37.38	262.55
20	0.024	0.094	7.32	0.258	10.59	77.51	3.87	41.00	299.93
21	0.022	0.092	7.60	0.242	10.84	82.34	4.14	44.86	340.93
22	0.020	0.090	7.87	0.226	11.06	87.08	4.43	49.01	385.79
23	0.019	0.089	8.14	0.211	11.27	91.72	4.74	53.44	434.80
24	0.017	0.087	8.39	0.197	11.47	96.25	5.07	58.18	488.23
25	0.016	0.086	8.64	0.184	11.65	100.68	5.43	63.25	546.41
26	0.015	0.085	8.88	0.172	11.83	104.98	5.81	68.68	609.66
27	0.013	0.083	9.11	0.161	11.99	109.17	6.21	74.48	678.33
28	0.012	0.082	9.33	0.150	12.14	113.23	6.65	80.70	752.81
29	0.011	0.081	9.54	0.141	12.28	117.16	7.11	87.35	833.51
30	0.011	0.081	9.75	0.131	12.41	120.97	7.61	94.46	920.86
35	0.007	0.077	10.67	0.094	12.95	138.13	10.68	138.24	1474.79
40	0.005	0.075	11.42	0.067	13.33	152.29	14.97	199.63	2280.47
45	0.003	0.073	12.04	0.048	13.61	163.76	21.00	285.75	3439.23
50	0.002	0.072	12.53	0.034	13.80	172.91	29.46	406.52	5093.19
500	0.000	0.070	14.29	0.000	14.29	204.08	********	********	*********

Table A3.8 $i = 8\%$

n	(a/f)	(a/p)	(a/g)	(p/f)	(p/a)	(p/g)	(f/p)	(f/a)	(f/g)
1	1.000	1.080	0.0	0.926	0.93	0.00	1.08	1.00	0.0
2	0.481	0.561	0.48	0.857	1.78	0.86	1.17	2.08	1.00
3	0.308	0.388	0.95	0.794	2.58	2.44	1.26	3.25	3.08
4	0.222	0.302	1.40	0.735	3.31	4.65	1.36	4.51	6.33
5	0.170	0.250	1.85	0.681	3.99	7.37	1.47	5.87	10.83
6	0.136	0.216	2.28	0.630	4.62	10.52	1.59	7.34	16.70
7	0.112	0.192	2.69	0.583	5.21	14.02	1.71	8.92	24.03
8	0.094	0.174	3.10	0.540	5.75	17.81	1.85	10.64	32.96
9	0.080	0.160	3.49	0.500	6.25	21.81	2.00	12.49	43.59
10	0.069	0.149	3.87	0.463	6.71	25.98	2.16	14.49	56.08
11	0.060	0.140	4.24	0.429	7.14	30.27	2.33	16.65	70.57
12	0.053	0.133	4.60	0.397	7.54	34.63	2.52	18.98	87.21
13	0.047	0.127	4.94	0.368	7.90	39.05	2.72	21.50	106.19
14	0.041	0.121	5.27	0.340	8.24	43.47	2.94	24.21	127.69
15	0.037	0.117	5.59	0.315	8.56	47.89	3.17	27.15	151.90
16	0.033	0.113	5.90	0.292	8.85	52.26	3.43	30.32	179.05
17	0.030	0.110	6.20	0.270	9.12	56.59	3.70	33.75	209.38
18	0.027	0.107	6.49	0.250	9.37	60.84	4.00	37.45	243.13
19	0.024	0.104	6.77	0.232	9.60	65.01	4.32	41.45	280.58
20	0.022	0.102	7.04	0.215	9.82	69.09	4.66	45.76	322.02
21	0.020	0.100	7.29	0.199	10.02	73.06	5.03	50.42	367.78
22	0.018	0.098	7.54	0.184	10.20	76.93	5.44	55.46	418.21
23	0.016	0.096	7.78	0.170	10.37	80.67	5.87	60.89	473.66
24	0.015	0.095	8.01	0.158	10.53	84.30	6.34	66.76	534.56
25	0.014	0.094	8.23	0.146	10.67	87.80	6.85	73.11	601.32
26	0.013	0.093	8.44	0.135	10.81	91.18	7.40	79.95	674.43
27	0.011	0.091	8.64	0.125	10.94	94.44	7.99	87.35	754.38
28	0.010	0.090	8.83	0.116	11.05	97.57	8.63	95.34	841.73
29	0.010	0.090	9.01	0.107	11.16	100.57	9.32	103.97	937.07
30	0.009	0.089	9.19	0.099	11.26	103.46	10.06	113.28	1041.04
35	0.006	0.086	9.96	0.068	11.65	116.09	14.79	172.32	1716.45
40	0.004	0.084	10.57	0.046	11.92	126.04	21.72	259.06	2738.19
45	0.003	0.083	11.04	0.031	12.11	133.73	31.92	386.50	4268.79
50	0.002	0.082	11.41	0.021	12.23	139.59	46.90	573.77	6547.09
500	0.000	0.080	12.50	0.000	12.50	156.25	********	********	*********

Table A3.9 $i = 9\%$

n	(a/f)	(a/p)	(a/g)	(p/f)	(p/a)	(p/g)	(f/p)	(f/a)	(f/g)
1	1.000	1.090	0.00	0.917	0.92	0.00	1.09	1.00	0.00
2	0.478	0.568	0.48	0.842	1.76	0.84	1.19	2.09	1.00
3	0.305	0.395	0.94	0.772	2.53	2.39	1.30	3.28	3.09
4	0.219	0.309	1.39	0.708	3.24	4.51	1.41	4.57	6.37
5	0.167	0.257	1.83	0.650	3.89	7.11	1.54	5.98	10.94
6	0.133	0.223	2.25	0.596	4.49	10.09	1.68	7.52	16.93
7	0.109	0.199	2.66	0.547	5.03	13.37	1.83	9.20	24.45
8	0.091	0.181	3.05	0.502	5.53	16.89	1.99	11.03	33.65
9	0.077	0.167	3.43	0.460	6.00	20.57	2.17	13.02	44.68
10	0.066	0.156	3.80	0.422	6.42	24.37	2.37	15.19	57.70
11	0.057	0.147	4.15	0.388	6.81	28.25	2.58	17.56	72.89
12	0.050	0.140	4.49	0.356	7.16	32.16	2.81	20.14	90.45
13	0.044	0.134	4.82	0.326	7.49	36.07	3.07	22.95	110.59
14	0.038	0.128	5.13	0.299	7.79	39.96	3.34	26.02	133.54
15	0.034	0.124	5.43	0.275	8.06	43.81	3.64	29.36	159.56
16	0.030	0.120	5.72	0.252	8.31	47.58	3.97	33.00	188.92
17	0.027	0.117	6.00	0.231	8.54	51.28	4.33	36.97	221.92
18	0.024	0.114	6.27	0.212	8.76	54.89	4.72	41.30	258.90
19	0.022	0.112	6.52	0.194	8.95	58.39	5.14	46.02	300.20
20	0.020	0.110	6.77	0.178	9.13	61.78	5.60	51.16	346.21
21	0.018	0.108	7.00	0.164	9.29	65.05	6.11	56.76	397.37
22	0.016	0.106	7.22	0.150	9.44	68.20	6.66	62.87	454.14
23	0.014	0.104	7.44	0.138	9.58	71.24	7.26	69.53	517.01
24	0.013	0.103	7.64	0.126	9.71	74.14	7.91	76.79	586.54
25	0.012	0.102	7.83	0.116	9.82	76.93	8.62	84.70	663.33
26	0.011	0.101	8.02	0.106	9.93	79.59	9.40	93.32	748.02
27	0.010	0.100	8.19	0.098	10.03	82.12	10.24	102.72	841.35
28	0.009	0.099	8.36	0.090	10.12	84.54	11.17	112.97	944.07
29	0.008	0.098	8.52	0.082	10.20	86.84	12.17	124.13	1057.03
30	0.007	0.097	8.67	0.075	10.27	89.03	13.27	136.30	1181.16
35	0.005	0.095	9.31	0.049	10.57	98.36	20.41	215.71	2007.84
40	0.003	0.093	9.80	0.032	10.76	105.38	31.41	337.87	3309.70
45	0.002	0.092	10.16	0.021	10.88	110.56	48.33	525.84	5342.69
50	0.001	0.091	10.43	0.013	10.96	114.32	74.35	815.05	8500.62
500	0.000	0.090	11.11	0.000	11.11	123.46	********	********	********

Table A3.10 $i = 10\%$

n	(a/f)	(a/p)	(a/g)	(p/f)	(p/a)	(p/g)	(f/p)	(f/a)	(f/g)
1	1.000	1.100	0.00	0.909	0.91	0.00	1.10	1.00	0.00
2	0.476	0.576	0.48	0.826	1.74	0.83	1.21	2.10	1.00
3	0.302	0.402	0.94	0.751	2.49	2.33	1.33	3.31	3.10
4	0.215	0.315	1.38	0.683	3.17	4.38	1.46	4.64	6.41
5	0.164	0.264	1.81	0.621	3.79	6.86	1.61	6.11	11.05
6	0.130	0.230	2.22	0.564	4.36	9.68	1.77	7.72	17.16
7	0.105	0.205	2.62	0.513	4.87	12.76	1.95	9.49	24.87
8	0.087	0.187	3.00	0.467	5.33	16.03	2.14	11.44	34.36
9	0.074	0.174	3.37	0.424	5.76	19.42	2.36	13.58	45.79
10	0.063	0.163	3.73	0.386	6.14	22.89	2.59	15.94	59.37
11	0.054	0.154	4.06	0.350	6.50	26.40	2.85	18.53	75.31
12	0.047	0.147	4.39	0.319	6.81	29.90	3.14	21.38	93.84
13	0.041	0.141	4.70	0.290	7.10	33.38	3.45	24.52	115.22
14	0.036	0.136	5.00	0.263	7.37	36.80	3.80	27.97	139.75
15	0.031	0.131	5.28	0.239	7.61	40.15	4.18	31.77	167.72
16	0.028	0.128	5.55	0.218	7.82	43.42	4.59	35.95	199.49
17	0.025	0.125	5.81	0.198	8.02	46.58	5.05	40.54	235.44
18	0.022	0.122	6.05	0.180	8.20	49.64	5.56	45.60	275.99
19	0.020	0.120	6.29	0.164	8.36	52.58	6.12	51.16	321.59
20	0.017	0.117	6.51	0.149	8.51	55.41	6.73	57.27	372.74
21	0.016	0.116	6.72	0.135	8.65	58.11	7.40	64.00	430.02
22	0.014	0.114	6.92	0.123	8.77	60.69	8.14	71.40	494.02
23	0.013	0.113	7.11	0.112	8.88	63.15	8.95	79.54	565.42
24	0.011	0.111	7.29	0.102	8.98	65.48	9.85	88.50	644.96
25	0.010	0.110	7.46	0.092	9.08	67.70	10.83	98.35	733.46
26	0.009	0.109	7.62	0.084	9.16	69.79	11.92	109.18	831.80
27	0.008	0.108	7.77	0.076	9.24	71.78	13.11	121.10	940.98
28	0.007	0.107	7.91	0.069	9.31	73.65	14.42	134.21	1062.08
29	0.007	0.107	8.05	0.063	9.37	75.41	15.86	148.63	1196.29
30	0.006	0.106	8.18	0.057	9.43	77.08	17.45	164.49	1344.92
35	0.004	0.104	8.71	0.036	9.64	83.99	28.10	271.02	2360.20
40	0.002	0.102	9.10	0.022	9.78	88.95	45.26	442.58	4025.84
45	0.001	0.101	9.37	0.014	9.86	92.45	72.89	718.89	6738.89
50	0.001	0.101	9.57	0.009	9.91	94.89	117.39	1163.88	11138.80
500	0.000	0.100	10.00	0.000	10.00	100.00	********	********	*********

Table A3.11 $i = 11\%$

n	(a/f)	(a/p)	(a/g)	(p/f)	(p/a)	(p/g)	(f/p)	(f/a)	(f/g)
1	1.000	1.110	0.00	0.901	0.90	0.00	1.11	1.00	0.00
2	0.474	0.584	0.47	0.812	1.71	0.81	1.23	2.11	1.00
3	0.299	0.409	0.93	0.731	2.44	2.27	1.37	3.34	3.11
4	0.212	0.322	1.37	0.659	3.10	4.25	1.52	4.71	6.45
5	0.161	0.271	1.79	0.593	3.70	6.62	1.69	6.23	11.16
6	0.126	0.236	2.20	0.535	4.23	9.30	1.87	7.91	17.39
7	0.102	0.212	2.59	0.482	4.71	12.19	2.08	9.78	25.30
8	0.084	0.194	2.96	0.434	5.15	15.22	2.30	11.86	35.09
9	0.071	0.181	3.31	0.391	5.54	18.35	2.56	14.16	46.94
10	0.060	0.170	3.65	0.352	5.89	21.52	2.84	16.72	61.11
11	0.051	0.161	3.98	0.317	6.21	24.69	3.15	19.56	77.83
12	0.044	0.154	4.29	0.286	6.49	27.84	3.50	22.71	97.39
13	0.038	0.148	4.58	0.258	6.75	30.93	3.88	26.21	120.10
14	0.033	0.143	4.86	0.232	6.98	33.94	4.31	30.09	146.32
15	0.029	0.139	5.13	0.209	7.19	36.87	4.78	34.41	176.41
16	0.026	0.136	5.38	0.188	7.38	39.70	5.31	39.19	210.82
17	0.022	0.132	5.62	0.170	7.55	42.41	5.90	44.50	250.01
18	0.020	0.130	5.84	0.153	7.70	45.01	6.54	50.40	294.51
19	0.018	0.128	6.06	0.138	7.84	47.49	7.26	56.94	344.90
20	0.016	0.126	6.26	0.124	7.96	49.84	8.06	64.20	401.84
21	0.014	0.124	6.45	0.112	8.08	52.08	8.95	72.26	466.04
22	0.012	0.122	6.63	0.101	8.18	54.19	9.93	81.21	538.31
23	0.011	0.121	6.80	0.091	8.27	56.19	11.03	91.15	619.52
24	0.010	0.120	6.96	0.082	8.35	58.07	12.24	102.17	710.67
25	0.009	0.119	7.10	0.074	8.42	59.83	13.59	114.41	812.84
26	0.008	0.118	7.24	0.066	8.49	61.49	15.08	128.00	927.25
27	0.007	0.117	7.38	0.060	8.55	63.04	16.74	143.08	1055.25
28	0.006	0.116	7.50	0.054	8.60	64.50	18.58	159.82	1198.33
29	0.006	0.116	7.61	0.048	8.65	65.85	20.62	178.40	1358.14
30	0.005	0.115	7.72	0.044	8.69	67.12	22.89	199.02	1536.54
35	0.003	0.113	8.16	0.026	8.86	72.25	38.57	341.59	2787.15
40	0.002	0.112	8.47	0.015	8.95	75.78	65.00	581.82	4925.64
45	0.001	0.111	8.68	0.009	9.01	78.16	109.53	986.63	8560.25
50	0.001	0.111	8.82	0.005	9.04	79.73	184.56	1668.75	14715.92
500	0.000	0.110	9.09	0.000	9.09	82.64	********	********	*********

Table A3.12 $i = 12\%$

n	(a/f)	(a/p)	(a/g)	(p/f)	(p/a)	(p/g)	(f/p)	(f/a)	(f/g)
1	1.000	1.120	0.0	0.893	0.89	0.00	1.12	1.00	0.0
2	0.472	0.592	0.47	0.797	1.69	0.80	1.25	2.12	1.00
3	0.296	0.416	0.92	0.712	2.40	2.22	1.40	3.37	3.12
4	0.209	0.329	1.36	0.636	3.04	4.13	1.57	4.78	6.49
5	0.157	0.277	1.77	0.567	3.60	6.40	1.76	6.35	11.27
6	0.123	0.243	2.17	0.507	4.11	8.93	1.97	8.12	17.63
7	0.099	0.219	2.55	0.452	4.56	11.64	2.21	10.09	25.74
8	0.081	0.201	2.91	0.404	4.97	14.47	2.48	12.30	35.83
9	0.068	0.188	3.26	0.361	5.33	17.36	2.77	14.78	48.13
10	0.057	0.177	3.58	0.322	5.65	20.25	3.11	17.55	62.91
11	0.048	0.168	3.90	0.287	5.94	23.13	3.48	20.65	80.45
12	0.041	0.161	4.19	0.257	6.19	25.95	3.90	24.13	101.11
13	0.036	0.156	4.47	0.229	6.42	28.70	4.36	28.03	125.24
14	0.031	0.151	4.73	0.205	6.63	31.36	4.89	32.39	153.27
15	0.027	0.147	4.98	0.183	6.81	33.92	5.47	37.28	185.66
16	0.023	0.143	5.21	0.163	6.97	36.37	6.13	42.75	222.94
17	0.020	0.140	5.44	0.146	7.12	38.70	6.87	48.88	265.70
18	0.018	0.138	5.64	0.130	7.25	40.91	7.69	55.75	314.58
19	0.016	0.136	5.84	0.116	7.37	43.00	8.61	63.44	370.33
20	0.014	0.134	6.02	0.104	7.47	44.97	9.65	72.05	433.77
21	0.012	0.132	6.19	0.093	7.56	46.82	10.80	81.70	505.82
22	0.011	0.131	6.35	0.083	7.64	48.55	12.10	92.50	587.52
23	0.010	0.130	6.50	0.074	7.72	50.18	13.55	104.60	680.02
24	0.008	0.128	6.64	0.066	7.78	51.69	15.18	118.16	784.63
25	0.007	0.127	6.77	0.059	7.84	53.10	17.00	133.33	902.78
26	0.007	0.127	6.89	0.053	7.90	54.42	19.04	150.33	1036.11
27	0.006	0.126	7.00	0.047	7.94	55.64	21.32	169.37	1186.45
28	0.005	0.125	7.11	0.042	7.98	56.77	23.88	190.70	1355.82
29	0.005	0.125	7.21	0.037	8.02	57.81	26.75	214.58	1546.52
30	0.004	0.124	7.30	0.033	8.06	58.78	29.96	241.33	1761.10
35	0.002	0.122	7.66	0.019	8.18	62.61	52.80	431.66	3305.52
40	0.001	0.121	7.90	0.011	8.24	65.12	93.05	767.09	6059.07
45	0.001	0.121	8.06	0.006	8.28	66.73	163.99	1358.22	10943.54
50	0.000	0.120	8.16	0.003	8.30	67.76	289.00	2400.01	19583.40
500	0.000	0.120	8.33	0.000	8.33	69.44	********	********	*********

Table A3.13 $i = 13\%$

n	(a/f)	(a/p)	(a/g)	(p/f)	(p/a)	(p/g)	(f/p)	(f/a)	(f/g)
1	1.000	1.130	0.00	0.885	0.88	0.00	1.13	1.00	0.00
2	0.469	0.599	0.47	0.783	1.67	0.78	1.28	2.13	1.00
3	0.294	0.424	0.92	0.693	2.36	2.17	1.44	3.41	3.13
4	0.206	0.336	1.35	0.613	2.97	4.01	1.63	4.85	6.54
5	0.154	0.284	1.76	0.543	3.52	6.18	1.84	6.48	11.39
6	0.120	0.250	2.15	0.480	4.00	8.58	2.08	8.32	17.87
7	0.096	0.226	2.52	0.425	4.42	11.13	2.35	10.40	26.19
8	0.078	0.208	2.87	0.376	4.80	13.77	2.66	12.76	36.59
9	0.065	0.195	3.20	0.333	5.13	16.43	3.00	15.42	49.35
10	0.054	0.184	3.52	0.295	5.43	19.08	3.39	18.42	64.77
11	0.046	0.176	3.81	0.261	5.69	21.69	3.84	21.81	83.19
12	0.039	0.169	4.09	0.231	5.92	24.22	4.33	25.65	105.00
13	0.033	0.163	4.36	0.204	6.12	26.67	4.90	29.98	130.65
14	0.029	0.159	4.61	0.181	6.30	29.02	5.53	34.88	160.63
15	0.025	0.155	4.84	0.160	6.46	31.26	6.25	40.42	195.52
16	0.021	0.151	5.06	0.141	6.60	33.38	7.07	46.67	235.93
17	0.019	0.149	5.26	0.125	6.73	35.39	7.99	53.74	282.60
18	0.016	0.146	5.45	0.111	6.84	37.27	9.02	61.72	336.34
19	0.014	0.144	5.63	0.098	6.94	39.04	10.20	70.75	398.06
20	0.012	0.142	5.79	0.087	7.02	40.69	11.52	80.95	468.81
21	0.011	0.141	5.95	0.077	7.10	42.22	13.02	92.47	549.76
22	0.009	0.139	6.09	0.068	7.17	43.65	14.71	105.49	642.23
23	0.008	0.138	6.22	0.060	7.23	44.97	16.63	120.20	747.71
24	0.007	0.137	6.34	0.053	7.28	46.20	18.79	136.83	867.91
25	0.006	0.136	6.46	0.047	7.33	47.33	21.23	155.62	1004.74
26	0.006	0.136	6.56	0.042	7.37	48.37	23.99	176.85	1160.36
27	0.005	0.135	6.66	0.037	7.41	49.33	27.11	200.84	1337.20
28	0.004	0.134	6.75	0.033	7.44	50.21	30.63	227.95	1538.04
29	0.004	0.134	6.83	0.029	7.47	51.02	34.62	258.58	1765.98
30	0.003	0.133	6.91	0.026	7.50	51.76	39.11	293.19	2024.56
35	0.002	0.132	7.20	0.014	7.59	54.61	72.07	546.67	3935.90
40	0.001	0.131	7.39	0.008	7.63	56.41	132.78	1013.67	7489.80
45	0.001	0.131	7.51	0.004	7.66	57.51	244.63	1874.10	14070.02
50	0.000	0.130	7.58	0.002	7.68	58.19	450.72	3459.38	26226.01
500	0.000	0.130	7.69	0.000	7.69	59.17	********	********	*********

Table A3.14 $i = 14\%$

n	(a/f)	(a/p)	(a/g)	(p/f)	(p/a)	(p/g)	(f/p)	(f/a)	(f/g)
1	1.000	1.140	0.00	0.877	0.88	0.00	1.14	1.00	0.00
2	0.467	0.607	0.47	0.769	1.65	0.77	1.30	2.14	1.00
3	0.291	0.431	0.91	0.675	2.32	2.12	1.48	3.44	3.14
4	0.203	0.343	1.34	0.592	2.91	3.90	1.69	4.92	6.58
5	0.151	0.291	1.74	0.519	3.43	5.97	1.93	6.61	11.50
6	0.117	0.257	2.12	0.456	3.89	8.25	2.19	8.54	18.11
7	0.093	0.233	2.48	0.400	4.29	10.65	2.50	10.73	26.65
8	0.076	0.216	2.82	0.351	4.64	13.10	2.85	13.23	37.38
9	0.062	0.202	3.15	0.308	4.95	15.56	3.25	16.09	50.61
10	0.052	0.192	3.45	0.270	5.22	17.99	3.71	19.34	66.69
11	0.043	0.183	3.73	0.237	5.45	20.36	4.23	23.04	86.03
12	0.037	0.177	4.00	0.208	5.66	22.64	4.82	27.27	109.08
13	0.031	0.171	4.25	0.182	5.84	24.82	5.49	32.09	136.35
14	0.027	0.167	4.48	0.160	6.00	26.90	6.26	37.58	168.43
15	0.023	0.163	4.70	0.140	6.14	28.86	7.14	43.84	206.01
16	0.020	0.160	4.90	0.123	6.27	30.71	8.14	50.98	249.86
17	0.017	0.157	5.09	0.108	6.37	32.43	9.28	59.12	300.84
18	0.015	0.155	5.26	0.095	6.47	34.04	10.58	68.39	359.95
19	0.013	0.153	5.42	0.083	6.55	35.53	12.06	78.97	428.35
20	0.011	0.151	5.57	0.073	6.62	36.91	13.74	91.02	507.31
21	0.010	0.150	5.71	0.064	6.69	38.19	15.67	104.77	598.34
22	0.008	0.148	5.84	0.056	6.74	39.37	17.86	120.43	703.10
23	0.007	0.147	5.95	0.049	6.79	40.45	20.36	138.30	823.54
24	0.006	0.146	6.06	0.043	6.84	41.44	23.21	158.66	961.83
25	0.005	0.145	6.16	0.038	6.87	42.34	26.46	181.87	1120.49
26	0.005	0.145	6.25	0.033	6.91	43.17	30.17	208.33	1302.36
27	0.004	0.144	6.33	0.029	6.94	43.93	34.39	238.50	1510.68
28	0.004	0.144	6.41	0.026	6.96	44.62	39.20	272.89	1749.18
29	0.003	0.143	6.48	0.022	6.98	45.24	44.69	312.09	2022.06
30	0.003	0.143	6.54	0.020	7.00	45.81	50.95	356.78	2334.15
35	0.001	0.141	6.78	0.010	7.07	47.95	98.10	693.56	4704.00
40	0.001	0.141	6.93	0.005	7.11	49.24	188.88	1342.00	9299.98
45	0.000	0.140	7.02	0.003	7.12	50.00	363.67	2590.50	18182.16
50	0.000	0.140	7.07	0.001	7.13	50.44	700.21	4994.38	35317.05
500	0.000	0.140	7.14	0.000	7.14	51.02	********	********	********

Table A3.15 $i = 15\%$

n	(a/f)	(a/p)	(a/g)	(p/f)	(p/a)	(p/g)	(f/p)	(f/a)	(f/g)
1	1.000	1.150	0.00	0.870	0.87	0.00	1.15	1.00	0.00
2	0.465	0.615	0.47	0.756	1.63	0.76	1.32	2.15	1.00
3	0.288	0.438	0.91	0.658	2.28	2.07	1.52	3.47	3.15
4	0.200	0.350	1.33	0.572	2.85	3.79	1.75	4.99	6.62
5	0.148	0.298	1.72	0.497	3.35	5.78	2.01	6.74	11.62
6	0.114	0.264	2.10	0.432	3.78	7.94	2.31	8.75	18.36
7	0.090	0.240	2.45	0.376	4.16	10.19	2.66	11.07	27.11
8	0.073	0.223	2.78	0.327	4.49	12.48	3.06	13.73	38.18
9	0.060	0.210	3.09	0.284	4.77	14.75	3.52	16.79	51.91
10	0.049	0.199	3.38	0.247	5.02	16.98	4.05	20.30	68.69
11	0.041	0.191	3.65	0.215	5.23	19.13	4.65	24.35	88.99
12	0.034	0.184	3.91	0.187	5.42	21.18	5.35	29.00	113.34
13	0.029	0.179	4.14	0.163	5.58	23.14	6.15	34.35	142.34
14	0.025	0.175	4.36	0.141	5.72	24.97	7.08	40.50	176.70
15	0.021	0.171	4.56	0.123	5.85	26.69	8.14	47.58	217.20
16	0.018	0.168	4.75	0.107	5.95	28.30	9.36	55.72	264.78
17	0.015	0.165	4.93	0.093	6.05	29.78	10.76	65.07	320.50
18	0.013	0.163	5.08	0.081	6.13	31.16	12.38	75.84	385.57
19	0.011	0.161	5.23	0.070	6.20	32.42	14.23	88.21	461.41
20	0.010	0.160	5.37	0.061	6.26	33.58	16.37	102.44	549.62
21	0.008	0.158	5.49	0.053	6.31	34.64	18.82	118.81	652.06
22	0.007	0.157	5.60	0.046	6.36	35.62	21.64	137.63	770.87
23	0.006	0.156	5.70	0.040	6.40	36.50	24.89	159.28	908.50
24	0.005	0.155	5.80	0.035	6.43	37.30	28.62	184.17	1067.78
25	0.005	0.155	5.88	0.030	6.46	38.03	32.92	212.79	1251.94
26	0.004	0.154	5.96	0.026	6.49	38.69	37.86	245.71	1464.73
27	0.004	0.154	6.03	0.023	6.51	39.29	43.53	283.57	1710.44
28	0.003	0.153	6.10	0.020	6.53	39.83	50.07	327.10	1994.01
29	0.003	0.153	6.15	0.017	6.55	40.31	57.57	377.17	2321.11
30	0.002	0.152	6.21	0.015	6.57	40.75	66.21	434.74	2698.28
35	0.001	0.151	6.40	0.008	6.62	42.36	133.17	881.16	5641.07
40	0.001	0.151	6.52	0.004	6.64	43.28	267.86	1779.07	11593.80
45	0.000	0.150	6.58	0.002	6.65	43.81	538.76	3585.08	23600.54
50	0.000	0.150	6.62	0.001	6.66	44.10	1083.64	7217.61	47784.08
500	0.000	0.150	6.67	0.000	6.67	44.44	********	********	*********

Table A3.16 $i = 17\%$

n	(a/f)	(a/p)	(a/g)	(p/f)	(p/a)	(p/g)	(f/p)	(f/a)	(f/g)
1	1.000	1.170	0.00	0.855	0.85	0.00	1.17	1.00	0.00
2	0.461	0.631	0.46	0.731	1.59	0.73	1.37	2.17	1.00
3	0.283	0.453	0.90	0.624	2.21	1.98	1.60	3.54	3.17
4	0.195	0.365	1.31	0.534	2.74	3.58	1.87	5.14	6.71
5	0.143	0.313	1.69	0.456	3.20	5.40	2.19	7.01	11.85
6	0.109	0.279	2.05	0.390	3.59	7.35	2.57	9.21	18.86
7	0.085	0.255	2.38	0.333	3.92	9.35	3.00	11.77	28.07
8	0.068	0.238	2.70	0.285	4.21	11.35	3.51	14.77	39.84
9	0.055	0.225	2.99	0.243	4.45	13.29	4.11	18.28	54.62
10	0.045	0.215	3.26	0.208	4.66	15.17	4.81	22.39	72.90
11	0.037	0.207	3.50	0.178	4.84	16.94	5.62	27.20	95.29
12	0.030	0.200	3.73	0.152	4.99	18.62	6.58	32.82	122.49
13	0.025	0.195	3.94	0.130	5.12	20.17	7.70	39.40	155.32
14	0.021	0.191	4.13	0.111	5.23	21.62	9.01	47.10	194.72
15	0.018	0.188	4.31	0.095	5.32	22.95	10.54	56.11	241.82
16	0.015	0.185	4.47	0.081	5.41	24.16	12.33	66.65	297.93
17	0.013	0.183	4.62	0.069	5.47	25.27	14.43	78.98	364.58
18	0.011	0.181	4.75	0.059	5.53	26.28	16.88	93.40	443.56
19	0.009	0.179	4.87	0.051	5.58	27.19	19.75	110.28	536.96
20	0.008	0.178	4.98	0.043	5.63	28.01	23.11	130.03	647.24
21	0.007	0.177	5.08	0.037	5.66	28.75	27.03	153.14	777.27
22	0.006	0.176	5.16	0.032	5.70	29.42	31.63	180.17	930.41
23	0.005	0.175	5.24	0.027	5.72	30.01	37.01	211.80	1110.57
24	0.004	0.174	5.31	0.023	5.75	30.54	43.30	248.80	1322.37
25	0.003	0.173	5.38	0.020	5.77	31.02	50.66	292.10	1571.17
26	0.003	0.173	5.44	0.017	5.78	31.44	59.27	342.76	1863.27
27	0.002	0.172	5.49	0.014	5.80	31.81	69.34	402.02	2206.03
28	0.002	0.172	5.53	0.012	5.81	32.15	81.13	471.37	2608.05
29	0.002	0.172	5.57	0.011	5.82	32.44	94.92	552.50	3079.41
30	0.002	0.172	5.61	0.009	5.83	32.70	111.06	647.42	3631.91
500	0.000	0.170	5.88	0.000	5.88	34.60	********	********	********

Table A3.17 $i = 20\%$

n	(a/f)	(a/p)	(a/g)	(p/f)	(p/a)	(p/g)	(f/p)	(f/a)	(f/g)
1	1.000	1.200	0.0	0.833	0.83	0.00	1.20	1.00	0.0
2	0.455	0.655	0.45	0.694	1.53	0.69	1.44	2.20	1.00
3	0.275	0.475	0.88	0.579	2.11	1.85	1.73	3.64	3.20
4	0.186	0.386	1.27	0.482	2.59	3.30	2.07	5.37	6.84
5	0.134	0.334	1.64	0.402	2.99	4.91	2.49	7.44	12.21
6	0.101	0.301	1.98	0.335	3.33	6.58	2.99	9.93	19.65
7	0.077	0.277	2.29	0.279	3.60	8.26	3.58	12.92	29.58
8	0.061	0.261	2.58	0.233	3.84	9.88	4.30	16.50	42.50
9	0.048	0.248	2.84	0.194	4.03	11.43	5.16	20.80	58.99
10	0.039	0.239	3.07	0.162	4.19	12.89	6.19	25.96	79.79
11	0.031	0.231	3.29	0.135	4.33	14.23	7.43	32.15	105.75
12	0.025	0.225	3.48	0.112	4.44	15.47	8.92	39.58	137.90
13	0.021	0.221	3.66	0.093	4.53	16.59	10.70	48.50	177.48
14	0.017	0.217	3.82	0.078	4.61	17.60	12.84	59.20	225.98
15	0.014	0.214	3.96	0.065	4.68	18.51	15.41	72.04	285.18
16	0.011	0.211	4.09	0.054	4.73	19.32	18.49	87.44	357.21
17	0.009	0.209	4.20	0.045	4.77	20.04	22.19	105.93	444.65
18	0.008	0.208	4.30	0.038	4.81	20.68	26.62	128.12	550.58
19	0.006	0.206	4.39	0.031	4.84	21.24	31.95	154.74	678.70
20	0.005	0.205	4.46	0.026	4.87	21.74	38.34	186.69	833.44
21	0.004	0.204	4.53	0.022	4.89	22.17	46.00	225.03	1020.13
22	0.004	0.204	4.59	0.018	4.91	22.55	55.21	271.03	1245.15
23	0.003	0.203	4.65	0.015	4.92	22.89	66.25	326.24	1516.18
24	0.003	0.203	4.69	0.013	4.94	23.18	79.50	392.48	1842.42
25	0.002	0.202	4.74	0.010	4.95	23.43	95.40	471.98	2234.90
26	0.002	0.202	4.77	0.009	4.96	23.65	114.48	567.38	2706.88
27	0.001	0.201	4.80	0.007	4.96	23.84	137.37	681.85	3274.26
28	0.001	0.201	4.83	0.006	4.97	24.00	164.84	819.22	3956.11
29	0.001	0.201	4.85	0.005	4.97	24.14	197.81	984.06	4775.32
30	0.001	0.201	4.87	0.004	4.98	24.26	237.38	1181.88	5759.39
500	0.000	0.200	5.00	0.000	5.00	25.00	********	********	*********

Table A3.18 $i = 23\%$

n	(a/f)	(a/p)	(a/g)	(p/f)	(p/a)	(p/g)	(f/p)	(f/a)	(f/g)
1	1.000	1.230	0.00	0.813	0.81	0.00	1.23	1.00	0.00
2	0.448	0.678	0.45	0.661	1.47	0.66	1.51	2.23	1.00
3	0.267	0.497	0.86	0.537	2.01	1.74	1.86	3.74	3.23
4	0.178	0.408	1.24	0.437	2.45	3.05	2.29	5.60	6.97
5	0.127	0.357	1.59	0.355	2.80	4.47	2.82	7.89	12.58
6	0.093	0.323	1.91	0.289	3.09	5.91	3.46	10.71	20.47
7	0.071	0.301	2.20	0.235	3.33	7.32	4.26	14.17	31.18
8	0.054	0.284	2.46	0.191	3.52	8.66	5.24	18.43	45.35
9	0.042	0.272	2.69	0.155	3.67	9.90	6.44	23.67	63.78
10	0.033	0.263	2.90	0.126	3.80	11.03	7.93	30.11	87.45
11	0.026	0.256	3.09	0.103	3.90	12.06	9.75	38.04	117.56
12	0.021	0.251	3.26	0.083	3.99	12.98	11.99	47.79	155.60
13	0.017	0.247	3.40	0.068	4.05	13.79	14.75	59.78	203.38
14	0.013	0.243	3.53	0.055	4.11	14.51	18.14	74.53	263.16
15	0.011	0.241	3.64	0.045	4.15	15.13	22.31	92.67	337.69
16	0.009	0.239	3.74	0.036	4.19	15.68	27.45	114.98	430.36
17	0.007	0.237	3.83	0.030	4.22	16.15	33.76	142.43	545.34
18	0.006	0.236	3.90	0.024	4.24	16.56	41.52	176.19	687.77
19	0.005	0.235	3.97	0.020	4.26	16.92	51.07	217.71	863.96
20	0.004	0.234	4.02	0.016	4.28	17.22	62.82	268.78	1081.67
21	0.003	0.233	4.07	0.013	4.29	17.48	77.27	331.60	1350.45
22	0.002	0.232	4.11	0.011	4.30	17.70	95.04	408.87	1682.05
23	0.002	0.232	4.15	0.009	4.31	17.89	116.90	503.91	2090.92
24	0.002	0.232	4.18	0.007	4.32	18.05	143.79	620.81	2594.84
25	0.001	0.231	4.21	0.006	4.32	18.18	176.86	764.60	3215.65
26	0.001	0.231	4.23	0.005	4.33	18.30	217.53	941.46	3980.24
27	0.001	0.231	4.25	0.004	4.33	18.39	267.57	1158.99	4921.70
28	0.001	0.231	4.26	0.003	4.33	18.48	329.11	1426.56	6080.69
29	0.001	0.231	4.28	0.002	4.34	18.55	404.80	1755.66	7507.24
30	0.000	0.230	4.29	0.002	4.34	18.60	497.91	2160.47	9262.90
500	0.000	0.230	4.35	0.000	4.35	18.90	********	********	*********

Table A3.19 $i = 25\%$

n	(a/f)	(a/p)	(a/g)	(p/f)	(p/a)	(p/g)	(f/p)	(f/a)	(f/g)
1	1.000	1.250	0.00	0.800	0.80	0.00	1.25	1.00	0.00
2	0.444	0.694	0.44	0.640	1.44	0.64	1.56	2.25	1.00
3	0.262	0.512	0.85	0.512	1.95	1.66	1.95	3.81	3.25
4	0.173	0.423	1.22	0.410	2.36	2.89	2.44	5.77	7.06
5	0.122	0.372	1.56	0.328	2.69	4.20	3.05	8.21	12.83
6	0.089	0.339	1.87	0.262	2.95	5.51	3.81	11.26	21.03
7	0.066	0.316	2.14	0.210	3.16	6.77	4.77	15.07	32.29
8	0.050	0.300	2.39	0.168	3.33	7.95	5.96	19.84	47.37
9	0.039	0.289	2.60	0.134	3.46	9.02	7.45	25.80	67.21
10	0.030	0.280	2.80	0.107	3.57	9.99	9.31	33.25	93.01
11	0.023	0.273	2.97	0.086	3.66	10.85	11.64	42.57	126.26
12	0.018	0.268	3.11	0.069	3.73	11.60	14.55	54.21	168.83
13	0.015	0.265	3.24	0.055	3.78	12.26	18.19	68.76	223.04
14	0.012	0.262	3.36	0.044	3.82	12.83	22.74	86.95	291.79
15	0.009	0.259	3.45	0.035	3.86	13.33	28.42	109.69	378.74
16	0.007	0.257	3.54	0.028	3.89	13.75	35.53	138.11	488.43
17	0.006	0.256	3.61	0.023	3.91	14.11	44.41	173.63	626.53
18	0.005	0.255	3.67	0.018	3.93	14.41	55.51	218.04	800.17
19	0.004	0.254	3.72	0.014	3.94	14.67	69.39	273.55	1018.21
20	0.003	0.253	3.77	0.012	3.95	14.89	86.73	342.94	1291.76
21	0.002	0.252	3.80	0.009	3.96	15.08	108.42	429.67	1634.70
22	0.002	0.252	3.84	0.007	3.97	15.23	135.52	538.09	2064.37
23	0.001	0.251	3.86	0.006	3.98	15.36	169.40	673.61	2602.46
24	0.001	0.251	3.89	0.005	3.98	15.47	211.75	843.02	3276.08
25	0.001	0.251	3.91	0.004	3.98	15.56	264.69	1054.77	4119.09
26	0.001	0.251	3.92	0.003	3.99	15.64	330.87	1319.46	5173.86
27	0.001	0.251	3.93	0.002	3.99	15.70	413.58	1650.33	6493.32
28	0.000	0.250	3.95	0.002	3.99	15.75	516.98	2063.91	8143.64
29	0.000	0.250	3.96	0.002	3.99	15.80	646.22	2580.88	10207.54
30	0.000	0.250	3.96	0.001	4.00	15.83	807.77	3227.10	12788.42
500	0.000	0.250	4.00	0.000	4.00	16.00	********	********	********

Table A3.20 $i = 30\%$

n	(a/f)	(a/p)	(a/g)	(p/f)	(p/a)	(p/g)	(f/p)	(f/a)	(f/g)
1	1.000	1.300	0.00	0.769	0.77	0.00	1.30	1.00	0.00
2	0.435	0.735	0.43	0.592	1.36	0.59	1.69	2.30	1.00
3	0.251	0.551	0.83	0.455	1.82	1.50	2.20	3.99	3.30
4	0.162	0.462	1.18	0.350	2.17	2.55	2.86	6.19	7.29
5	0.111	0.411	1.49	0.269	2.44	3.63	3.71	9.04	13.48
6	0.078	0.378	1.77	0.207	2.64	4.67	4.83	12.76	22.52
7	0.057	0.357	2.01	0.159	2.80	5.62	6.27	17.58	35.28
8	0.042	0.342	2.22	0.123	2.92	6.48	8.16	23.86	52.86
9	0.031	0.331	2.40	0.094	3.02	7.23	10.60	32.01	76.72
10	0.023	0.323	2.55	0.073	3.09	7.89	13.79	42.62	108.73
11	0.018	0.318	2.68	0.056	3.15	8.45	17.92	56.40	151.35
12	0.013	0.313	2.80	0.043	3.19	8.92	23.30	74.33	207.75
13	0.010	0.310	2.89	0.033	3.22	9.31	30.29	97.62	282.08
14	0.008	0.308	2.97	0.025	3.25	9.64	39.37	127.91	379.71
15	0.006	0.306	3.03	0.020	3.27	9.92	51.19	167.28	507.62
16	0.005	0.305	3.09	0.015	3.28	10.14	66.54	218.47	674.90
17	0.004	0.304	3.13	0.012	3.29	10.33	86.50	285.01	893.37
18	0.003	0.303	3.17	0.009	3.30	10.48	112.45	371.51	1178.38
19	0.002	0.302	3.20	0.007	3.31	10.60	146.19	483.97	1549.89
20	0.002	0.302	3.23	0.005	3.32	10.70	190.05	630.16	2033.86
21	0.001	0.301	3.25	0.004	3.32	10.78	247.06	820.21	2664.02
22	0.001	0.301	3.26	0.003	3.32	10.85	321.18	1067.27	3484.22
23	0.001	0.301	3.28	0.002	3.33	10.90	417.53	1388.44	4551.48
24	0.001	0.301	3.29	0.002	3.33	10.94	542.79	1805.98	5939.93
25	0.000	0.300	3.30	0.001	3.33	10.98	705.63	2348.77	7745.90
26	0.000	0.300	3.30	0.001	3.33	11.00	917.32	3054.40	10094.67
27	0.000	0.300	3.31	0.001	3.33	11.03	1192.51	3971.72	13149.06
28	0.000	0.300	3.32	0.001	3.33	11.04	1550.27	5164.23	17120.77
29	0.000	0.300	3.32	0.000	3.33	11.06	2015.35	6714.49	22284.99
30	0.000	0.300	3.32	0.000	3.33	11.07	2619.95	8729.83	28999.46
500	0.000	0.300	3.33	0.000	3.33	11.11	********	********	*********

Table A3.21 $i = 35\%$

n	(a/f)	(a/p)	(a/g)	(p/f)	(p/a)	(p/g)	(f/p)	(f/a)	(f/g)
1	1.000	1.350	0.00	0.741	0.74	0.00	1.35	1.00	0.00
2	0.426	0.776	0.43	0.549	1.29	0.55	1.82	2.35	1.00
3	0.240	0.590	0.80	0.406	1.70	1.36	2.46	4.17	3.35
4	0.151	0.501	1.13	0.301	2.00	2.26	3.32	6.63	7.52
5	0.100	0.450	1.42	0.223	2.22	3.16	4.48	9.95	14.16
6	0.069	0.419	1.67	0.165	2.39	3.98	6.05	14.44	24.11
7	0.049	0.399	1.88	0.122	2.51	4.72	8.17	20.49	38.55
8	0.035	0.385	2.06	0.091	2.60	5.35	11.03	28.66	59.04
9	0.025	0.375	2.21	0.067	2.67	5.89	14.89	39.70	87.70
10	0.018	0.368	2.33	0.050	2.72	6.34	20.11	54.59	127.40
11	0.013	0.363	2.44	0.037	2.75	6.70	27.14	74.70	181.99
12	0.010	0.360	2.52	0.027	2.78	7.00	36.64	101.84	256.69
13	0.007	0.357	2.59	0.020	2.80	7.25	49.47	138.48	358.53
14	0.005	0.355	2.64	0.015	2.81	7.44	66.78	187.95	497.01
15	0.004	0.354	2.69	0.011	2.83	7.60	90.16	254.74	684.96
16	0.003	0.353	2.72	0.008	2.83	7.72	121.71	344.89	939.70
17	0.002	0.352	2.75	0.006	2.84	7.82	164.31	466.61	1284.59
18	0.002	0.352	2.78	0.005	2.84	7.89	221.82	630.92	1751.20
19	0.001	0.351	2.79	0.003	2.85	7.95	299.46	852.74	2382.12
20	0.001	0.351	2.81	0.002	2.85	8.00	404.27	1152.20	3234.86
500	0.000	0.350	2.86	0.000	2.86	8.16	********	********	*********

Table A3.22 $i = 40\%$

n	(a/f)	(a/p)	(a/g)	(p/f)	(p/a)	(p/g)	(f/p)	(f/a)	(f/g)
1	1.000	1.400	0.0	0.714	0.71	0.00	1.40	1.00	0.0
2	0.417	0.817	0.42	0.510	1.22	0.51	1.96	2.40	1.00
3	0.229	0.629	0.78	0.364	1.59	1.24	2.74	4.36	3.40
4	0.141	0.541	1.09	0.260	1.85	2.02	3.84	7.10	7.76
5	0.091	0.491	1.36	0.186	2.04	2.76	5.38	10.95	14.86
6	0.061	0.461	1.58	0.133	2.17	3.43	7.53	16.32	25.81
7	0.042	0.442	1.77	0.095	2.26	4.00	10.54	23.85	42.13
8	0.029	0.429	1.92	0.068	2.33	4.47	14.76	34.39	65.99
9	0.020	0.420	2.04	0.048	2.38	4.86	20.66	49.15	100.38
10	0.014	0.414	2.14	0.035	2.41	5.17	28.93	69.81	149.53
11	0.010	0.410	2.22	0.025	2.44	5.42	40.50	98.74	219.35
12	0.007	0.407	2.28	0.018	2.46	5.61	56.69	139.23	318.09
13	0.005	0.405	2.33	0.013	2.47	5.76	79.37	195.93	457.32
14	0.004	0.404	2.37	0.009	2.48	5.88	111.12	275.30	653.25
15	0.003	0.403	2.40	0.006	2.48	5.97	155.57	386.42	928.55
16	0.002	0.402	2.43	0.005	2.49	6.04	217.79	541.99	1314.97
17	0.001	0.401	2.44	0.003	2.49	6.09	304.91	759.78	1856.95
18	0.001	0.401	2.46	0.002	2.49	6.13	426.88	1064.69	2616.73
19	0.001	0.401	2.47	0.002	2.50	6.16	597.63	1491.57	3681.42
20	0.000	0.400	2.48	0.001	2.50	6.18	836.68	2089.19	5172.99
500	0.000	0.400	2.50	0.000	2.50	6.25	********	********	*********

Table A3.23 $i = 50\%$

n	(a/f)	(a/p)	(a/g)	(p/f)	(p/a)	(p/g)	(f/p)	(f/a)	(f/g)
1	1.000	1.500	0.00	0.667	0.67	0.00	1.50	1.00	0.00
2	0.400	0.900	0.40	0.444	1.11	0.44	2.25	2.50	1.00
3	0.211	0.711	0.74	0.296	1.41	1.04	3.37	4.75	3.50
4	0.123	0.623	1.02	0.198	1.60	1.63	5.06	8.12	8.25
5	0.076	0.576	1.24	0.132	1.74	2.16	7.59	13.19	16.37
6	0.048	0.548	1.42	0.088	1.82	2.60	11.39	20.78	29.56
7	0.031	0.531	1.56	0.059	1.88	2.95	17.09	32.17	50.34
8	0.020	0.520	1.68	0.039	1.92	3.22	25.63	49.26	82.52
9	0.013	0.513	1.76	0.026	1.95	3.43	38.44	74.89	131.77
10	0.009	0.509	1.82	0.017	1.97	3.58	57.66	113.33	206.66
11	0.006	0.506	1.87	0.012	1.98	3.70	86.50	170.99	319.99
12	0.004	0.504	1.91	0.008	1.98	3.78	129.75	257.49	490.98
13	0.003	0.503	1.93	0.005	1.99	3.85	194.62	387.24	748.47
14	0.002	0.502	1.95	0.003	1.99	3.89	291.93	581.85	1135.71
15	0.001	0.501	1.97	0.002	2.00	3.92	437.89	873.78	1717.56
16	0.001	0.501	1.98	0.002	2.00	3.95	656.83	1311.67	2591.34
17	0.001	0.501	1.98	0.001	2.00	3.96	985.25	1968.50	3903.01
18	0.000	0.500	1.99	0.001	2.00	3.97	1477.87	2953.75	5871.50
19	0.000	0.500	1.99	0.000	2.00	3.98	2216.81	4431.62	8825.25
20	0.000	0.500	1.99	0.000	2.00	3.99	3325.21	6648.43	13256.87
500	0.000	0.500	2.00	0.000	2.00	4.00	———	———	———

INDEX

Accuracy, 83
Adjusted tableau, 288
Air pollution control for St. Louis, 322–326
Algebra, matrix, 278–308
Analysis of variance:
  and hypothesis testing, 110
  orthogonal design, 111
  with replication, 111
  two-variable, 109
  (*See also* Hypothesis tests)
Annual equivalent, 208
  (*See also* Benefit-cost)
Annuity, 186
Artificial variable, 275, 288
Autocorrelation, 156

Basis vector, 279, 308
Bayes, Thomas, 26
  and confidence intervals, 129, 131, 134, 135
  and hypothesis testing, 135, 169–171
  inference, 168–171
Bayes' rule, 27, 28, 119–120, 350, 393
Benefit-cost:
  cost-effectiveness, 179, 229
  multiattribute, 236–241
  (*See also* Sensitivity analysis)
  net present equivalent, 185, 205, 218
  rate of return, 201, 202, 211–215
    internal, 201, 202
    marginal, 201, 202
    return on investment, 201, 202
  ratio, 201, 202, 208–211
    marginal, 201, 202, 218

Benefit-cost:
  ratio: overall, 201, 202, 218
  (*See also* Decision rule; Information, value
    of; State of nature; Utility)
Benefits:
  economic, 202, 205
  equity, 241–243
  multifaceted, 236–241
  nonmonetary, 235, 236
  (*See also* Benefit-cost)
Bernoulli, J., 18
Bernoulli model, 65, 76, 78, 125
Bernoulli trials, 65, 66
Beta model, 75, 76, 122
Big M method, 287–295
Binomial coefficients, 66
Binomial model, 65, 76, 78
Boffey, Philip, 29
Break-even analysis, 228–231
Breast cancer detection, 385–398

Central limit theorem, 70, 78, 94
Certainty equivalent, 368
Chi-square model, 72, 76
Coefficient of determination, 146
Computer analysis packages, 113, 117, 317
Conditional probability, 25, 80
Confidence interval, 83
  (*See also* Estimation, interval, confidence)
Confidence level, 83
Consumer surplus, 203
Contingency analysis, 226–228

Contingency tables and cross-tabulated data, 108, 167
(*See also* Goodness of fit, chi-square)
Convex sets, 272−273
Corner point theorem, 273
Correlation:
covariance, 139
(*See also* Regression)
Correlation coefficient, 139
of determination, 146, 150
estimation of, 138−142
multiple, 150
partial, 150
significance of, 147−148
Cost, 205
capital, 209
(*See also* Benefit-cost; Net present equivalent)
Cost-effectiveness, 179, 221, 229, 249
Covariance, 139
Critical value, 95
Cumulative distribution function (CDF), 58−61
fractiles of, 122
(*See also* Probability)

Dantzig, George B., 274
Data summary:
histogram, 42−43
(*See also* Statistic)
Decision analysis (*see* Benefit-cost; Decision rule; Information, value of; State of nature; Utility)
Decision rule, 220, 335−337
expected utility, 337
expected value, 336
maximin, 335
minimax, 335
most likely outcome, 336
(*See also* Benefit-cost; Linear programming; Net present equivalent)
Decision strategy, 340−342, 390, 395−398
Decision tree:
extensive form, 337−340, 342−350
normal form, 340−350
Decision variable, 271, 314
Deductive reasoning, 15
De Finetti, B., 18
Degree:
of belief, 18, 19
of freedom, 72, 74, 107, 111*n*., 112
De Morgan, A., 18

Density function, probability, 58−61
Discount factor:
annuity, 186
arithmetic, 191−192
capital recovery factor, 187, 188
future worth: of present sum, 185, 188
of uniform series, 188
geometric, 192
present worth: of future sum, 192, 186, 188
of uniform series, 187, 188
uniform series: of a future sum, 188
of a present sum, 187, 188
Discount rate (*see* Interest rate)
Distribution (*see* Probability, distributions)

Earth Resources Survey System (ERS), 251−261
Efficiency, 202−205
(*See also* Benefit-cost)
Endogenous, 142
Equity, 241−243
Error:
of estimate, 83, 85
interval, 83
standard error of estimate, 146
Type I, 93
Type II, 93
Estimation:
comparison of estimates: means, 99−102, 104, 130
multiple means, 108−113
multiple proportions, 105−108
proportions, 93−99, 104, 124
variances, 102−104
interval, confidence, 82−92
average or mean, 88−91, 131
correlation coefficient, 138−142
proportion, 84−88, 129
variance, 91−92
sample size and confidence, 83, 84, 87, 89
single values (*see* Statistic)
Evidence:
conditional probability, 25
inference, 15
necessary, 21
sufficient, 21
(*See also* Degree, of belief)
Exogenous, 142
Expected value, 61−62
for sum of variables, 64
Exponential model, 69, 76

F model, 74, 76
False positive (FP) rates, 388, 391, 394
Feasible solution, 271
Federal statistical system, 411–413
Fisher's exact test, 99, 167
Free variable, 296–298
Frequency distribution (histogram), 43
Future equivalent, 208
  (*See also* Benefit-cost)

Gauss, K. F., 55
Geometric model, 66, 76
Goodness of fit, chi-square, 105, 108
  (*See also* Hypothesis tests)

Health decisions:
  adolescent growth, 49–50
  breast cancer detection, 385–398
  cost effectiveness, 221, 270–272, 284, 352,
    363
  diagnosis, 30–31, 381–382
  environment, 231, 301, 321, 322, 326
  epidemiology, 33, 232
  EPSDT Medicaid, 400–408
  food additives, 164, 171
  income, 134, 135
  insurance, 116
  life expectancy, 117, 231
  new drug, 96–98
  pancreatic surgery, 347–348
  physician attitudes, 115
  polio vaccine case, 261–266
  saccharin, 99, 164–171
  sex selection, 78, 94
  swine flu, 29–30
  tumorogenesis, 164–171
  Uniformed Services University of the Health
    Sciences, 247–251
Hicks, J. R., 180
Histogram, 42–43
  relative frequency, 43
Hypothesis tests:
  Fisher's exact test, 99, 167
  for paired data, 101–102
  in regression, 147, 148, 150*n*.
  (*See also* Estimation)

Independence, stochastic, 24
Inductive reasoning, 15
Inflation, 195–196

Information, value of: opportunity loss, 356,
    357, 361
  perfect information, 359, 361
  sample (imperfect) information, 358, 360
Interest rate:
  compound, 185
  compound frequency, 193–194
  continuous, 193–194
  equivalent annual, 193
  nominal, 193
  period, 193
  simple, 184
  social discount rate, 185, 198
Interest tables, 434–456

Jeffreys, H., 18

Kaldor, N., 180
Keynes, J. M., 18
Kolb, David, 2

Laplace, P. S. de, 18
Laws of large numbers, 83
Least squares fit (*see* Regression)
Likelihood:
  as probabillity, 16, 350
  prospective ratio, 33
  retrospective ratio, 33
Linear programming:
  artificial variable, 275, 288
  basis inverse, 308
  capital budgeting, 292–293
  convex sets, 272–273
  corner point theorem, 273
  decision variable, 271
  degenerate solution, 275
  feasible solution, 271, 349
  free variable, 296–298
  graphic solution, 269–272, 349
  objective function, 273–275
  optimal solution, 271, 279
  personnel assignment, 293–295
  piecewise linear approximation, 295, 296
  postoptimality analysis, 298, 308
  sensitivity, 298, 308–317, 349
  shadow price, 311
  simplex method, 274–281
  simplex tableau, 279–281
  slack variable, 274–275
  starting solution, 275–276

Linear programming:
  surplus variable, 274 – 275, 287
  transportation, 290 – 292
  unbounded solution, 275
Little, I. M. D., 180
Logical connectors:
  and (conjunction), 19
  not, 19
  or (disjunction), 19

Marginal analysis (*see* Benefit-cost)
Marginal probability, 64
Marglin, Stephen A., 196
Mass function, probability, 58 – 59
Matrix:
  addition, 304 – 305
  dimension, 303, 304
  form, 303
  multiplication, 303 – 304, 306 – 308
  row manipulation, 278
Maximin decision rule, 335
Mean:
  arithmetic, 36, 61
  estimation of, 88 – 91
  geometric, 36, 37
  harmonic, 36, 38
Median, 36, 39
Minimax decision rule. 335
Mode, 36, 39
Morgenstern, Oscar, 333
Mosteller, Frederick, 29

Necessary evidence, 21
Net present equivalent (NPE), 205 – 208
  (*See also* Benefit-cost, net present
    equivalent)
Normal model, 70, 76
NPE (net present equivalent), 205 – 208
  (*See also* Benefit-cost, net present equivalent)

Objective function, 273 – 275
Operating characteristic curve, 116
Opportunity loss, 183, 184
Optimal, 218, 271, 279, 280
  (*See also* Linear programming)

Paired data test, 101 – 102
Pareto, Vilfredo, 180
Partial correlation coefficient, 150
Pascal model, 68, 76

Perfect information, 357, 359, 361
Poisson model, 68, 76
Polio vaccine, development of, 261 – 266
Portfolio analysis, 215 – 217
Power, statistical: calculation of, 116
  of hypothesis test, 93
  and significance, 93
Present equivalent, 208
  (*See also* Benefit-cost; Net present
    equivalent)
Probability:
  a posteriori, 27, 393
  a priori, 19, 26, 134
  calibration, 391
  CDF, 58 – 61
  conditional, 25, 80
  confirmatory value, 393
  cumulative distribution function (CDF),
    58 – 61
  density function (PDF), 58 – 61
  distributions: Bernoulli, 65, 76, 78, 125
    beta, 75, 76, 122
    binomial, 65, 76, 78
    chi-square, χ^2, 72, 76
    exponential, 69, 76
    F, 74, 76
    Gauss, 70, 76, 130
    geometric, 66, 76, 169
    negative binomial, 68, 76
    normal, 70, 76, 130
    Pascal, 68, 76
    Poisson, 68, 76
    Student's-t, 73, 76
    t, 73, 76
    triangular, 60, 62
  false positive, 388, 391, 394
  interpretations: degree of belief, 18 – 19
    judgmental, 18 – 19, 29, 120 – 124, 134,
      407
    objective, 18 – 19
    relative frequency, 18 – 19
  likelihood, 16, 350
    posterior, 27, 120 – 124, 363
    prior, 19, 26, 120 – 124, 363
    ratio, 33
  mass function (PMF), 58 – 59
  nomogram, 258, 394
  rules, 21 – 28
    Bayes', 27 – 28, 119 – 120
    conditional, 25
    disjunctive, 23
    independence stochastic, 24
    multiplicative, 25
  tables, 77, 413 – 433
  true positive, 388, 391, 394

Probability density function (PDF), 58—61
Probability mass function (PMF), 58—59
Project evalation (*see* Benefit-cost; Portfolio
 analysis)
Project lifetime, 187, 188, 223

Queuing, 68, 69, 76

r-square (*see* Regression)
Raiffa, Howard, 18, 123, 136
Ramsey, F. P., 18
Random variable, 56—58
 assignment, 57—58
 continuous, 59
 integer (discrete), 58
 standardized, 71, 77
 sums, 64
Rate of return, 211—215
 (*See also* Efficiency)
Regression:
 auto-, 155—156
 multivariate model, 138, 148—151
 residual of, 151
 simple linear model, 138, 143, 148
 stepwise, 138
 on time, 155
Reliability, 30
Risk, 366—368
 and certainty equivalent, 368
 expected loss, 385n.
 and insurance, 368, 369
 in medicine, 385n.
Risk aversion, 369
Risk neutral, 369
Risk premium, 368, 369
Risk prone, 369
Rules of probability, 21—28
 Bayes', 27, 28
 conditioning, 25
 disjunctive, 23
 independence stochastic, 24
 multiplicative, 25

Sample:
 and estimation: average, 88—89
  correlation, 140
  proportion, 87—88
  skewness, 41—42
  standard deviation, 40—41
  variance, 41
 random, 81—82
 stratified, 81

Savage, L., 18
Selvidge, J., 29
Sensitivity analysis, 223, 235, 236, 250, 298,
 302, 309—317, 349—350
Scatter diagram, 138
Shadow price, 219, 220, 311
Significance, statistical: of hypothesis test, 93,
 95, 166
 relation to confidence, 94
 (*See also* Estimation)
Simplex method, 274—281
Skewness coefficient, 41
Slack variable, 274—275
Sludge management case, 326—331
Standard deviation, 40—41, 62
Standard error of estimate, 146
State of nature, 338
Statements:
 complimentary, 20
 contingency, 19
 exhaustive, 58
 independent, 24
 logically equivalent, 20
 mutually exclusive, 20, 58
 self-contradictory, 19
 tautology, 19
Statistic:
 average, 35
  arithmetic mean, 36
  geometric mean, 36, 37
  harmonic mean, 36, 38
  median, 36, 39
 correlation coefficient, 139
 mode, 36, 39
 proportion, 37
 relative frequency, 43
 skewness coefficient, 41
 standard deviation, 40—41
 variance, 41
Statistical Analysis System (SAS), 113
Statistical Package for Social Sciences (SPSS),
 113, 117
Stochastic independence (*see* Probability)
Strategy (*see* Decision strategy)
Student's-t model, 73, 76
Sufficient evidence, 20—21
Surplus variable, 274—275, 287
Swine flu decision of 1976, 29

Tradeoff analysis, 1, 367, 373
 (*See also* Sensitivity analysis)
Transportation problem, 290—292, 300
 balanced, 290

Trends:
 cyclic, 46
 extrapolation, 44
 geometric, 44
 insulin demand, 160–164
 moving average, 47
 regression, 155
True positive (TP) rates, 388, 391, 394

Uncertainty, 6, 223, 333, 367
 (*See also* Degree, of belief; Probability)
Uniformed Services University of the Health
  Sciences, 247–251
Utility, 365
 and certainty equivalent, 368
 independence, 374
 isopreference, 366, 367
 marginal, 378, 384
 for money, 369–370
 multiattribute, 370, 378
 product form, 375

Utility:
 and pumped storage site selection, 398–400
  quasi-separable, 374–378, 399
 and risk, 369
  separable, 370–374, 399

Variability:
 average deviation, 40
 range, 40
 standard deviation, 40–41
 variance, 41, 62, 64
Variance, 41, 62–64
Vector:
 dimension, 303, 304
 form, 303
 unit basis, 279, 308
Vector addition, 304–305
Vector multiplication, 303–304
Von Neumann, John, 333

Westoff, Charles F., 50